MW01632024

Configural Frequency Analysis

Methods, Models, and Applications

Configural Frequency Analysis

Methods, Models, and Applications

Alexander von Eye
Michigan State University

2002

LAWRENCE ERLBAUM ASSOCIATES, PUBLISHERS
Mahwah, New Jersey London

Lawrence Erlbaum Associates, Inc., Publishers
10 Industrial Avenue
Mahwah, NJ 07430

Cover design by Kathryn Houghtaling Lacey

Library of Congress Cataloging-in-Publication Data

Eye, Alexander von.
Configural frequency analysis : methods, models, and applications / Alexander von Eye.
p. cm.
Includes bibliographical references and index.

ISBN 0-8058-4323-X (cloth : alk. paper)
ISBN 0-8058-4324-8 (pbk. : alk. paper)

1. Psychometrics. 2. Discriminant analysis. I. Title.

BF39 .E94 2002
150'.1'519532 — dc21 2002016979
CIP

Books published by Lawrence Erlbaum Associates are printed on acid-free paper, and their bindings are chosen for strength and durability.

Printed in the United States of America
10 9 8 7 6 5 4 3 2 1

List of contents

Configural Frequency Analysis - Methods, Models, and Applications

Preface

Events that occur as expected are rarely deemed worth mentioning. In contrast, events that are surprising, unexpected, unusual, shocking, or colossal appear in the news. Examples of such events include terrorist attacks, when we are informed about the events in New York, Washington, and Pennsylvania on September 11, 2001; or on the more peaceful side, the weather, when we hear that there is a drought in the otherwise rainy Michigan; accident statistics, when we note that the number of deaths from traffic accidents that involved alcohol is smaller in the year 2001 than expected from earlier years; or health, when we learn that smoking and lack of exercise in the population does not prevent the life expectancy in France from being one of the highest among all industrial countries.

Configural Frequency Analysis (CFA) is a statistical method that allows one to determine whether events that are unexpected in the sense exemplified above are significantly discrepant from expectancy. The idea is that for each event, an expected frequency is determined. Then, one asks whether the observed frequency differs from the expected more than just randomly.

As was indicated in the examples, discrepancies come in two forms. First, events occur more often than expected. For example, there may be more sunny days in Michigan than expected from the weather patterns usually observed in the Great Lakes region. If such events occur *significantly more often than expected*, the pattern under study constitutes a *CFA type*. Other events occur less often than expected. For example, one can ask whether the number of alcohol-related deaths in traffic accidents is *significantly below expectation*. If this is the case, the pattern under study constitutes a *CFA antitype*.

According to Lehmacher (2000), questions similar to the ones answered using CFA, were asked already in 1922 by Pfaundler and von Sehr. The authors asked whether symptoms of medical diseases can be shown to co-occur above expectancy. Lange and Vogel (1965) suggested that the term *syndrom* be used only if individual symptoms co-occurred above expectancy. Lienert, who is credited with the development of the concepts and principles of CFA, proposed in 1968 (see Lienert, 1969) to test for each cell in a cross-classification whether it constitutes a type or an antitype.

The present text introduces readers to the method of Configural Frequency Analysis. It provides an almost complete overview of approaches, ideas, and techniques. The first part of this text covers concepts and methods of CFA. This part introduces the goals of CFA, discusses the base models that are used to test event patterns against, describes and compares statistical tests, presents descriptive measures, and explains methods to protect the significance level α.

The second part introduces CFA base models in more detail. Models that assign the same status to all variables are distinguished from models that discriminate between variables that differ in status, for instance, predictors and criteria. Methods for the comparison of two or more groups are discussed in detail, including specific significance tests and descriptive measures.

The third part of this book focuses on CFA methods for longitudinal data. It is shown how differences between time-adjacent observations can be analyzed using CFA. It is also shown that the analysis of differences can require special probability models. This part of the book also illustrates the analysis of shifts in location, and the analysis of series of measures that are represented by polynomials, autocorrelations, or auto-distances.

The fourth part of this book contains the *CFA Specialty File*. Methods are discussed that allow one to deal with such problems as structural zeros, and that allow one to include covariates into CFA. The graphical representation of CFA results is discussed, and the configural analysis of groups of cells is introduced. It is shown how CFA results can be simplified (aggregated). Finally, this part presents two powerful alternatives to standard CFA. The first of these alternatives, proposed by Kieser and Victor (1999), uses the more general log-linear models of quasi-independence as base models. Using these models, certain artifacts can be prevented. The second alternative, proposed by Wood, Sher and von Eye (1994) and by Gutiérrez-Peña and von Eye (2000), is *Bayesian CFA*. This method (a) allows one to consider a priori existing information, (b) provides a natural way to analyzing groups of cells, and (c) does not require one to adjust the significance level α.

Computational issues are discussed in the fifth part. This part shows how CFA can be performed using standard general purpose statistical software such as SYSTAT. In addition, this part shows how Bayesian CFA can be performed using Splus. The features of a specialized CFA program are illustrated in detail.

There are several audiences for a book like this. First. students in the behavioral, social, biological, and medical sciences, or students in empirical sciences in general, may benefit from the possibility to pursue questions that arise from taking the cell-oriented (Lehmacher, 2000) or person-oriented perspectives (Bergman & Magnusson, 1997). CFA can be used either as the only method to answer questions concerning individual cells of cross-classifications, or it can be used in tandem with such methods as discriminant analysis, logistic regression, or log-linear modeling.

The level of statistical expertise needed to benefit most from this book is that of a junior or senior in the empirical behavioral and social sciences. At this level, students have completed introductory statistics courses and know such methods as χ^2-tests. In addition, they may have taken courses in categorical data analysis or log-linear modeling, both of which would make it easier to work with this book on CFA. To perform CFA, no more than a general purpose software package such as SAS, SPSS, Splus, or SYSTAT is needed. However, specialized CFA programs as illustrated in Part 5 of this book are more flexible, and they are available free (for details see Chapter 12).

Acknowledgments. When I wrote this book, I benefitted greatly from a number of individuals' support, encouragement, and help. First of all, Donata, Maxine, Valerie, and Julian tolerate my lengthy escapades in my study, and provide me with the human environment that keeps me up when I happen to venture out of this room. My friends Eduardo Gutiérrez-Peña, Eun-Young Mun, Mike Rovine, and Christof Schuster read the entire first draft of the manuscript and provided me with a plethora of good-willing, detailed, and insightful comments. They found the mistakes that are not in this manuscript any more. I am responsible for the ones still in the text. The publishers at Lawrence Erlbaum, most notably Larry Erlbaum himself, Debra Riegert, and Jason Planer expressed their interest in this project and encouraged me from the first day of our collaboration. I am deeply grateful for all their support.

Gustav A. Lienert, who initiated CFA, read and comment on almost the entire manuscript in the last days of his life. I feel honored by this effort. This text reflects the changes he proposed. This book is dedicated to his memory.

Alexander von Eye
Okemos, April 2002

Configural Frequency Analysis

Methods, Models, and Applications

Part 1: Concepts and Methods of CFA

"The best methods are capable of uncovering patterns which are unexpected."

Diggle, Liang, & Zeger, 1996, p. 33

1. Introduction: The Goals and Steps of Configural Frequency Analysis

This first chapter consists of three parts. First, it introduces readers to the basic concepts of Configural Frequency Analysis (CFA). It begins by describing the questions that can be answered with CFA. Second, it embeds CFA in the context of *Person Orientation*, that is, a particular research perspective that emerged in the 1990s. Third, it discusses the five steps involved in the application of CFA. The chapter concludes with a first complete data example of CFA.

1.1 Questions that can be answered with CFA

Configural Frequency Analysis (CFA; Lienert, 1968, 1971a) allows researchers to identify those patterns of categories that were observed more often or less often than expected based on chance. Consider, for example, the contingency table that can be created by crossing the three psychiatric symptoms Narrowed Consciousness (C), Thought Disturbance (T), and Affective Disturbance (A; Lienert, 1964, 1969, 1970; von Eye, 1990). In a sample of 65 students who participated in a study on the effects of LSD 50, each of these symptoms was scaled as 1 = present or 2 = absent. The cross-classification C x T x A, which has been used repeatedly in illustrations of CFA (see, e.g., Heilmann & Schütt, 1985; Lehmacher, 1981; Lindner, 1984; Ludwig, Gottlieb, & Lienert, 1986), appears in Table 1.

Table 1: Cross-classification of the three variables Narrowed Consciousness (C), Thought Disturbance (T), and Affective Disturbance (A); N = 65

Pattern CTA	Observed Frequency
111	20
112	1
121	4
122	12
211	3
212	10
221	15
222	0

In the context of CFA, the patterns denoted by the cell indices 111, 112, ..., 222 are termed *Configurations*. If d variables are under study, each configuration consists of d elements. The configurations differ from each other in at least one and maximally in all d elements. For instance, the first configuration, 111, describes the 20 students who experienced all three disturbances. The second configuration, 112, differs from the first in the last digit. This configuration describes the sole student who experiences narrowed consciousness and thought disturbances, but no affective disturbance. The last configuration, 222, differs from the first in all $d = 3$ elements. It suggests that no student was found unaffected by LSD 50. A complete CFA of the data in Table 1 follows in Section 3.7.2.2.

The observed frequencies in Table 1 indicate that the eight configurations do not appear at equal rates. Rather, it seems that experiencing no effects is unlikely, experiencing all three effects is most likely, and experiencing only two effects is relatively unlikely. To make these descriptive statements, one needs no further statistical analysis. However, there may be questions beyond the purely descriptive. Given a cross-classification of two or more variables. CFA can be used to answer questions of the following types:

(1) *How do the observed frequencies compare with the expected frequencies*? As interesting and important as it may be to interpret observed frequencies, one often wonders whether the extremely high or low numbers are still that extreme when we compare them with their expected counterparts. The same applies to the less extreme frequencies. Are they still about average when compared to what could have been expected? To answer these questions, one needs to estimate expected cell frequencies. The expected cell frequencies conform to the specifications made in so-called *base models*. These are models that reflect the assumptions concerning the relationships among the variables under study. Base models are discussed in Sections 2.1 - 2.3. It goes without saying that *different base models can lead to different expected cell frequencies* (Mellenbergh, 1996). As a consequence, the answer to this first question depends on the base model selected for frequency comparison, and the interpretation of discrepancies between observed and expected cell frequencies must always consider the characteristics of the base model specified for the estimation of the expected frequencies. The selection of base models is not arbitrary (see Chapter 2 for the definition of a valid CFA base model). The comparison of observed with expected cell frequencies allows one to identify those configurations that were *observed as often as expected*. It allows one also to identify those configurations that were *observed more often than expected* and those configurations that were *observed less often than expected*. Configurations that are observed at different frequencies than expected are of particular interest in CFA applications.

(2) *Are the discrepancies between observed and expected cell frequencies statistically significant*? It is rarely the case that observed and expected cell frequencies are identical. In most instances, there will be numerical differences. CFA allows one to answer the question whether a numerical difference is random or too large to be considered random. If an observed cell frequency is significantly larger than the expected cell frequency, the respective configuration is said to constitute a *CFA type*. If an observed frequency is significantly smaller than its expected counterpart, the configuration is said to constitute a *CFA antitype*. Configurations

with observed frequencies that differ from their expectancies only randomly, constitute neither a type nor an antitype. In most CFA applications, researchers will find both, that is, cells that constitute neither a type nor an antitype, and cells that deviate significantly from expectation.

(3) *Do two or more groups of respondents differ in their frequency distributions*? In the analysis of cross-classifications, this question typically is answered using some form of the χ^2-test, some log-linear model, or logistic regression. Variants of χ^2-tests can be employed in CFA too (for statistical tests employed in CFA, see Chapter 2). However, CFA focuses on individual configurations rather than on overall goodness-of-fit. CFA indicates the configurations in which groups differ. If the difference is statistically significant, the respective configuration is said to constitute a *discrimination type*.

(4) *Do frequency distributions change over time and what are the characteristics of such changes*? There is a large number of CFA methods available for the investigation of change and patterns of change. For example, one can ask whether shifts from one category to some other category occur as often as expected from some chance model. This is of importance, for instance, in investigations of treatment effects, therapy outcome, or voter movements. Part III of this book covers methods of longitudinal CFA.

(5) *Do groups differ in their change patterns*? In developmental research, in research concerning changes in consumer behavior, in research on changes in voting preferences, or in research on the effects of medicinal or leisure drugs, it is one issue of concern whether groups differ in the changes that occur over time. What are the differences in the processes that lead some customers to purchase holiday presents on the web and others in the stores? CFA allows one to describe these groups, to describe the change processes, and to determine whether differences in change are greater than expected.

(6) *Are there predictor-criterion relationships*? In educational research, in studies on therapy effects, in investigations on the effects of drugs, and in many other contexts, researchers ask

whether events or configurations of events allow one to predict other configurations of events. CFA allows one to identify those configurations for which one can predict that other configurations occur more often than expected, and those configurations for which one can predict that other configurations occur less often than expected based on chance.

This book presents methods of CFA that enable researchers to answer these and more questions.

1.2 CFA and the person perspective[1]

William Stern introduced in 1911 the distinction between *variability* and *psychography*. Variability is the focus when many individuals are observed in one characteristic with the goal to describe the distribution of this characteristic in the population. Psychographic methods aim at describing one individual in many characteristics. Stern also states that these two methods can be combined.

When describing an individual in a psychographic effort, results are often presented in the form of a *profile*. For example, test results of the MMPI personality test typically are presented in the form of individual profiles, and individuals are compared to reference profiles. For example, a profile may resemble the pattern typical of schizophrenics. A profile describes the position of an individual on standardized, continuous scales. Thus, one can also compare the individual's relative standing across several variables. Longitudinally, one can study an individual's relative standing and/or the correlation with some reference change. Individuals can be grouped based on profile similarity.

In contrast to profiles, *configurations* are not based on continuous but on categorical variables. As was explained in Section 1.1, the ensemble of categories that describes a cell of a cross-classification is called configuration (Lienert, 1969). Configurational analysis using CFA investigates such configurations from several perspectives. First, CFA identifies configurations (see Table 1). This involves creating cross-classifications or, when variables are originally continuous, categorization

---

[1]The following section borrows heavily from von Eye (2002b; see also von Eye, Indurkhya, & Kreppner, 2000).

and then creating cross-classifications. Second, CFA asks, whether the number of times a configuration was observed could have been expected from some a priori specified model, the base model. Significant deviations will then be studied in more detail. Third, researchers often ask in a step that goes beyond CFA, whether the cases described by different configurations also differ in their mean and covariance structures in variables not used for the cross-classification. This question concerns the external validity of configurational statements (Aksan et al., 1999; see Section 10.11). Other questions that can be answered using CFA have been listed above. In the following paragraphs, CFA will be embedded in Differential Psychology and the Person-Oriented Approach.

This section covers two roots of CFA, *Differential Psychology* and the *Person-Oriented Approach.* The fundamental tenet of Differential Psychology is that "individual differences are worthy of study in their own right" (Anastasi, 1994, p. ix). This is often seen in contrast to General Psychology where it is the main goal to create statements that are valid for an entire population. General Psychology is chiefly interested in variables, their variability, and their covariation (see Stern, 1911). The data carriers themselves, for example, humans, play the role of replaceable random events. They are not of interest per se. In contrast, Differential Psychology considers the data carriers units of analysis. The smallest unit would be the individual at a given point in time. However, larger units are often considered, for example, all individuals that meet the criteria of geniuses, alcoholics, and basketball players.

Differential Psychology as both a scientific method and an applied concept presupposes that the data carriers' characteristics are measurable. In addition, it must be assumed that the scales used for measurement have the same meaning for every data carrier. Third, it must be assumed that the differences between individuals are measurable. In other words, it must be assumed that data carriers are indeed different when they differ in their location on some scale. When applying CFA, researchers make the same assumptions.

The *Person-Oriented Approach* (Bergman & Magnusson, 1991, 1997; Magnusson, 1998; Magnusson & Bergman, 2000; von Eye et al., 2000) is a relative of Differential Psychology. It is based on five propositions (Bergman & Magnusson, 1997; von Eye et al., 1999a):

(1) Functioning, process, and development (FPD) are, at least in part, specific to the individual.

(2) FPD are complex and necessitate including many factors and their interactions.
(3) There is lawfulness and structure in (a) individual growth and (b) interindividual differences in FPD.
(4) Processes are organized and function as *patterns* of the involved factors. The meaning of the involved factors is given by the factors' interactions with other factors.
(5) Some patterns will be observed more frequently than other patterns, or more frequently than expected based on prior knowledge or assumptions. These patterns can be called *common types*. Examples of common types include the *types identified by CFA*. Accordingly, there will be patterns that are observed less frequently than expected from some chance model. CFA terms these the *antitypical* patterns or *antitypes*.

Two consequences of these five propositions are of importance for the discussion and application of CFA. The first is that, in order to describe human functioning and development, differential statements can be fruitful in addition to statements that generalize to variable populations, person populations, or both. Subgroups, characterized by group-specific patterns, can be described more precisely. This is the reason why methods of CFA (and cluster analysis) are positioned so prominently in person-oriented research. Each of these methods of analysis focuses on groups of individuals that share in common a particular pattern and differ in at least one, but possibly in all characteristics (see Table 1, above).

The second consequence is that functioning needs to be described at an individual-specific basis. If it is a goal to compare individuals based on their characteristics of FPD, one needs a valid description of each individual. Consider, for example, Proposition 5, above. It states that some patterns will occur more frequently and others less frequently than expected based on chance or prior knowledge. An empirical basis for such a proposition can be provided only if intra-individual functioning and development is known.

Thus, the person-oriented approach and CFA meet where (a) patterns of scores or categories are investigated, and (b) where the tenet of differential psychology is employed according to which it is worth the effort to investigate individuals and groups of individuals. The methodology employed for studies within the framework of the person-oriented approach is typically that of CFA. The five steps involved in this methodology are presented in the next section.

1.3 The five steps of CFA

This section introduces readers to the five steps that a typical CFA application involves. This introduction is brief and provides no more than an overview. The remainder of this book provides the details for each of these steps. These steps are

(1) Selection of a CFA base model and estimation of expected cell frequencies; the base model (i) reflects theoretical assumptions concerning the nature of the variables as either of equal status or grouped into predictors and criteria, and (ii) considers the sampling scheme under which the data were collected;
(2) Selection of a concept of deviation from independence;
(3) Selection of a significance test;
(4) Performance of significance tests and identification of configurations that constitute types or antitypes;
(5) Interpretation of types and antitypes.

The following paragraphs give an overview of these five steps. The following sections provide details, illustrations, and examples. Readers already conversant with CFA will notice the many new facets that have been developed to increase the number of models and options of CFA. Readers new to CFA will realize the multifaceted nature of the method.

(1) *Selection of a CFA base model and estimation of expected cell frequencies*. Expected cell frequencies for most CFA models[2] can be estimated using the log-frequency model

$$\log E = X\lambda ,$$

where E is the array of model frequencies, that is, frequencies that conform to the model specifications. X is the *design matrix*, also called *indicator matrix*. Its vectors reflect the CFA base model or, in other contexts, the log-frequency model under study. λ is the vector of model parameters. These parameters are not of interest per se in frequentist CFA. Rather, CFA focuses on the discrepancies between the expected and the observed cell frequencies. In contrast to log-linear modeling, CFA is not applied with the

---

[2]Exceptions are presented, for instance, in the section on CFA for repeated observations (see Section 8.2.3; cf. von Eye & Niedermeier, 1999).

goal of identifying a model that describes the data sufficiently and parsimoniously (for a brief introduction to log-linear modeling, see Appendix A). Rather, *a CFA base model takes into account all effects that are NOT of interest to the researchers*, and it is assumed that the base model fails to describe the data well. If types and antitypes emerge, they indicate where the most prominent discrepancies between the base model and the data are.

Consider the following example of specifying a base model. In Prediction CFA, the effects that are NOT of interest concern the relationships among the predictors and the relationships among the criteria. Thus, the indicator matrix X for the Prediction CFA base model includes all relationships among the predictors and all relationships among the criteria. In other words, the typical base model for Prediction CFA is saturated in the predictors and the criteria. However, the base model must not include any effect that links predictors to criteria. If types and antitypes emerge, they reflect relationships between predictors and criteria, but not among the predictors or among the criteria. These predictor-criterion relationships manifest in configurations that were observed more often than expected from the base model or in configurations that were observed less often than expected from the base model. A type suggests that a particular predictor configuration allows one to predict the occurrence of a particular criterion configuration. An antitype allows one to predict that a particular predictor configuration is *not* followed by a particular criterion configuration.

In addition to considering the nature of variables as either all belonging to one group, or as predictors and criteria as in the example with Prediction CFA, the sampling scheme must be considered when specifying the base model. Typically, the sampling scheme is *multinomial*. Under this scheme, respondents (or responses; in general, the units of analysis) are randomly assigned to the cells of the entire cross-tabulation. When the sampling scheme is multinomial, any CFA base model is admissible. Please notice that this statement does not imply that any log-frequency model is admissible as a CFA base model (see Section 2.2). However, the multinomial sampling scheme itself does not place any particular constraints on the selection of a base model.

An example of a cross-classification that can be formed for configurational analysis involves the variables, Preference for type of car (P; 1 = minivan; 2 = sedan; 3 = sports utility vehicle; 4 = convertible; 5 = other) and number of miles driven per year (M; 1 = 0 - 10,000; 2 = 10,001 - 15,000; 3 = 15,001 - 20,000; 4 = more). Suppose a sample of 200 respondents indicated their car preference and the number of miles they typically drive in a year. Then, each respondent can be randomly assigned to the 20 cells of the entire 5 x 4 cross-classification of P and M, and there is no constraint concerning the specification of base models.

In other instances, the sampling scheme may be *product-multinomial*. Under this scheme, the units of analysis can be assigned only to a selection of cells in a cross-classification. For instance, suppose the above sample of 200 respondents includes 120 women and 80 men, and the gender comparison is part of the aims of the study. Then, the number of cells in the cross-tabulation increases from 5 x 4 to 2 x 5 x 4, and the sampling scheme becomes product-multinomial in the gender variable. Each respondent can be assigned only to that part of the table that is reserved for his or her gender group. From a CFA perspective, the most important consequence of selecting the product-multinomial sampling scheme is that the marginals of variables that are sampled product-multinomially must always be reproduced. Thus, base models that do not reproduce these marginals are excluded by definition. This applies accordingly to multivariate product-multinomial sampling, that is, sampling schemes with more than one fixed marginal. In the present example, including the gender variable precludes zero-order CFA from consideration. Zero-order CFA, also called *Configural Cluster Analysis*, uses the no effect model for a base model, that is, the log-linear model $\log E = \mathbf{1}\lambda$, where $\mathbf{1}$ is a vector of ones and λ is the intercept parameter. This model may not reproduce the sizes of the female and male samples and is therefore not admissible.

(2) *Selection of a concept of deviation from independence and Selection of a significance test*. In all CFA base models, types and antitypes emerge when the discrepancy between an observed and an expected cell frequency is statistically significant. However, the measures that are available to describe the discrepancies use different definitions of discrepancy, and differ in the assumptions that must be made for proper application. The χ^2-based measures and their normal approximations assess the magnitude of the discrepancy relative to the expected frequency. This group of measures differs mostly in statistical power, and can be employed regardless of sampling scheme. The hypergeometric test and its normal approximations, and the binomial test also assess the magnitude of the discrepancy, but they presuppose product-multinomial sampling. The relative risk, RR_i, is defined as the ratio N_i/E_i where i indexes the configurations. This measure indicates the frequency with which an event was observed, relative to the frequency

with which it was expected. RR_i is a descriptive measure (see Section 4.1; DuMouchel, 1999). There exists an equivalent measure, I_i, that results from a logarithmic transformation, that is, $I_i = \log_2(RR_i$; cf. Church & Hanks, 1991). This measure was termed *mutual information*. RR_i and I_i do not require any specific sampling scheme. The measure log P (for a formal definition see DuMouchel, 1999, or Section 4.2) has been used descriptively and also to test CFA null hypotheses. If used for statistical inference, the measure is similar to the binomial and other tests used in CFA, although the rank order of the assessed extremity of the discrepancy between the observed and the expected cell frequencies can differ dramatically (see Section 4.2; DuMouchel, 1999; von Eye & Gutiérrez-Peña, in preparation). In the present context of CFA, we use logP as a descriptive measure.

In two-sample CFA, two groups of respondents are compared. The comparison uses information from two sources. The first source consists of the frequencies with which Configuration i was observed in both samples. The second source consists of the sizes of the comparison samples. The statistics can be classified based on whether they are *marginal-dependent* or *marginal-free*. Marginal-dependent measures indicate the magnitude of an association that also takes the marginal distribution of responses into account. Marginal-free measures only consider the association. It is very likely that marginal-dependent tests suggest a different appraisal of data than marginal-free tests (von Eye, Spiel, & Rovine, 1995).

(3) *Selection of significance test.* Four criteria are put forth that can guide researchers in the selection of measures for one-sample CFA: *exact versus approximative* test, *statistical power*, *sampling scheme*, and use for *descriptive versus inferential purposes*. In addition, the tests employed in CFA differ in their sensitivity to types and antitypes. More specifically, when sample sizes are small, most tests identify more types than antitypes. In contrast when sample sizes are large, most tests are more sensitive to antitypes than types. one consistent exception is Anscombe's (1953) z-approximation which always tends to find more antitypes than types, even when sample sizes are small. Section 3.8 provides more detail and comparisons of these and other tests, and presents arguments for the selection of significance tests for CFA.

(4) *Performing significance tests and identifying configurations as types or antitypes*. This fourth step of performing a CFA is routine to the extent that significance tests come with tail probabilities that allow one to determine

immediately whether a configuration constitutes a type, an antitype, or supports the null hypothesis. It is important, however, to keep in mind that exploratory CFA involves employing significance tests to each cell in a cross-classification. This procedure can lead to wrong statistical decisions first because of capitalizing of chance. Each test comes with the nominal error margin α. Therefore, α% of the decisions can be expected to be incorrect. In large tables, this percentage can amount to large numbers of possibly wrong conclusions about the existence of types and antitypes. Second, the cell-wise tests can be dependent upon each other. Consider, for example, the case of two-sample CFA. If one of the two groups displays more cases than expected, the other, by necessity, will display fewer cases than expected. The results of the two tests are completely dependent upon each other. The result of the second test is determined by the result of the first, because the null hypothesis of the second test stands no chance of surviving if the null hypothesis of the first test was rejected.

Therefore, after performing the cell-wise significance tests, and before labeling configurations as type/antitype constituting, measures must be taken to protect the test-wise α. A selection of such measures is presented in Section 3.10.

(5) *Interpretation of types and antitypes*. The interpretation of types and antitypes is fueled by five kinds of information. The first is the *meaning of the configuration* itself (see Table 1, above). The meaning of a configuration can often be seen in tandem with its nature as a type or antitype. For instance, it may not be a surprise that there exist no toothbrushes with brushes made of steel. Therefore, in the space of dental care equipment, steel-brushed brushes may meaningfully define an antitype. Inversely, one may entertain the hypothesis that couples that stay together for a long time are happy. Thus, in the space of couples, happy, long lasting relationships may form a type.

The second source of information is the *CFA base model*. The base model determines the nature of types and antitypes. Consider, for example, classical CFA which has a base model that proposes independence among all variables. Only main effects are taken into account. If this model yields types or antitypes, they can be interpreted as local associations (Havránek & Lienert, 1984) among variables. Another example is Prediction CFA (P-CFA). As was explained above, P-CFA has a base model that is saturated both in the predictors and the criteria. The relationships among predictors and criteria are not taken into account, thus constituting the only possible reason for the emergence of types and antitypes. If P-CFA yields types or antitypes, they are reflective of predictive relationships among predictors and criteria, not just of any association.

The third kind of information is the *sampling scheme*. In multinomial sampling, types and antitypes describe the entire population from which the sample was drawn. In product-multinomial sampling, types and antitypes describe the particular population in which they were found. Consider again the above example where men and women are compared in the types of car they prefer and the number of miles they drive annually. Suppose a type emerges for men who prefer sport utility vehicles and drive them more than 20,000 miles a year. This type only describes the male population, not the female population, nor the human population in general.

The fourth kind of information is the *nature of the statistical measure* that was employed for the search for types and antitypes. As was indicated above and will be illustrated in detail in Sections 3.8 and 7.2, different measures can yield different harvests of types and antitypes. Therefore, interpretation must consider the nature of the measure, and results from different studies can be compared only if the same measures were employed.

The fifth kind of information is external in the sense of *external validity*. Often, researchers are interested in whether types and antitypes also differ in other variables than the ones used in CFA. Methods of discriminant analysis, logistic regression, MANOVA, or CFA can be used to compare configurations in other variables. Two examples shall be cited here. First, (Görtelmeyer, 1988) identified six types of sleep problems using CFA. Then, he used analysis of variance methods to compare these six types in the space of psychological personality variables. The second example is a study in which researchers first used CFA to identify temperamental types among preschoolers (Aksan et al., 1999). In a subsequent step, the authors used correlational methods to discriminate their types and antitypes in the space of parental evaluation variables. An example of CFA with subsequent discriminant analysis appears in Section 10.9.2.

1.4 A first complete CFA data example

In this section, we present a first complete data analysis using CFA. We introduce methods "on the fly" and explain details in later sections. The first example is meant to provide the reader with a glimpse of the statements that can be created using CFA. The data example is taken from von Eye and Niedermeier (1999).

In a study on the development of elementary school children, 86 students participated in a program for elementary mathematics skills. Each student took three consecutive courses. At the end of each course the students took a comprehensive test, on the basis of which they obtained a 1 for reaching the learning criterion and a 2 for missing the criterion. Thus, for each student, information on three variables was created: Test 1 (T1), Test 2 (T2), and Test 3 (T3). Crossed, these three dichotomous variables span the 2 x 2 x 2 table that appears in Table 2, below. We now analyze these data using exploratory CFA. The question that we ask is whether any of the eight configurations that describe the development of the students' performance in mathematics occurred more often or less often than expected based on the CFA base model of independence of the three tests. To illustrate the procedure, we explicitly take each of the five steps listed above.

Step 1: Selection of a CFA base model and estimation of expected cell frequencies. In the present example we opt for a log-linear main effect model as the CFA base model (for a brief introduction to log-linear modeling, see Appendix A). This can be explained as follows.

(1) The main effect model takes the main effects of all variables into account. As a consequence, emerging types and antitypes will not reflect the varying numbers of students who reach the criterion. (Readers are invited to confirm from the data in Table 2 that the number of students who pass increases from Test 1 to Test 2, and then again from Test 2 to Test 3). Rather, types and antitypes will reflect the development of students (see Point 2).

(2) The main effect model proposes that the variables T1, T2, and T3 are independent of each other. As a consequence, types and antitypes can emerge only if there are local associations between the variables. These associations indicate that the performance measures for the three tests are related to each other, which manifests in configurations that occurred more often (types) or less often (antitypes) than could be expected from the assumption of independence of the three tests.

It is important to note that many statistical methods require strong assumptions about the nature of the longitudinal variables (remember, e.g., the discussion of compound symmetry in analysis of variance; see Neter, Kutner, Nachtsheim, & Wasserman, 1996). The assumption of independence of repeatedly observed variables made in the second proposition of the present CFA base model seems to contradict these assumptions. However, when applying CFA, researchers do not simply assume that repeatedly observed variables are autocorrelated. Rather, they propose in the base model that the variables are independent. Types and antitypes will then provide detailed information about the nature of the autocorrelation, if it exists.

It is also important to realize that other base models may make sense too. For instance, one could ask whether the information provided by the first test allows one to predict the outcomes in the second and third tests. Alternatively, one could ask whether the results in the first two tests allow one to predict the results of the third test. Another model that can be discussed is that of randomness of change. One can estimate the expected cell frequencies under the assumption of random change and employ CFA to identify those instances where change is not random.

The expected cell frequencies can be estimated by hand calculation, or by using any of the log-linear modeling programs available in the general purpose statistical software packages such as SAS, SPSS, or SYSTAT. Alternatively, one can use a specialized CFA program (von Eye, 2001). Table 2 displays the estimated expected cell frequencies for the main effect base model. These frequencies were calculated using von Eye's CFA program (see Section 12.3.1). In many instances, in particular when simple base models are employed, the expected cell frequencies can be hand-calculated. This is shown for the example in Table 2 below the table.

Step 2: Selection of a concept of deviation. Thus far, the characteristics of the statistical tests available for CFA have only been mentioned. The tests will be explained in more detail in Sections 3.2 - 3.6, and criteria for selecting tests will be introduced in Sections 3.7 - 3.9. Therefore, we use here a concept that is widely known. It is the concept of the difference between the observed and the expected cell frequency, relative to the standard error of this difference. This concept is known from Pearson's X^2-test (see Step 4).

Step 3: Selection of a significance test. From the many tests that can be used and will be discussed in Sections 3.2 - 3.9, we select the Pearson X^2 for the present example, because we suppose that this test is well known to most readers. The X^2 component that is calculated for each configuration is

$$X^2 = \frac{(N_i - \hat{E}_i)^2}{\hat{E}_i},$$

where i indexes the configurations. Summed, the X^2-components yield the Pearson X^2-test statistic. In the present case, we focus on the X^2-components which serve as test statistics for the cell-specific CFA H_0. Each of the X^2 statistics can be compared to the χ^2-distribution under 1 degree of freedom.

Step 4: Performing significance tests and identifying types and antitypes. The results from employing the X^2-component test and the tail probabilities for each test appear in Table 2. To protect the nominal significance threshold α against possible test-wise errors, we invoke the Bonferroni method. This method adjusts the nominal α by taking into consideration the total number of tests performed. In the present example, we have eight tests, that is, one test for each of the eight configurations. Setting α to the usual 0.05, we obtain an adjusted $\alpha^* = \alpha/8 = 0.00625$. The tail probability of a CFA test is now required to be less than α^* for a configuration to constitute a type or an antitype.

Table 2 is structured in a format that we will use throughout this book. The left-most column contains the cell indices, that is, the labels for the configurations. The second column displays the observed cell frequencies. The third column contains the expected cell frequencies. The fourth column presents the values of the test statistic, the fifth column displays the tail probabilities, and the last column shows the characterization of a configuration as a type, **T**, or an antitype, **A**.

The unidimensional marginal frequencies are $T1_1 = 31$, $T1_2 = 55$, $T2_1 = 46$, $T2_2 = 40$, $T3_1 = 47$, $T3_2 = 39$. We now illustrate how the expected cell frequencies in this example can be hand-calculated. For three variables, the equation is

$$\hat{E}_{ijk} = \frac{N_{i..} N_{.j.} N_{..k}}{N^2},$$

where N indicates the sample size, $N_{i..}$ are the marginal frequencies of the first variable, $N_{.j.}$ are the marginal frequencies of the second variable, $N_{..k}$ are the marginal frequencies of the third variable, and i, j, and k are the indices for the cell categories. In the present example, i, j, k, = {1, 2}.

Table 2: CFA of results in three consecutive mathematics courses

Cell Indices	Frequencies		Significance Tests		Type/ Antitype ?
T1 T2 T3	observed	expected	X^2	p^a	
1 1 1	20	9.06	13.20	0.0003	**T**
1 1 2	4	7.52	1.65	0.1993	
1 2 1	2	7.88	4.39	0.0362	
1 2 2	5	6.54	0.36	0.5474	
2 1 1	19	16.08	0.53	0.4661	
2 1 2	3	13.34	8.02	0.0046	**A**
2 2 1	6	13.98	4.56	0.0328	
2 2 2	27	11.60	20.44	$< \alpha^*$	**T**

[a] $< \alpha^*$ indicates that this tail probability is smaller than can be expressed with 4 decimal places.

Inserting, for example, the values for Configuration 111, we calculate

$$\hat{E}_{111} = \frac{31 \cdot 46 \cdot 47}{86^2} = 9.062.$$

This is the first value in Column 3 of Table 2. The values for the remaining expected cell frequencies are calculated accordingly.

The value of the test statistic for the first configuration is calculated as

$$X^2_{111} = \frac{(20 - 9.062)^2}{9.062} = 13.202.$$

This is the first value in Column 4 of Table 2. The tail probability for this value is $p = 0.0002796$ (Column 5). This probability is smaller than the critical adjusted α^* which is 0.00625. We thus reject the null hypothesis according to which the deviation of the observed cell frequency from the frequency that was estimated based on the main effect model of variable independence is random.

Step 5: Interpretation of types and antitypes. We conclude that there exists a local association which manifests in a *type of success in mathematics.* Configuration 111 describes those students who pass the final examination in each of the three mathematics courses. Twenty students were found to display this pattern, but only about 9 were expected based on the model of independence. Configuration 212 constitutes an antitype. This configuration describes those students who fail the first and the third course but pass the second. Over 13 students were expected to show this profile, but only 3 did show it. Configuration 222 constitutes a second type. These are the students who consistently fail the mathematics classes. 27 students failed all three finals, but less than 12 were expected to do so. Together, the two types suggest that students' success is very stable, and so is lack of success. The antitype suggests that at least one pattern of instability was significantly less frequently observed than expected based on chance alone.

As was indicated above, one method of establishing the external validity of these types and the antitype could involve a MANOVA or discriminant analysis. We will illustrate this step in Section 10.11.2 (see also Aksan et al., 1999). As was also indicated above, CFA results are typically *non-exhaustive.* That is, only a selection of the eight configurations in this example stand out as types and antitypes. Thus, because CFA results are non-exhaustive, one can call the variable relationships that result in types and antitypes *local* associations. Only a non-exhaustive number of sectors in the data space reflects a relationship. The remaining sectors show data that conform with the base model of no association.

It should also be noticed that Table 2 contains two configurations for which the values of the test statistic had tail probabilities less than the nominal, non-adjusted $\alpha = 0.05$. These are Configurations 121 and 221. For both configurations we found fewer cases than expected from the base model. However, because we opted to protect our statistical decisions against the possibly inflated α-error, we are not in a situation in which we can interpret these two configurations as antitypes. In Section 10.3, we present CFA methods that allow one to answer the question whether the group of configurations that describe varying performance constitutes a composite antitype.

The next chapter introduces log-linear models for CFA that can be used to estimate expected cell frequencies. In addition, the chapter defines CFA base models. Other CFA base models that are not log-linear will be introduced in the chapter on longitudinal CFA (Section 8.2.3).

2. Log-linear Base Models for CFA

The main effect and interaction structure of the variables that span a cross-classification can be described in terms of log-linear models (a brief introduction into the method of log-linear modeling is provided in Appendix A). The general log-linear model is

$$\log E = X\lambda ,$$

where E is an array of model frequencies, X is the design matrix, also called indicator matrix, and λ is a parameter vector (Christensen, 1997; Evers & Namboodiri, 1978; von Eye, Kreppner, & Weßels, 1994). The design matrix contains column vectors that express the main effects and interactions specified for a model. There exist several ways to express the main effects and interactions. Most popular are dummy coding and effect coding. Dummy coding uses only the values of 0 and 1. Effect coding typically uses the values of -1, 0, and 1. However, for purposes of weighting, other values are occasionally used also. Dummy coding and effect coding are equivalent. In this book, we use effect coding because a design matrix specified in effect coding terms is easier for many researchers to interpret than a matrix specified using dummy coding.

The parameters are related to the design matrix by

$$\lambda = (X'X)^{-1}X'\mu ,$$

where $\mu = \log E$, and the $'$ sign indicates a transposed matrix. In CFA applications, the parameters of a base model are typically not of interest because it is assumed that the base model does not describe the data well.

Types and antitypes describe deviations from the base model. If the base model fits, there can be no types or antitypes. Accordingly, the goodness-of-fit X^2 values of the base model are typically not interpreted in CFA.

In general, log-linear modeling provides researchers with the following three options (Goodman, 1984; von Eye et al., 1994):

(1) *Analysis of the joint frequency distribution of the variables that span a cross-classification.* The results of this kind of analysis can be expressed in terms of a distribution jointly displayed by the variables. For example, two variables can be symmetrically distributed such that the transpose of their cross-classification, say A', equals the original matrix, A.

(2) *Analysis of the association pattern of response variables.* The results of this kind of analysis are typically expressed in terms of first and higher order interactions between the variables that were crossed. For instance, two variables can be associated with each other. This can be expressed as a significant deviation from independence using the classical Pearson X^2-test. Typically, and in particular when the association (interaction) between these two variables is studied in the context of other variables, researchers interpret an association based on the parameters that are significantly different than zero.

(3) *Assessment of the possible dependence of a response variable on explanatory or predictor variables.* The results of this kind of analysis can be expressed in terms of conditional probabilities of the states of the dependent variable, given the levels of the predictors. In a most elementary case, one can assume that the states of the dependent variable are conditionally equiprobable, given the predictor states.

Considering these three options and the status of CFA as a prime method in the domain of person-oriented research (see Section 1.2), one can make the different goals of log-linear modeling and CFA explicit. As indicated in the formulation of the three above options, log-linear modeling focuses on variables. Results are expressed in terms of parameters that represent the relationships among variables, or in terms of distributional parameters. Log-linear parameters can be interpreted only if a model fits.

In contrast, CFA focuses on the discrepancies between some base model and the data. These discrepancies appear in the form of types and antitypes. If types and antitypes emerge, the base model is contradicted and does not describe the data well. Because types and antitypes are interpreted at the level of configurations rather than variables, they indicate *local* associations (Havránek & Lienert, 1984) rather than standard, global associations among variables. It should be noticed, however, that local associations often result in the description of a variable association as existing.

Although the goals of log-linear modeling and CFA are fundamentally different, the two methodologies share two important characteristics in common. First, both methodologies allow the user to consider all variables under study as response variables (see Option 2, above). Thus, unlike in regression analysis or analysis of variance, there is no need to always think in terms of predictive or dependency structures. However, it is also possible to distinguish between independent and dependent variables or between predictors and criteria, as will be demonstrated in Section 6.2 on Prediction CFA (cf. Option 3, above). Second, because most CFA base models can be specified in terms of log-linear models, the two methodologies use the same algorithms for estimating expected cell frequencies. For instance, the CFA program that is introduced in Section 12.3 uses the same Newton-Raphson methods to estimate expected cell frequencies as some log-linear modeling programs. It should be emphasized again, however, that (1) not all CFA base models are log-linear models, and (2) not all log-linear models qualify as CFA base models. The chapters on repeated observations (Part III of this book) and on Bayesian CFA (Section 11.12) will give examples of such base models.

Section 2.1 presents sample CFA base models and their assumptions. These assumptions are important because the interpretation of types and antitypes rests on them. For each of the sample base models, a design matrix will be presented. Section 2.2 discusses admissibility of log-linear models as CFA base models. Section 2.3 discusses the role played by sampling schemes, Section 2.4 presents a grouping of CFA base models, and Section 2.5 summarizes the decisions that must be made when selecting a CFA base model.

2.1 Sample CFA base models and their design matrices

For the following examples we use models of the form log $E = X\lambda$, where E is the array of expected cell frequencies, X is the design matrix, and λ is the parameter vector. In the present section, we focus on the design matrix X, because the base model is specified in X. The following paragraphs present the base models for three sample CFA base models: classical CFA of three dichotomous variables; Prediction CFA with two dichotomous predictors and two dichotomous criterion variables; and classical CFA of two variables with more than two categories. More examples follow throughout this text.

The base model of classical CFA for a cross-classification of three variables. Consider a cross-classification that is spanned by three dichotomous variables and thus has 2 x 2 x 2 = 8 cells. Table 2 is an example of such a table. In "classical" CFA (Lienert, 1969), the base model is the log-linear main effect model of variable independence. When estimating expected cell frequencies, this model takes into account

(1) The *main effects of all variables* that are crossed. When main effects are taken into account, types and antitypes cannot emerge just because the probabilities of the categories of the variables in the cross-classification differ;

(2) *None of the first or higher order interactions*. If types and antitypes emerge, they indicate that (local) interactions exist because these were not part of the base model.

Consider the data example in Table 2. The emergence of two types and one antitype suggests that the three test results are associated such that consistent passing or failing occurs more often than expected under the independence model, and that one pattern of inconsistent performance occurs less often than expected.

Based on the two assumptions of the main effect model, the design matrix contains two kinds of vectors. The first is the vector for the intercept, that is, the constant vector. The second kind includes the vectors for the main effects of all variables. Thus, the design matrix for this 2 x 2 x 2 table is

$$X = \begin{bmatrix} 1 & 1 & 1 & 1 \\ 1 & 1 & 1 & -1 \\ 1 & 1 & -1 & 1 \\ 1 & 1 & -1 & -1 \\ 1 & -1 & 1 & 1 \\ 1 & -1 & 1 & -1 \\ 1 & -1 & -1 & 1 \\ 1 & -1 & -1 & -1 \end{bmatrix}.$$

The first column in matrix X is the constant vector. This vector is part of all log-linear models considered for CFA. It plays a role comparable to the constant vector in analysis of variance and regression which yields the estimate of the intercept. Accordingly, the first parameter in the vector λ, that is, λ_0, can be called the intercept of the log-linear model (for more detail see, e.g., Agresti, 1990; Christensen, 1997). The second vector in X contrasts the first category of the first variable with the second category. The third vector in X contrasts the first category of the second variable with the second category. The last vector in X contrasts the two categories of the third variable. The order of variables and the order of categories has no effect on the magnitude of the estimated parameters or expected cell frequencies.

The base model for Prediction CFA with two predictors and two criteria. This section presents a base model that goes beyond the standard main effect model. Specifically, we show the design matrix for a model with two predictors and two criteria. All four variables in this example are dichotomous. The base model takes into account the following effects:

(1) *Main effects of all variables.* The main effects are taken into account to prevent types and antitypes from emerging that would be caused by discrepancies from a uniform distribution rather than predictor-criterion relationships.

(2) *The interaction between the two predictors.* If types and antitypes are of interest that reflect local relationships between predictors and criterion variables, types and antitypes that are caused by relationships among the predictors must be prevented. This can be

done by making the interaction between the two predictors part of the base model. This applies accordingly when an analysis contains more than two predictors.

(3) *The interaction between the two criterion variables*. The same rationale applies as for the interaction between the two predictors.

If types and antitypes emerge for this base model, they can only be caused by predictor-criteria relationships, but not by any main effect, interaction among predictors, or interaction among criteria. The reason for this conclusion is that none of the possible interactions between predictors and criteria are considered in the base model, and these interactions are the only terms not considered. Based on the effects proposed in this base model, the design matrix contains three kinds of vectors. The first is the vector for the intercept, that is, the constant vector. The second kind includes the vectors for the main effects of all variables. The third kind of vector includes the interaction between the two predictors and the interaction between the two criterion variables. Thus, the design matrix for this 2 x 2 x 2 x 2 table is

$$X = \begin{bmatrix} 1 & 1 & 1 & 1 & 1 & 1 & 1 \\ 1 & 1 & 1 & 1 & -1 & 1 & -1 \\ 1 & 1 & 1 & -1 & 1 & 1 & -1 \\ 1 & 1 & 1 & -1 & -1 & 1 & 1 \\ 1 & 1 & -1 & 1 & 1 & -1 & 1 \\ 1 & 1 & -1 & 1 & -1 & -1 & -1 \\ 1 & 1 & -1 & -1 & 1 & -1 & -1 \\ 1 & 1 & -1 & -1 & -1 & -1 & 1 \\ 1 & -1 & 1 & 1 & 1 & -1 & 1 \\ 1 & -1 & 1 & 1 & -1 & -1 & -1 \\ 1 & -1 & 1 & -1 & 1 & -1 & -1 \\ 1 & -1 & 1 & -1 & -1 & -1 & 1 \\ 1 & -1 & -1 & 1 & 1 & 1 & 1 \\ 1 & -1 & -1 & 1 & -1 & 1 & -1 \\ 1 & -1 & -1 & -1 & 1 & 1 & -1 \\ 1 & -1 & -1 & -1 & -1 & 1 & 1 \end{bmatrix}.$$

This design matrix displays the constant vector in its first column. The vectors for the four main effects follow. The last two column vectors represent the interactions between the two predictors and the two criteria. The first interaction vector results from element-wise multiplication of the second with the third column in X. The second interaction vector results from element-wise multiplication of the fourth with the fifth column vector in X.

The base model for a CFA of two variables with more than two categories. In this third example, we create the design matrix for the base model of a CFA for two variables. The model will only take main effects into account, so that types and antitypes can emerge only from (local) associations between these two variables. The goal pursued with this example is to illustrate CFA for a variable A which has three and variable B which has four categories. The design matrix for the log-linear main effect model for this cross-classification is

$$X = \begin{bmatrix} 1 & 1 & 0 & 1 & 0 & 0 \\ 1 & 1 & 0 & 0 & 1 & 0 \\ 1 & 1 & 0 & 0 & 0 & 1 \\ 1 & 1 & 0 & -1 & -1 & -1 \\ 1 & 0 & 1 & 1 & 0 & 0 \\ 1 & 0 & 1 & 0 & 1 & 0 \\ 1 & 0 & 1 & 0 & 0 & 1 \\ 1 & 0 & 1 & -1 & -1 & -1 \\ 1 & -1 & -1 & 1 & 0 & 0 \\ 1 & -1 & -1 & 0 & 1 & 0 \\ 1 & -1 & -1 & 0 & 0 & 1 \\ 1 & -1 & -1 & -1 & -1 & -1 \end{bmatrix}.$$

The first vector in this design matrix is the constant column, for the intercept. The second and third vectors represent the main effects of variable A. The first of these vectors contrasts the first category of variable A with the third category. The second of these vectors contrasts the second category of variable A with the third category. The last three column vectors of X represent the main effects of variable B. The three vectors contrast the first, second, and third categories of variable B with the fourth category.

Notation. In the following sections, we use the explicit form of the design matrices only occasionally, to illustrate the meaning of a base model. In most other instances, we use a more convenient form to express the same model. This form is $\log E = X\lambda$. Because each column of X is linked to one λ, the model can uniquely be represented by only referring to its parameters. The form of this representation is

$$\log E = \lambda_0 + \sum_{\text{main effects}} \lambda_i + \sum_{\text{first order interactions}} \lambda_{ij} + \sum_{\text{second order interactions}} \lambda_{ijk} + \ldots,$$

where λ_0 is the intercept and subscripts i, j, and k index variables. For a completely written-out example, consider the four variables A, B, C, and D. The saturated model, that is, the model that contains all possible effects for these four variables is

$$\log E = \lambda_0 + \lambda_i^A + \lambda_j^B + \lambda_k^C + \lambda_l^D + \lambda_{ij}^{AB} + \lambda_{ik}^{AC} + \lambda_{il}^{AD} + \lambda_{jk}^{BC} + \lambda_{jl}^{BD} + \lambda_{kl}^{CD} + \lambda_{ijk}^{ABC} + \lambda_{ijl}^{ABD} + \lambda_{ikl}^{ACD} + \lambda_{jkl}^{BCD} + \lambda_{ijkl}^{ABCD},$$

where the subscripts index the parameters estimated for each effect, and the superscripts indicate the variables involved. For CFA base models, the parameters not estimated are set equal to zero, that is, are not included in the model. This implies that the respective columns are not included in the design matrix.

To illustrate, we now reformulate the three above examples, for which we provided the design matrices, in terms of this notation. The first model included three variables for which the base model was a main effect model. This model includes only the intercept parameter and the parameters for the main effects of the three variables. Labeling the three variables A, B, and C, this model can be formulated as

$$\log E = \lambda_0 + \lambda_i^A + \lambda_j^B + \lambda_k^C.$$

The second model involved the four variables A, B, C, and D, and the interactions between A and B and between C and D. This model can be formulated as

$$\log E = \lambda_0 + \lambda_i^A + \lambda_j^B + \lambda_k^C + \lambda_l^D + \lambda_{ij}^{AB} + \lambda_{kl}^{CD}.$$

The third model involved the two variables A and B. The base model for these two variables was

$$\log E = \lambda_0 + \lambda_i^A + \lambda_j^B .$$

This last expression shows that the λ-terms have the same form for dichotomous and polytomous variables.

2.2 Admissibility of log-linear models as CFA base models

The issue of admissibility of log-linear models as CFA base models is covered in two sections. In the present section, admissibility is treated from the perspective of interpretability. In the next section, we introduce the implications from employing particular sampling schemes.

With the exception of saturated models which cannot yield types or antitypes by definition, every log-linear model can be considered as a CFA base model. However, the interpretation of types and antitypes is straightforward in particular when certain admissibility criteria are fulfilled. The following four criteria have been put forth (von Eye & Schuster, 1998):

(1) *Uniqueness of interpretation of types and antitypes*. This criterion requires that there be only one reason for discrepancies between observed and expected cell frequencies. Examples of such reasons include the existence of effects beyond the main effects, the existence of predictor-criterion relationships, and the existence of effects on the criterion side.

Consider, for instance, a cross-classification that is spanned by the three variables A, B, and C. For this table, a number of log-linear models can serve as base models. Three of these are discussed here. The first of these models is the so-called null model. This is the model that takes into account no effect at all (the constant is usually not considered an effect). This model has the form $\log E = \mathbf{1}\lambda$, where **1** is a vector of ones, and λ contains only the intercept parameter. If this base model yields types and antitypes, there must be non-negligible effects that allow one to describe the data. Without further analysis, the nature of these effects remains unknown. However, the CFA types and antitypes indicate where "the action is," that is, where these effects manifest. This interpretation is unique in the sense that all variables have the same status and effects can be of any nature, be they main effects or interactions. No variable has a status such that effects are a priori excluded. Types from this model are always constituted by the configurations with the largest frequencies, and antitypes are always constituted by the configurations with the smallest frequencies. This is the reason why this base model of CFA has also been called the base model of *Configural Cluster Analysis* (Krüger, Lienert, Gebert, & von Eye, 1979; Lienert & von Eye, 1985; see Section 5.1).

The second admissible model for the three variables A, B, and C is the main effect model $\log E = \lambda_0 + \lambda_i^A + \lambda_j^B + \lambda_k^C$. This model also assigns all variables the same status. However, in contrast to CCA, types and antitypes can emerge here only if variables interact. No particular interaction is excluded, and interactions can be of any order. Main effects are part of the base model and cannot, therefore, be the reason for the emergence of types or antitypes.

Consider the following example of Configural Cluster Analysis (CCA) and Configural Frequency Analysis (CFA). In its first issue of the year 2000, the magazine *Popular Photography* published the 70 winners and honorable mentions of an international photography contest (Schneider, 2000). The information provided in this article about the photographs can be analyzed using the variables *Type of Camera* (C; 1 = medium format; 2 = Canon; 3 = Nikon; 4 = other), *Type of Film* used (F; 1 = positive film (slides); 2 = other (negative film, black and white, sheet film, etc.)), and *Price Level* (P; 1 = Grand or First Prize; 2 = Second Prize; 3 = Third Prize; 4 = honorable mention). We now analyze the 4 x 2 x 4 cross-tabulation of C, F, and P using the null model of CCA and the model of variable independence, that is, the main effect base model of CFA. Table 3 displays the cell indices and the observed cell frequencies along with the results from these two base models. For both analyses we used an approximation of the standard normal z-test (this test will be explained in detail in Section 3.3), and we Bonferroni-adjusted $\alpha = 0.05$ which led to $\alpha^* = 0.05/32 = 0.0015625$.

The results in the fourth column of Table 3 suggest that three configural clusters and no configural anticlusters exist. The first cluster, constituted by configuration 224 suggests that more pictures that were taken with Canon cameras on negative film were awarded honorable mentions than expected based on the null model. The second cluster, constituted by Configuration 314, suggests that more pictures that were taken with Nikon

Table 3: CFA of contestwinning pictures based on null model and independence model

Cell Indices	Observed frequencies	Null model		Independence model	
CFP		$\hat{e}_{ijk}$	p_{ijk}	$\hat{e}_{ijk}$	p_{ijk}
111	1	2.188	.2110	.456	.21
112	1	2.188	.2110	.414	.18
113	1	2.188	.2110	.414	.18
114	0	2.188	.0670	1.616	.10
121	2	2.188	.4500	.644	.05
122	0	2.188	.0670	.586	.22
123	0	2.188	.0670	.586	.22
124	2	2.188	.4500	2.284	.43
211	0	2.188	.0670	1.367	.42
212	0	2.188	.0670	1.243	.13
213	2	2.188	.4500	1.243	.25
214	4	2.188	.1102	4.847	.35
221	3	2.188	.2914	1.933	.22
222	2	2.188	.4500	1.757	.43
223	2	2.188	.4500	1.757	.43
224	8	2.188	$<\alpha^{*a}$ **T**	6.853	.33
311	1	2.188	.2110	1.758	.28
312	2	2.188	.4500	1.589	.38

/cont.

Cell Indices	Observed frequencies	Null model		Independence model	
CFP		$\hat{e}_{ijk}$	p_{ijk}	$\hat{e}_{ijk}$	p_{ijk}
313	2	2.188	.4500	1.598	.38
314	7	2.188	.0006**T**	6.232	.38
321	2	2.188	.4500	2.485	.38
322	1	2.188	.2110	2.259	.20
323	1	2.188	.2110	2.259	.20
324	11	2.188	$<\alpha^{*a}$ **T**	8.811	.23
411	0	2.188	.0670	0.977	.16
412	2	2.188	.4500	0.888	.12
413	2	2.188	.4500	0.888	.12
414	4	2.188	.1102	3.462	.39
421	2	2.188	.4500	1.381	.30
422	2	2.188	.4500	1.255	.25
423	0	2.188	.0670	1.255	.13
424	3	2.188	.2914	4.895	.20

[a] $< \alpha^*$ indicates that the tail probability is smaller than can be expressed with four decimal places.

cameras on slide film won honorable mentions than expected from the null model. The third cluster, constituted by Configuration 324, indicates that more picture that were taken with Nikon cameras on negative film won honorable mentions than expected from the null model. None of the other configurations appeared more often or less often than expected from the null model. Notice that the small expected frequencies prevented antitypes from emerging (Indurkhya & von Eye, 2000).

While these results are interesting in themselves, they do not indicate whether the three types resulted from main effects (e.g., the different frequencies with which camera types or film types had been used) or interactions among the three variables, C, F, and P. To determine whether main effects or interactions caused the three types, we also performed a CFA using the main effect model of variable independence as the base model. The overall goodness-of-fit Pearson $X^2 = 21.27$ ($df = 24$; $p = 0.62$) suggests that the main effect model describes the data well. Accordingly, no types or antitypes appeared. We thus conclude that the three types were caused by main effects. After taking into account the main effects in the base model, the types disappeared. We therefore conclude that there exists no association between type of camera used, type of film, and type of prize awarded that could result in types or antitypes.

A third base model that may be of interest when analyzing the three variables A, B, and C is that of Prediction CFA (P-CFA). Suppose that A and B are predictors and C is the criterion. The P-CFA base model for this design is saturated in the predictors and proposes independence between A and B on the one side and C on the other side. Specifically, the base model is $\log E = \lambda_o + \lambda_i^A + \lambda_j^B + \lambda_k^C + \lambda_{ij}^{AB}$. This model assigns variables to the two groups of predictors and criteria. Thus, variable status is no longer the same for all variables. Nevertheless, this model has a unique interpretation. Only one group of variable relationships is left out of consideration in the base model. These are the predictor-criterion relationships. Therefore, the model is admissible as a CFA base model.

(2) *Parsimony*. Parsimony is a generally valid criterion for quality scientific research. This criterion requires that a base model be as simple as possible; that is, a base model must include as few terms as possible and as simple terms as possible. The topic of parsimony will be taken up in more detail in Section 10.2.

(3) *Consideration of sampling scheme*. This criterion requires that the sampling schemes of all variables be considered (see Section 2.3).

2.3 Sampling schemes and admissibility of CFA base models

Data can be collected under a large number of sampling schemes. The best known and most frequently employed are the multinomial and the product multinomial sampling schemes (Christensen, 1997; Jobson, 1992; von Eye & Schuster, 1998). These two schemes are discussed here. Before discussing the consequences of selecting a particular sampling scheme for the selection of CFA base models, it should be made explicit that employing either sampling scheme considered here does not impact the usefulness of log-linear base models for analyzing the data. In addition, parameter estimates will stay the same and so will overall goodness-of-fit of base models and log-linear models in general. However, the selection of possible base models may be constrained by the use of a particular sampling scheme.

2.3.1 Multinomial sampling

Multinomial sampling is performed when a random sample of individuals is classified according to categorical variables. When there is only one categorical variable, such as gender or kind of disease, the sampling is *multinomial*. When the classification categories result from crossing two or more variables, the sampling is *cross-classified multinomial*. To shorten and simplify presentation, we consider in the following sections only cross-classified multinomial sampling, because CFA is virtually always used to analyze cross-classifications of two or more variables. Cross-classified multinomial sampling allows for random assignment of individuals to any cell of the entire cross-classification. Suppose a two-dimensional table is created with R rows and C columns, and $i = 1, ..., R$ and $j = 1, ..., C$. Then the joint density of the sample cell frequencies is

$$f(N_{11}, N_{12}, ..., N_{RC}) = \frac{N!}{\prod_{i=1}^{R}\prod_{j=1}^{C} N_{ij}!} \prod_{i=1}^{R}\prod_{j=1}^{C} \pi_{ij}^{N_{ij}} ,$$

where π_{ij} indicates the probability for Cell ij, $\sum_{i=1}^{R}\sum_{j=1}^{C} \pi_{ij} = 1$, and $\sum_{i=1}^{R}\sum_{j=1}^{C} N_{ij} = N$. The expectancies of the N_{ij} are $E[N_{ij}] = N\pi_{ij}$. The variances of the N_{ij} are $V[N_{ij}] = N\pi_{ij}(1 - \pi_{ij})$ for $i = 1, ..., R$ and $j = 1, ..., C$. The covariances are $Cov[N_{ij}, N_{kl}] = -N\pi_{ij}\,\pi_{kl}$, for $i \neq k; j \neq l$; $i, k = 1, ..., R$; and $j, l = 1, ..., C$. Because the assignment of cases is to the cells in the entire table, there is no constraint on the expected frequencies other than $\sum_i \sum_j N_{ij} = N.$

2.3.2 Product-multinomial sampling

The *product-multinomial* distribution describes the joint distribution of two or more independent multinomial distributions. Consider an R x C cross-classification with *fixed row marginals* $N_{i.}$ for $i = 1, ..., R$. Row marginals are fixed when the number of cases in the rows is determined a priori. This can be the case by design, or when individuals in each row are members of subpopulations, for instance, females and males, or smokers and non-smokers. The joint density of the R rows results from multiplying the row-specific multinomials. In an R x C table this product is

$$f(N_{11}, N_{12}, ..., N_{RC}) = \prod_{i=1}^{R} \left[\frac{N_{i.}!}{\prod_{j=1}^{C} N_{ij}!} \prod_{j=1}^{C} \left[\frac{\pi_{ij}}{\pi_{i.}} \right]^{N_{ij}} \right].$$

This equation indicates that the probability of observing the contingency table with cell frequencies $N_{11}, N_{12}, ..., N_{RC}$ is given as the product of probabilities of observing each of the R independent vectors of row probabilities $(N_{11}, ..., N_{1C}), ..., (N_{R1}, ..., N_{RC})$. This applies accordingly if column marginals are fixed, or if the marginals are fixed for more than one variable (*cross-classified product-multinomial*).

While the estimation of parameters is the same for these two sampling schemes, kind and number of models that can be considered, differ. Consider the following example (von Eye & Schuster, 1998): Researchers design a study on the effects of drinking in which they include two independent classification variables, Drinking (D; yes - no) and Gender (G; female - male), and one dependent variable, Liver Cancer (C; shows signs of liver cancer - does not show signs of liver cancer). Together, these three variables form a 2 x 2 x 2 cross-classification. Drinking and Gender are the independent variables, and Liver Cancer is the dependent variable. Now, the researchers decide to fix the margins of the two independent variables. Specifically, they fix variable Drinking determining the number of alcohol consumers and resisters to be included in the sample a priori. The number of male and female respondents was also determined a priori. In addition, the numbers of alcohol consumers and resisters were fixed per gender. Therefore, any model of these three variables must include a provision to reproduce the bivariate Gender - Drinking marginals, $m_{ij.}$. All models that include the (hierarchical) term D x G, which we express in this context as λ^{DG}, fulfill this condition. These are the five models that include the terms λ^{DCG}; λ^{DG}, λ^{DC}, λ^{GC}; λ^{DG}, λ^{DC}; λ^{DG}, λ^{GC}; and λ^{DG}, λ^{C}. All models without the D x G term are not admissible. The inadmissible models include, for instance, the main effect model with the terms λ^{D}, λ^{G}, λ^{C}, and the model λ^{DC}, λ^{GC}. We illustrate the implications of sampling schemes in the context of CFA in the following section for standard, main effect CFA. The implications for prediction CFA and discriminant CFA are discussed in Section 6.2 and Chapter 7 (k-sample CFA).

2.3.3 Sampling schemes and their implications for CFA

The most routinely applied CFA base model is that of variable independence. Indeed, for many years, this model was the only one considered for CFA. It is one of the thus far rarely discussed conditions for proper application of the main effect base model that the sampling scheme NOT be cross-classified product-multinomial. The reason for this constraint is that the cross-classified product-multinomial sampling creates two-, three-, or higher-dimensional margins that must be reproduced by the base model. These margins are not automatically reproduced by the main effect model.

To illustrate, consider the case where researchers study 50 female and 50 male smokers, and their responses to physical exercise. Each of the subsamples is subdivided in groups of 25 based on the rigor of their exercise. The design for this study can be depicted as in Table 4.

This table displays four cells with 25 respondents each. If Gender and Exercise regimen are crossed with one or more response variables, these cells turn into the bivariate marginals of a larger design. If data from this design are analyzed using the main effect model base model, the expected cell frequencies may not sum up to 25 for the four bivariate marginals any more. For instance, the base model may predict that only 17 respondents are female and participate in the rigorous exercise program. This would be incorrect, and spurious types and antitypes could emerge just because of this error. In the following paragraphs, we give a real data example, and illustrate the effects of a wrong choice of base model.

In the 1999 indictment, the U.S. senate voted on whether President Clinton was guilty of perjury and of obstruction of justice. A total of 100 senators voted, 55 of whom were Republicans and 45 of whom were Democrats. In addition, 62 senators had been senators for two or more terms, and 38 senators were freshmen. On both accusations, the voting was either

Table 4: Design for smoking and exercise study

		Gender	
		female	male
Exercise	rigorous	25	25
	less rigorous	25	25

guilty or *not guilty*. Together, these four variables form the 2 x 2 x 2 x 2 cross-classification of the variables Party Membership (M; 1 = Democrat, 2 = Republican), Number of Terms (T; 1 = two or more terms, 2 = freshman), judgment on Perjury (P; 1 = not guilty, 2 = guilty), and judgment on Obstruction of Justice (O; 1 = not guilty, 2 = guilty). Table 5 displays this table, along with results from standard CFA using the main effect base model of variable independence. For the CFA we employed the standard normal *z*-test and we Bonferroni-protected the nominal significance level which led to $\alpha^* = 0.003125$.

The results in Table 5 indicate the existence of four types and two antitypes. The first type, constituted by Configuration 1111, suggests that more seasoned Democrat senators than expected from the base model voted not guilty on both accounts. The second type, constituted by Configuration 1211, suggests that more freshman Democrat senators than expected from the base model voted not guilty on both accounts. The third type, constituted by Configuration 2122, indicates that more seasoned Republican senators than expected from chance voted guilty on both accounts, and the fourth type, constituted by Configuration 2222 suggests that more freshman Republicans than expected voted guilty on both accounts.

The two antitypes can be interpreted as follows. The first antitype, constituted by Configuration 1112, suggests that fewer seasoned Democrats than expected voted Clinton not guilty on the Perjury account but guilty on the Obstruction of Justice account. The second antitype, constituted by Configuration 2121, indicates that fewer seasoned Republicans than expected voted Clinton guilty on the Perjury account but not guilty on the obstruction of justice account.

Table 5: CFA of the variables Party Membership (M), Number of Terms (T), Judgment on Perjury (P), and Judgment on Obstruction of Justice (O) (main effect model)

Cell indices	Frequencies		Statistics		Type/ Antitype ?
MTPO	observed	expected	z	$p(z)$	
1111	32	7.67	8.78	< α*	**T**
1112	0	7.67	-2.77	.0028	**A**
1121	0	6.28	-2.51	.0061	
1122	0	6.28	-2.51	.0061	
1211	13	4.70	3.83	< α*	**T**
1212	0	4.70	-2.17	.0151	
1221	0	3.85	-1.96	.0249	
1222	0	3.85	-1.96	.0249	
2111	3	9.38	-2.08	.0186	
2112	4	9.38	-1.76	.0395	
2121	0	7.67	-2.77	.0028	**A**
2122	23	7.67	5.53	< α*	**T**
2211	2	5.75	-1.56	.0590	
2212	1	5.75	-1.98	.0238	
2221	0	4.70	-2.17	.0151	
2222	22	4.70	7.98	< α*	**T**

[a] < α* indicates that the tail probability is smaller than can be expressed with four decimal places.

These results seem to describe the voting according to party lines nicely. They do not describe the jumping-of-party-lines of 10 Republicans (see Configurations 2111, 2112, 2211, and 2212) for which no antitypes could be established for lack of statistical power (the topic of differential statistical power for types and antitypes will be taken up again in Section 3.9). However, **these results may be based on a wrong choice of base model**, and may therefore be invalid. More specifically, the M x T x P x O cross-classification contains two cross-classified variables that can be considered sampled according to a bivariate product-multinomial sampling scheme. These are the variables Party Membership (M) and Number of Terms (T). The M x T bivariate marginals of this design must then be reproduced. The main effect-only base model that was used for Table 5 is unable to achieve this. Collapsing over the response variables P and O, we create the 2 x 2 cross-classification of the two product-multinomial variables, M and T. Table 6 displays the bivariate frequencies in this cross-classification in regular type face. The frequencies according to the base model in Table 5 are included in *italics*.

Table 6: Bivariate marginals of the variables P and O from Table 5

		T		Totals
		1	2	
M	1	32 *27.898*	13 *17.102*	45 *45*
	2	30 *34.102*	25 *20.808*	55 *55*
Totals		62 *62*	38 *38*	100

Table 6 shows clearly two of the consequences of mis-specification of a base model: the bi- or multivariate marginals can be mis-estimated. For instance, according to the base model there were 17.1 neophyte Democrats in the senate in 1999. However in reality, there were 13. The second consequence of mis-specification of base models is that types and antitypes can emerge just because of this mis-specification. Such types and antitypes reflect the specification error rather than data characteristics. To determine whether the pattern of types and antitypes in Table 5 changes when the base model is correctly specified, we re-calculated the CFA under a different base model.

For the results in Table 5, the base model was $\log E = \lambda_0 + \lambda_i^M + \lambda_j^T + \lambda_k^P + \lambda_l^O$, that is, the main effects model. We now re-calculate this analysis under the base model
$\log E = \lambda_0 + \lambda_i^M + \lambda_j^T + \lambda_k^P + \lambda_l^O + \lambda_{ij}^{MT}$. This model considers the interaction between Number of Terms and Party Membership. The results for this analysis appear in Table 7. To create results that are comparable to the ones presented in Table 4, the test-wise α was protected using the Bonferroni method and the z-test was employed.

Table 7 suggests that the consideration of the bivariate product-multinomial nature of the variables Party Membership and Number of Terms changes the harvest of antitypes. Configuration 2121 no longer constitutes an antitype. Thus, the knowledge about the number of senators in their first terms and in their second or later terms in both parties allows one to expect a smaller number of Republican neophyte votes of guilty perjury and not guilty of obstruction of justice than based on the main effect model. As a result, the observed zero is not significantly different than the expected 6.75[3]. In addition, none of the expected cell frequencies is the same under both models.

While losing one antitype may not be considered a major change by all researchers (more dramatic base model-related changes in type/antitype patterns are presented in Section 6.2, on Prediction CFA), one important result of this comparison is that the expected cell frequencies in Table 7 now add up to the correct uni- and bivariate marginal frequencies. For instance, summing the first four expected frequencies in Table 7 yields $N_{11..} = 32$. This is exactly the required value (see Table 5). Readers are invited to confirm that the remaining three expected cell frequencies reproduce the bivariate marginals of the M x T subtable exactly.

This example suggests that mis-specification of the base model can result in (a) patterns of types and antitypes that reflect the discrepancies

---

[3]Note that, from a log-linear modeling perspective, the added M x T interaction failed to improve the model fit significantly. The likelihood ratio goodness-of-fit for the main effect model is $LR\text{-}X^2 = 214.85$ ($df = 11$; $p < 0.01$). The likelihood ratio goodness-of-fit for the model in Table 7 is $LR\text{-}X^2 = 211.93$ ($df = 10$; $p < 0.01$). The difference between these two nested models is non-significant ($\Delta X^2 = 2.92$; $\Delta df = 1$; $p = 0.088$).

Table 7: CFA of the variables Party Membership (M), Number of Terms (T), Judgment on Perjury (P), and Judgment on Obstruction of Justice (O)

Cell indices	Frequencies		Statistics		Type/ Antitype ?
MTPO	observed	expected	z	$p(z)$	
1111	32	8.80	7.82	< α*	**T**
1112	0	8.80	-2.97	.0015	**A**
1121	0	7.20	-2.68	.0036	
1122	0	7.20	-2.68	.0036	
1211	13	3.58	4.96	< α*	**T**
1212	0	3.58	-1.89	.0293	
1221	0	2.93	-1.71	.0436	
1222	0	2.93	-1.71	.0436	
2111	3	8.25	-1.83	.0338	
2112	4	8.25	-1.48	.0694	
2121	0	6.75	-2.60	.0047	
2122	23	6.75	6.26	< α*	**T**
2211	2	6.88	-1.86	.0315	
2212	1	6.88	-2.24	.0125	
2221	0	5.63	-2.37	.0088	
2222	22	5.63	6.90	< α*	**T**

[a] < α* indicates that the tail probability is smaller than can be expressed with four decimal places.

from the design and sampling characteristics that should have been considered in the base model, and (b) mis-estimation of uni-, bi- or multivariate marginal frequencies. We thus conclude that

(1) when variables are observed under a product-multinomial sampling scheme, their marginals must be exactly reproduced. The CFA base model must therefore include the main effects of these variables.

(2) When variables are observed under a cross-classified product-multinomial sampling scheme, their bivariate or multivariate marginals must also be exactly reproduced. The CFA base model must therefore include the main effects and the interactions of these variables. Specifically, the CFA base model must be saturated in the variables that are cross-classified product-multinomial.

2.4 A grouping of CFA base models

This section presents a heuristic scheme that allows researchers to group CFA base models. The heuristic uses the two classification criteria *order of relationships* and *status of variables*. The order of relationships is defined by the order of terms included in a base model. For d variables, the order varies from 0 to d - 1. When the order is 0, no effects are taken into account at all (see Table 3). If first-order effects are taken into account, the base model considers main effects. If second-order effects are taken into account, the base model considers pair-wise interactions, and so on. CFA base models typically are hierarchical. When higher order effects are taken into account, all lower order effects of the variables included in the higher order effect terms are implied.

If the status of all variables is the same, we can use the order of effects to create a first group of CFA models, the group of *global CFA base models*. These base models (a) consider the same effects for each variable. Exceptions are made only if variables are cross-classified product-multinomial (see Tables 4 through 6); and (b) are classified according to the order of relationships taken into account when estimating the expected cell frequencies.

The following three global CFA base models have been explicitly discussed in the literature and are covered in this volume:

(1) *Zero Order CFA*, also called *Configural Cluster Analysis* (see

Table 3, above; Lienert & von Eye, 1984; Lienert & von Eye, 1985). This base model considers no effect at all (see Section 5.1);

(2) *First Order CFA*, that is, the standard main effect CFA (see Table 2; Krauth & Lienert, 1973a; Lienert, 1969; von Eye, 1988). This base model takes into account all main effects of all variables (see Section 5.2);

(3) *Second Order CFA* (von Eye & Lienert, 1984). This base model takes into account all main effects and all two-way interactions of all pairs of variables (an example of second order CFA is given in Section 5.3). CFA base models higher than second order have, to the best of our knowledge, not been employed in the analysis of empirical data.

The second group of base models, called *regional CFA base models*, assigns variables to different groups. These groups are exhaustive and non-overlapping. Four regional models have been discussed in the literature:

(1) *Interaction Structure Analysis* (ISA; Krauth & Lienert, 1973a; Krauth & Lienert, 1974). Consider a study in which researchers investigate the relationships between two groups of variables, and in which the two groups of variables have the same status. In other words, the groups are not considered predictor or criterion variables, or independent or dependent variables. Then, ISA links configurations in one group of variables to configurations in the second group of variables. The base model for ISA is (a) saturated in the first group of variables, (b) saturated in the second group of variables, and (c) proposes independence across the groups. That is, the ISA base model proposes that there exists no interaction among variables from the two groups. Types and antitypes can thus emerge only if such interactions exist.

To illustrate the concept of ISA, consider the four variables, A, B, C, and D. When two groups are created, two or more of four variables can be used. Table 8 displays the possible groups for each of these situations. If d variables are given, the total number of groupings, g, that can be considered for ISA is

$$g = \frac{3^d + 1}{2} - 2^d.$$

Table 8: ISA groupings for the four variables A, B, C, and D

Number of variables in grouping	Number of groupings	Groupings[a]
2	$\binom{4}{2} = 6$	A.B, A.C, A.D, B.C, B.D, C.D
3	$\binom{4}{3}\binom{3}{2} = 12$	A.BC, A.BD, A.CD, B.AC, B.AD, B.CD, C.AB, C.AD, C.BD, D.AB, D.AC, D.BC
4	$\binom{4}{1} + \frac{\binom{4}{2}}{2} = 7$	A.BCD, B.ACD, C.ABD, D.ABC, AB.CD, AC.BD, AD.BC

[a]Variables to the left of the period belong to the first group, variables to the right of the period belong to the second group. The order of the groups is of no importance.

(2) *Generalized ISA*. The original concept of ISA involved relating configurations from two groups of variables to each other. Types and antitypes can be interpreted without post hoc tests or hierarchical analyses. In generalized ISA (Lienert & von Eye, 1988), the number of groups is three or more. One problem that is hard to overcome in generalized ISA is that the number of significance tests can become very large. If the test-wise α is protected using a method such as Bonferroni's, the significance threshold can become prohibitively extreme. (For a strategy to reduce the number of tests, see von Eye, 1986.)

(3) *Prediction CFA* (P-CFA; Heilmann & Lienert, 1982; Heilmann, Lienert, & Maly, 1979; Lienert & Krauth, 1973a) has, at first look, the same structure as ISA. Configurations from two groups of variables are related to each other. However, there are two differences between these two models. The first difference is at the

interpretational level: one group of variables is interpreted as predictors, the other is interpreted as criterion variables. The second difference is that the predictors must be considered cross-classified product-multinomial, and in some instances criterion variables are cross–classified multinomial too. Based on this second difference, examples can be found in which there is a formal difference between ISA and P-CFA (von Eye, 1985; von Eye & Schuster, 1998). The base model for P-CFA is always saturated in the predictors. The order of interactions taken into account at the criterion variable side depends on sampling scheme and substantive hypotheses. The sampling scheme may constrain the selection of models that can meaningfully be estimated.

(4) *k-sample CFA* (Lienert, 1971, 1987) is a regional CFA model that also distinguishes between two groups of variables. In contrast to ISA and P-CFA, where the assignment of variables to variable groups is relatively flexible, k-sample CFA requires that one variable group is constituted by one or more grouping variables. These are variables that indicate the group a respondent belongs to.

2.5 The four steps of selecting a CFA base model

This section describes the four steps researchers go through when selecting a CFA base model. Taking these four steps will almost always lead to a correct base model. It should be noted that there is no one-to-one match between base models and substantive concepts. One reason for this is that CFA is typically employed in an exploratory context. Therefore, researchers often try out several base models as part of their exploratory efforts. A second reason is that in most applications, theory is flexible in the sense that more than one base model can be defended. However, some base models are incorrect and can lead to wrong interpretations of data. Therefore, taking the four steps is highly recommended.

(1) *Step 1: Consideration of the sampling scheme*. As was explained in Sections 2.3.1 - 2.3.3, multinomial sampling gives researchers the most liberty in the selection of CFA base models. If sampling is product-multinomial in single variables, zero-order CFA can no longer be considered, because the main effects of those variables need to be part of the base model that were sampled according to

a product-multinomial scheme. If sampling is cross-classified product-multinomial for a number of variables, the base model must be saturated in these variables. As a result, first- and higher order CFA can be out of consideration, and so may be base models that assume only main effects on the dependent variable side. In general, consideration of the sampling scheme is required for the selection of admissible CFA base models.

(2) *Step 2: Consideration of design*. The design determines whether variables can be grouped, whether variables are considered dependent versus independent, or have all the same status as response variables. If all variables have the same status, the sampling scheme and theory determine the possible base models which will all be global models. If variables are grouped, the grouping along with the sampling scheme and theory determine which base models are sensible. In CFA base models, each variable appears only once. Variable groups do not overlap.

(3) *Step 3: Making sure types and antitypes can be interpreted.* As was explained in Section 2.2, *uniqueness of interpretation of types and antitypes* is the foremost substantive admissibility criterion for CFA base models. For example, first order CFA types and antitypes can be interpreted as resulting from (local) associations rather than main effects. Second order CFA types and antitypes can be interpreted as resulting from (local) second or higher order associations. Accordingly, prediction types and antitypes reflect predictors-criteria relationships. If a first order CFA includes the interaction between two cross-classified product-multinomial variables, it still can be considered a first order CFA. The particular interaction must be included in the model, however, in order to prevent types and antitypes from emerging because of this interaction which reflects the sampling scheme rather than empirical relationships among variables.

(4) *Step 4: Reconsideration*. The first three steps usually lead to a defensible base model and possibly to types and antitypes. Because CFA is applied typically in exploratory contexts, we highly recommend reconsidering the base model after a first run. For instance, we recommend considering a second order CFA after a first order CFA, or we recommend considering a first order CFA

after a two-sample CFA (sampling scheme permitting). In each case, the new model will provide new insights. If a second order CFA yields the same types and antitypes as first order CFA, there must be second- or higher order associations among the variables that manifest in types and antitypes. Had the types and antitypes been the result of first order associations, second order CFA would have yielded neither types nor antitypes. This applies accordingly if only some types or antitypes disappear in second order CFA. In a similar fashion, the base model for two-sample CFA is typically saturated in the predictors. If types and antitypes emerge in a subsequent first order CFA that had not surfaced in two-sample CFA, they must be the results of interactions among the predictors which are taken into account in standard two-sample CFA (see Section 7). As a matter of course, all this can also be found out using log-linear models. However, CFA results are stated in terms of types and antitypes whereas log-linear results are stated in terms of main effects and interactions, and overall model goodness-of-fit. In other words, CFA is person-oriented, whereas log-linear modeling is variable-oriented. Therefore, the context of CFA makes it natural to perform the steps of data exploration also using CFA.

The next chapter introduces readers to the statistical testing procedures used to identify types and antitypes. These procedures consist of two parts. The first involves the statistical tests themselves. The second part involves protection of the test-wise α.

3. Statistical Testing in Global CFA

This chapter introduces readers to the statistical tests that can be used to come to a decision as to whether a configuration constitutes a type, an antitype, or remains inconspicuous. Interestingly, the selection of tests in CFA has more facets than usually involved in the selection of statistical tests. One facet is, as one might expect, the statistical power of the available tests. A second facet concerns the conditions (sampling scheme) under which tests can be employed. A third facet, most important in the context of CFA and only rarely discussed thus far, is the kind of type and antitype researchers are interested in. We present tests that cover a number of the ways to deviate from independence or, in general, a CFA base model. The present chapter covers tests for global base models that can also be used in most regional base models. We begin with the binomial test and its approximations. This is followed by the χ^2 and its approximations, and a section on tests that can be used when the margins are fixed as is the case in product-multinomial sampling. These tests will then be compared using results from simulation studies and an empirical data example. The presentation of significance tests is completed with a discussion of statistical power. There are additional tests that can be used only in two-sample CFA. These tests will be introduced in Section 7.2.

3.1 The null hypothesis in CFA

The questions that can be asked with CFA are all answered at the level of configurations, that is, single cells or groups of cells. The CFA null hypothesis is therefore formulated at the level of single configurations or groups of configurations. The former is of concern in this section. Methods for the analysis of groups of configurations are presented in Section 10.3 and in Section 11.2, on Bayesian CFA.

As was indicated in Section 1.3 the CFA null hypothesis is

$$H_0: \quad E[N_i] = E_i ,$$

where $E[...]$ indicates the expectancy, N_i is the observed frequency for configuration i, and i indexes the configurations (DuMouchel, 1999; Gutiérrez-Peña & von Eye et al., 2000). E_i is the expected cell frequency, estimated under some base model. Types emerge if $E[N_i] > E_i$. Antitypes emerge if $E[N_i] < E_i$. If the null hypothesis is tenable, a configuration constitutes neither a type nor an antitype. Typically, E_i is estimated using some log-linear model, called the CFA base model. However, E_i can also express prior knowledge, weights, an a priori probability, a distributional assumption, or can be derived from earlier results. The section on repeated measures CFA gives examples of prior probabilities that are not log-linear. The following sections are concerned with methods of testing the above CFA null hypothesis.

3.2 The binomial test

The first existing test that was proposed for use in CFA (Krauth & Lienert, 1973a) is the *binomial test*. This test allows one to estimate the probability B_i of the observed frequency, N_i, of Configuration i, given the probability, p, expected for this configuration from some base model, and the sample size, N. The probability can be calculated as follows. For an admissible CFA base model, the expected probability for Configuration i is estimated as $p_i = E_i/N$, where i indexes the configurations. Then, the exact tail probability is

$$B_i(N_i) = \sum_{j=a}^{l} \binom{N}{j} p^j q^{N-j},$$

where $q = 1 - p$, and

$$a = \begin{cases} 0 & \text{if } N_i < E_i \\ N_i & \text{if } N_i \geq E_i \end{cases},$$

and

$$l = \begin{cases} N_i & \text{if } a = 0 \\ N & \text{if } a = N_i \end{cases} .$$

The binomial test possesses a number of characteristics that make it particularly interesting for use in CFA. First, the test is nonparametric. Thus, there is no need to make distributional assumptions and there is no need to test such assumptions. Because of the nonparametric nature of the binomial test, CFA is considered a nonparametric method by many. Second, the test is exact. As can be seen in the formula for B_i, the point probabilities for the observed frequencies and each of the more extreme frequencies are calculated and summed. Thus, there is no need to assume that a test statistic is sufficiently near to some sampling distribution. B_i is the desired tail probability itself. Third and as a matter of course, the result of the test depends on the base model, because the estimated expected cell frequency E_i is determined by the base model.

One interesting characteristic of the binomial test is that it is both exact and conservative. That is, it suggests type/antitype decisions that tend to have a slight bias in favor of the null hypothesis. The reason for this characteristic is that the binomial test is fully valid only if the probability p is known. This is the case, for instance, in the urn examples in statistics textbooks. In CFA, however, the probability p typically needs to be estimated. Usually, researchers use maximum likelihood methods the estimate the expected cell frequencies, E_i, and thus the corresponding probabilities p_i. These estimates tend to describe the empirical data better than the true probabilities. As a consequence, the discrepancies between the estimated expected cell frequencies and the observed data are smaller than when the true probabilities were used, and types and antitypes are less likely to be found. Still, in spite of its slightly conservative nature, the binomial test is one of the better tests used in the context of CFA.

Data example: Meehl's Paradox. In the following paragraphs we present an example of the use of the binomial test. The example involves an artificial data set with a data structure according to Meehl's paradox (Meehl, 1950). Consider a psychiatrist who uses two items to diagnose schizophrenia in a sample of $N = 80$ inpatients. Based on the diagnoses at admission, the patients' status as schizophrenic versus not schizophrenic was known to the clinic yet unknown to the psychiatrist. The cross-tabulation of the three variables, Item 1 (1), Item 2 (2), and Patient Status (P) appears in Table 9, along with the observed frequencies, the expected frequencies, and the binomial tail probabilities. The patients' responses to both items were scored as T = true and F = false, and Patient Status was S = schizophrenic and N = not schizophrenic. The significance level α was Bonferroni-adjusted which yielded $\alpha^* = 0.05/8 = 0.00625$. The base model used to estimate the expected cell frequencies is a first order model, that is, the model of variable independence. Only main effects are considered. Thus, the log-linear base model is $\log \hat{E} = \lambda_0 + \lambda_i^1 + \lambda_j^2 + \lambda_k^P$, where the superscripts 1, 2, and P indicate the variables that span the cross-classification, and the subscripts i, j, and k indicate the main effect parameters in the model. The design matrix for this model is identical to the one in Section 2.1.

The results in Table 9 indicate four types and four antitypes. This is the rare CFA result where each cell constitutes either a type or an antitype. The types are constituted by the configurations TTS, TFN, FTN, and FFS. This pattern of types is of importance for the psychiatric diagnosis because it suggests that

(1) schizophrenics respond consistently to both items, that is, either TT or FF; and
(2) non-schizophrenics respond inconsistently to both items, that is, either TF or FT.

This result is complemented by the four antitypes which are constituted by the configurations TTN, TFS, FTS, and FFN. These antitypes suggest that

(3) schizophrenics are unlikely to respond inconsistently to both items; and
(4) non-schizophrenics are unlikely to respond consistently to both items.

These results are interesting first because they allow the psychiatrist to discriminate between schizophrenics and non-schizophrenics perfectly. All cases are concentrated on the types, and the frequencies for the antitypes are zero throughout.

Table 9: CFA binomial test results in an artificial data set that displays Meehl's paradox

Cell indices	Frequencies			
12P	observed	expected	*B*	Type/ Antitype?
TTS	20	10	.00168	**T**
TTN	0	10	.00002	**A**
TFS	0	10	.00002	**A**
TFN	20	10	.00168	**T**
FTS	0	10	.00002	**A**
FTN	20	10	.00168	**T**
FFS	20	10	.00168	**T**
FFN	0	10	.00002	**A**

But why is this paradoxical in any sense? From the perspective of a CFA user, there is nothing paradoxical at all in these results. To the contrary, these results suggest that the two items allow the psychiatrist to discriminate between schizophrenics and non-schizophrenics perfectly. However, from the perspective of classical test theory, these results are a major surprise. Two criteria that are used to decide in classical test theory whether to keep or to remove items are the inter-item correlations and the correlations between each item and the criterion, that is, the diagnosis. In the present example, we need to calculate three correlations, the one between Item 1 and Item 2, and the correlations between Item 1 and Patient Status, and between Item 2 and Patient Status. To correlate the two items, we create the Item 1 x Item 2 cross-classification by collapsing the tabulation in Table 9 over the categories of Patient Status. We obtain the cross-classification in Table 10.

Table 10: Item 1 x Item 2 cross-classification

Configurations 12	Observed frequencies
TT	20
TF	20
FT	20
FF	20

The Φ-correlation in a four-fold table can be estimated using

$$\Phi = \frac{N_{11} \cdot N_{22} - N_{12} \cdot N_{21}}{\sqrt{(N_{11} + N_{12})(N_{21} + N_{22})(N_{11} + N_{21})(N_{12} + N_{22})}}.$$

Inserting the frequencies from Table 8 yields

$$\Phi_{Item\ 1\ -\ Item\ 2} = \frac{20 \cdot 20 - 20 \cdot 20}{\sqrt{(20 + 20)(20 + 20)(20 + 20)(20 + 20)}} = 0.0 \ .$$

Using standard interpretation we conclude that these two items have nothing to do with each other. Ordinarily, this is a reason to eliminate these two items from consideration. But it gets worse. Collapsing Table 9 over Item 2 yields the cross-tabulation of Item 1 with the criterion, Patient Status. The frequency distribution for these two variables is identical with the distribution in Table 10, and we obtain $\Phi_{\text{Item 1 - Patient Status}} = 0.0$. The same holds true for $\Phi_{\text{Item 2 - Patient Status}}$. Thus, neither item fulfills the two criteria. From the perspective of classical test theory, they are useless. The results from CFA, however, suggest that these two items are perfect when it comes to discriminating between the two patient groups of schizophrenics and non-schizophrenics. This is the paradoxical aspect of Meehl's paradox.

This phenomenon can be explained using either log-linear modeling or CFA. We first use log-linear modeling. The data in Table 9 are constructed such that all main effects and all first order interaction terms are zero. The second order interaction is greater than zero and explains 100% of the variability in the cross-classification. The data structure thus is $\log E_i = \lambda_0 + \lambda_{ijk}^{Item\ 1,\ Item\ 2,\ Patient\ Status}$. The following printout which is part of a more detailed printout created using SYSTAT 10.0 (Wilkinson,

2000), illustrates this point[4]:

```
Observed Frequencies
====================
PSTATUS    ITEM2       |         ITEM1
                       | 1           2
---------+---------+------------------------
1          1           |       20.00         0.00
           2           |        0.00        20.00
                       +
2          1           |        0.00        20.00
           2           |       20.00         0.00
-------------------+------------------------
Pearson ChiSquare       0.0000   df      0   Probability
     LR ChiSquare      -0.0000   df      0   Probability
    Raftery's BIC      -0.0000
    Dissimilarity       0.0000

   Parameter    SE(Param)       Param/SE
   =========    =========       ========
        0.00         0.36           0.00    ITEM1
        0.00         0.36           0.00    ITEM2
        0.00         0.36           0.00    PSTATUS
        0.00         0.36           0.00    ITEM2      *ITEM1
        0.00         0.36           0.00    PSTATUS    *ITEM2
        0.00         0.36           0.00    PSTATUS    *ITEM1
        1.86         0.36           5.19    PSTATUS*ITEM2*ITEM1
        1.16         0.36           3.25    CONSTANT
```

The `Param/SE` values can be interpreted as z-scores. As can be seen, only the triple interaction describes an effect significantly different than zero. In the analyses based on classical test theory, only the first order correlations are considered. Therefore, it seems like the two items and the criterion variable have nothing to do with each other (for a discussion of classical test theory for categorical variables see Clogg & Manning, 1996).

The second way to explain Meehl's paradox uses CFA. The results in Table 9 were created with a first order CFA base model. As an alternative, we can use a second order CFA base model. The results from this analysis are identical to the ones displayed in Table 9. We thus conclude that the types and antitypes in Table 9 can result only from the second order interaction between all three variables.

---

[4]To avoid estimation problems, these results were created by invoking the Delta-option with $\Delta = 0.5$, that is, by adding the constant Δ to each cell frequency.

3.3 Three approximations of the binomial test

The following sections present three approximations of the binomial test. These approximations are useful because they require less numerical effort than the binomial test itself. The approximations are good, if certain conditions are met.

3.3.1 Approximation of the binomial test using Stirling's formula

To calculate the tail probability for a binomial test, the term $\binom{N}{j}$ needs to be calculated repeatedly. This calculation can be numerically intensive. The term can be written as

$$\binom{N}{j} = \frac{N!}{(N - j)!\, j!},$$

with $N! = N\cdot(N - 1)\cdot(N - 2)\cdot \ldots \cdot(N - (N - 2))$, and $j!$ accordingly. Stirling's formula (Feller, 1957; Hu, 1988; von Eye & Bergman, 1987) reduces the number of steps needed considerably by using

$$\lim_{N\to\infty} \frac{N!}{N^N e^{-N}\sqrt{2\pi N}} = 1,$$

with e = 2.71828182846..., and π = 3.14159265359... Using this expression, we can set

$$N! \approx N^N e^{-N}\sqrt{2\pi N}.$$

This relationship can be used to develop the products in $\binom{N}{j}$. Inserting into the binomial test equation yields

$$B_i^{Stirling}(N_i) = \sqrt{\frac{N}{2\pi N_i(N - N_i)}} \left(\frac{Np}{N_i}\right)^{N_i} \left(\frac{Nq}{N - N_i}\right)^{N - N_i}.$$

This expression is numerically less tedious than the expression that involves $N!$. To illustrate the accuracy of the approximation of the binomial terms that can be reached using Stirling's formula, suppose that $N = 80$, $N_i = 41$, $p = 0.4$, and $q = 0.6$. Inserting into the Stirling-approximated binomial formula. we obtain

$$B^{Stirling} = \sqrt{\frac{80}{2\pi 41\ (80 - 41)}} \left(\frac{80 \cdot 0.4}{41}\right)^{41} \left(\frac{80 \cdot 0.6}{39}\right)^{39} = 0.0113352.$$

Inserting into the original binomial formula yields

$$B = \frac{80!}{41!\ (80 - 41)!}\ 0.4^{41}\ 0.6^{(80 - 41)} = 0.0112998.$$

The difference between these two values is 0.0000354. In general, it has been shown (von Eye & Bergman, 1987) that the numerical approximation using Stirling's formula

(1) suggests statistical decisions that are generally slightly more conservative than the ones suggested by the exact test, that is, $B^{Stirling} > B$;

(2) deviates from the exact test the most when $N_i \approx E_i$; and

(3) is closest to the exact values when p is small, as is typical of most CFA applications, and when the discrepancy between N_i and E_i is large. For instance, when $p = 0.04$, $N = 50$, and $N_i = 49$, the difference between the values calculated by the exact binomial test and the Stirling-approximated binomial test is unequal to zero not before the 67th decimal.

3.3.2 Approximation of the binomial test using the DeMoivre-Laplace limit theorem

Using the DeMoivre-Laplace limit theorem, one can create a good approximation of the binomial test through

$$\hat{B}(N_i) = \Phi(z_{l + 0.5}) - \Phi(z_{a - 0.5}),$$

where $\Phi(z)$ is the area under the standard normal distribution that begins with z, and a and l are defined as for the binomial test, above (Feller, 1957). To estimate the z-values, we use the standard deviation of the binomial distribution which is $s = \sqrt{Npq}$. In large samples, the term $\Phi(z_{l + 0.5})$ approximates 1, and one can set

$$\hat{B}(N_i) = 1 - \Phi(z_{a - 0.5}).$$

Consider the same numerical example as before, that is, $N = 80$, $N_i = 41$, $p = 0.4$, and $q = 0.6$. Inserting in the formula for $\hat{B}$, one obtains

$$z_{a - 0.5} = \frac{41.5 - 80 \cdot 0.4}{\sqrt{80 \cdot 0.4 \cdot 0.6}} = 2.168069.$$

The one-sided tail probability for this z-value is $\hat{B}(40) = 0.01507674$, which is a value larger than the one suggested by the Stirling-approximated binomial test. This z-approximation, therefore, seems to suggest more conservative decisions than the binomial test. It is known, however, that this approximation can be less conservative at the extreme ends of the distribution, and when p assumes small values (Bergman & von Eye, 1987; von Eye & Bergman, 1987).

3.3.3 Standard normal approximation of the binomial test

The best known and most frequently used approximation of the binomial test is the standard normal distribution. This approximation is sufficiently accurate when N is large and p is not too extreme. It has been shown that the approximation does not create significant discrepancies between the normal and the binomial distributions if $np \geq 10$ (Osterkorn, 1975). If this condition is met, one can use the mean and the standard deviation of the binomial distribution and estimate the standard normal

$$z = \frac{N_i - Np}{\sqrt{Npq}}$$

instead of calculating the binomial probabilities. This approximation can suggest less conservative statistical decisions than the approximation presented in Section 3.3.2. If $5 \leq Np \leq 10$, the continuity-corrected

$$z = \frac{N_i - Np - 0.5}{\sqrt{Npq}}$$

is recommended (Krauth & Lienert, 1973a).

Without continuity correction, the z-approximation yields for the example with $N = 80$, $p = 0.4$, $N_i = 41$ the estimate $z = 2.05396$ and $p(z) = 0.0199898$. With continuity correction, one obtains $z = 1.939851$, and $p(z) = 0.0261990$.

3.3.4 Other approximations of the binomial test

A large number of approximations of the binomial test has been proposed (Molenaar, 1970; Naud, 1997). According to Molenaar and to Naud, there is *no single best* approximation. The accuracy of the approximation, that is, the nearness of the estimated probability to the binomial probability, depends on N and p. In addition, some approximations perform better for certain ranges of α while providing only rough estimates outside these ranges. Three of the approximations described in Molenaar (1970) and investigated by Naud (1997, 1999) are briefly reviewed here. The first of these three is the Camp-Paulson approximation of the binomial tail probability

$$\Phi\left[\frac{\left(9 - \frac{1}{k+1}\right)F^{\frac{1}{3}} - 9 + \frac{1}{N-k}}{3\left(\sqrt{F^{\frac{2}{3}}\frac{1}{k+1} + \frac{1}{N-k}}\right)}\right],$$

where $k = N_i$ and $F = \frac{(k+1)q}{(N-k)p}$. The second approximation to be briefly reviewed here is the Borges approximation,

$$\Phi\left[\sqrt{N + \frac{1}{3}}\,(pq)^{-\frac{1}{6}}\,J\left(\frac{k + \frac{2}{3}}{N + \frac{1}{3}}\right) - J(p)\right],$$

with $J(z) = \int_0^z t^{-\frac{1}{3}}(1 - t)^{-\frac{1}{3}}\,dt$. When $N_i < 20$, the Camp-Paulson approximation is slightly better than the Borges approximation. When $N_i > 50$, the Borges estimate is better in the tails, which is of importance for use in CFA. Numerically, the Borges is somewhat more labor-intensive because it requires numerical integration. As a substitution of the integral, one can use $J(z) = 1.5z^{\frac{2}{3}}\frac{60 - 17z}{60 - 25z}$, which supposedly does not reduce the accuracy of the approximation greatly.

The third approximation presented in this section is based on the Poisson distribution. This approximation is particularly useful when p is small, as is typically the case in CFA. Only when p is very small, will the tails of the distribution not be overestimated. If $N_i < E_i$, the Poisson approximation is

$$P = \sum_{i=0}^{N_i} \frac{e^{-\lambda}\lambda^i}{i!}.$$

If $N_i > E_i$, the Poisson approximation is

$$P = 1 - \sum_{i=0}^{N_i - 1} \frac{e^{-\lambda}\lambda^i}{i!}.$$

In both equations $\lambda = N_i\, p$. (For modifications of the λ parameter that prevent the approximation from overestimating the tails of the binomial distribution, see Molenaar, 1970; cf. Naud, 1997).

When p is small, the accuracy of the modified formulas is better than the accuracy of the Camp-Paulson approximation. Naud reports that these approximations perform even better than the best normal approximation. However, because these approximations have not been used in CFA except in Naud's simulations, they will not be discussed here any further. Other approximations will not be discussed either. For instance, the F-approximation (Heilmann & Schütt, 1985), equivalent to the binomial test, will not be discussed, for two reasons. First, this approximation provides no benefits over the binomial test. It is equivalent and numerically not much more parsimonious. Second, this approximation has been described in detail only for $N_i > E_i$, that is, for the search for types. Here we are interested in detecting both types and antitypes.

3.4 The χ^2 test and its normal approximation

The best known and most frequently used CFA test is the Pearson X^2-component test, commonly called the χ^2-test,

$$X_i^2 = \frac{(N_i - E_i)^2}{E_i},$$

with $df = 1$. Because for $df = 1$ the relation

$$z^2\left(\frac{\alpha}{2}\right) = \chi^2(\alpha)$$

holds, the z-distribution can be used to evaluate the X^2-components (Fienberg, 1980). For the example with $N = 80$, $p = 0.4$, $N_i = 41$ we calculate $X^2 = 2.53125$ and $p(X^2) = 0.111612$, a value larger by a factor of over 5 than the one calculated for the z-approximation of the binomial test. In this example, the X^2 component test suggests a more conservative decision than the binomial test and its normal approximations. This result carries over to almost all cases.

The normal approximation of the X^2-component test. The ordinate of standard normal scores x is given by

$$\Phi(x) = \frac{1}{\sqrt{2\pi}} e^{-\frac{x^2}{2}}.$$

The sum of the squared scores, Σx^2, is called χ^2, that is

$$\chi^2 = \sum_{i=1}^{t} x_i^2 ,$$

where i indexes the configurations and t is the number of configurations. The distribution of χ is known to be

$$\Phi(\chi) = Ce^{\frac{-\chi^2}{2}} \chi^{t-1},$$

with

$$C = \left[\frac{1}{\left(\frac{t-2}{2}\right)! \; 2}\right]^{\frac{1}{2}(t-2)}.$$

In exploratory CFA, only one cell is evaluated at a time. Therefore, the distribution of χ becomes

$$\Phi(\chi) = C\, e^{\frac{-\chi^2}{2}}.$$

This equation describes the positive values of a standard normally distributed variable. The χ^2 curve is positive by definition. Thus, the distribution of χ for $df = 1$ describes one half of the normal distribution. It follows from this relation that

$$z_i^2 = \frac{(N_i - Np_i)^2}{Np_i q_i}$$

is distributed as χ^2 with $df = 1$.

It should be noted that this expression is equivalent to the expression used for the normal approximation of the binomial test. These two expressions are equivalent, both in the numerator and in the denominator. However, these two expressions are not equivalent to the expression used for the X^2-component. The difference to the Pearson X^2-component is in the denominator, where the z-equations contain the term Npq and the X^2-component contains the term Np. This difference may by negligible as p approximates zero as can be the case for very large cross-tabulations. However, two consequences always follow:

(1) X^2 values will always be smaller than z^2-values.
(2) This discrepancy is more pronounced when $p > 0.5$.

To illustrate this discrepancy, we now present a little simulation that shows the behavior of the two test statistics for both $p > 0.5$ and $p < 0.5$. Specifically, a series of corresponding expected and observed frequencies was generated such that the observed frequencies varied from 20 to 1 in steps of one, and the expected frequencies varied from 1 to 20, also in steps of one. The sample size was set to 21 for each pattern of N_i and E_i. For each of the resulting 20 discrepancies, both X^2 and z^2 are depicted in Figure 1.

The left-hand side of Figure 1, from $N_i = 20$ to $N_i = 11$ displays the situation in which $p < 0.5$. The right-hand side of the figure, from $N_i = 10$ to $N_i = 1$ displays the situation in which $p > 0.5$. The curves suggest that

(1) the z^2-scores, displayed by the line with the diamonds, are always greater than the X^2-scores which are displayed by the line with the stars; thus the z-test will always have more power than the X^2-component test;
(2) the difference in power may be negligible when $N_i \approx E_i$; this, however, is the situation that is of least interest in CFA, because this is the situation for which no types or antitypes will emerge; the smallest difference was measured for $N_i = 11$ (and $E_i = 10$), where we calculated $X^2 = 0.091$ and $z^2 = 0.191$:

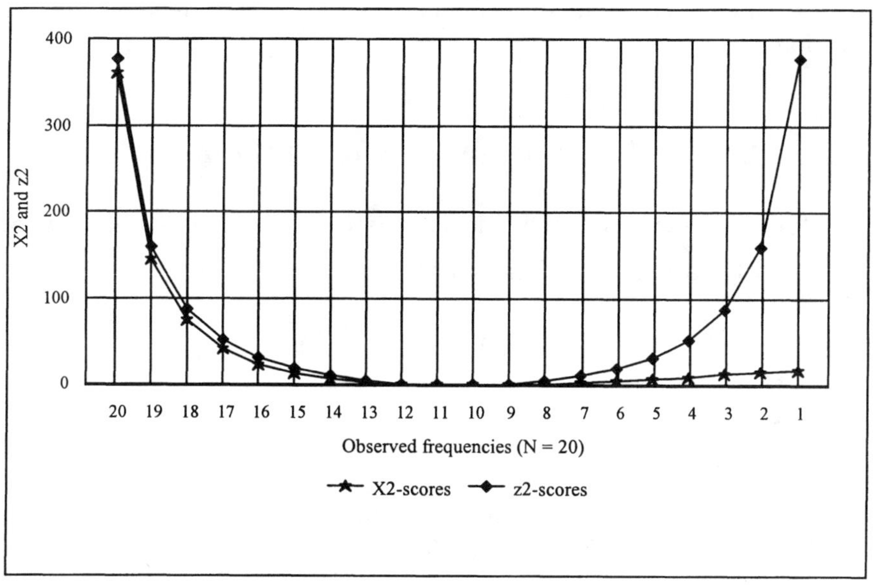

Figure 1: Comparison of z^2 and X^2 in CFA testing (for frequencies to the left of 11, p < 0.5, and for frequencies to the right of 11, p > 0.5)

(3) when $p < 0.5$, the difference between the two test statistics is still not large, but can be large enough for the z-test to identify a type or antitype and for the X^2-test not to identify it; the largest difference under this condition was calculated for $N_i = 20$ (and $E_i = 1$), where we calculated $X^2 = 361$ and $z^2 = 379.05$;

(4) when $p > 0.5$, the difference between the two test statistics can become colossal; the largest difference in this simulation was calculated for $N_i = 1$ (and $E_i = 20$), where we calculated $X^2 = 18.05$ and $z^2 = 379.05$.

Most interesting from the perspective of a researcher who looks for types and antitypes is that the right-hand side of Figure 1 also depicts the situation in which $N_i < E_i$, that is, where one can expect antitypes to emerge. Regardless of the size of p, the right-hand side of the figure shows that when $N_i < E_i$, the z-test is more likely to suggest the existence of antitypes than the X^2-test. As can be seen in the left-hand side of the figure, when $N_i > E_i$, the difference between these two tests still exists but is less overwhelming. The topic of differential power will be taken up again in Section 3.9.

3.5 Anscombe's normal approximation

The following definition of residuals (Anscombe, 1953) is supposed to be more nearly normally distributed than $\sqrt{X^2}$:

$$r_i^* = \frac{3[N_i^{\frac{2}{3}} - (E_i - \frac{1}{6})^{\frac{2}{3}}]}{2E_i^{\frac{1}{6}}}.$$

There are many other transformations and approximations that could be used, for instance, Haberman's (1973) adjusted residuals. However, we include here only those that have been used in CFA.

3.6 Hypergeometric tests and approximations

The CFA tests described in this section are asymptotic hypergeometric tests (Küchenhoff, 1986; Lehmacher, 1981). There exist exact hypergeometric tests (Lehmacher, 1981; Lindner, 1984). However, these tests suffer from a number of shortcomings that prevented them from being used in CFA applications. Lindner's test, for instance, was described for dichotomous variables only, and both Lindner's and Lehmacher's tests are tedious to calculate. Therefore, we focus, in the following two sections, on the more general and computationally simpler asymptotic versions of Lehmacher's test.

These tests can be used only if the following conditions are fulfilled:

(1) The margins are fixed as is the case in product-multinomial sampling;
(2) the sample size is very large, in particular when Lehmacher's original test is employed; and
(3) only a first order global base model is considered. Lehmacher's test will not work for any other base model, nor for models with

covariates.

3.6.1 Lehmacher's asymptotic hypergeometric test

Lehmacher (1981) proposed an asymptotic hypergeometric test for CFA (see also Lehmacher & Lienert, 1982). This test can be derived starting from the well known relation

$$X := \frac{N_i - E_i}{\sqrt{E_i}} \approx N(0, \sigma^2) ,$$

where $\sigma^2 < 1$ if the model fits (Christensen, 1997; Haberman, 1973). In words, this relation indicates that, under the null hypothesis, residuals are approximately normally distributed, but with a variance less than 1. To prevent researchers from making incorrect decisions concerning the existence of types or antitypes, Lehmacher derived the exact variance

$$\sigma_i^2 = Np_i[(1 - p_i - (N - 1)(p_i - \tilde{p}_i)],$$

where p_i is as for the binomial test estimated from the sample, i indexes the configurations, and N is the sample size. However, whereas any base model can be used for the binomial test, Lehmacher's test requires that a first order CFA base model, that is, a main effect model, be used. Suppose $d = 3$ variables are used. Then, p_{ijk} is estimated as

$$p_{ijk} = \frac{N_{i..}N_{.j.}N_{..k}}{N^2} ,$$

where i, j, and k index the categories of the three variables that span the cross-classification. Still for $d = 3$ variables, the probability $\tilde{p}_{ijk}$ is estimated as

$$\tilde{p}_{ijk} = \frac{(N_{i..} - 1)(N_{.j.} - 1)(N_{..k} - 1)}{(N - 1)^d}.$$

Using the exact variance, one can describe the asymptotically normally distributed test statistic

$$z_{L,i} = \frac{N_i - E_i}{\sigma_i},$$

where i indexes the cells of the cross-classification. Because $p > \tilde{p}$, Lehmacher's z_L will always assume larger values than the standard

z. The following inequality holds: $|X| < |z| < |z_L|$. In other words, of the X^2-component test, the standard z-test, and Lehmacher's z_L test, Lehmacher's z_L test has the most, and the X^2-component test the least, statistical power. Lehmacher's test requires that the sample size be very large.

3.6.2 Küchenhoff's continuity correction for Lehmacher's test

When selecting a significance test for CFA, researchers must take into account that large samples may be needed for Lehmacher's approximation of the exact hypergeometric test to be good. Küchenhoff (1986; cf. Perli, Hommel, & Lehmacher, 1987) performed simulations to determine the smallest difference N_i - E_i that suggests the existence of a type. The author created samples in size between N = 10 and N = 100 and found that the asymptotic test described in Section 3.6.1 often suggests overly liberal decisions, compared to the exact hypergeometric test. Simulations by Lindner (1984) suggest that there exist instances in which the tail probabilities that are estimated by the asymptotic test are smaller by a factor of 1: 730,000 than the tail probabilities estimated by the binomial test. Both Küchenhoff's and Lindner's results point to possibly nonconservative decisions suggested by Lehmacher's asymptotic test. In addition, Lindner showed that this nonconservative pattern occurs in particular when sample sizes are small.

To prevent researchers from making nonconservative decisions, Küchenhoff (1986; cf. Lautsch, Lienert, & von Eye, 1987) proposed using a continuity correction (Yates, 1934). This correction adjusts the observed cell frequency, N_i such that

$$N_i = \begin{cases} N_i - 0.5 & \text{if } N_i > E_i \\ N_i + 0.5 & \text{if } N_i \leq E_i \end{cases},$$

that is, for configurations that are candidates for types, one reduces the discrepancy between N_i and E_i by 0.5. For configurations that are candidates for antitypes, one increases N_i by 0.5 which has the effect that the discrepancy is also reduced by 0.5. Küchenhoff's (1986) continuity corrected version of Lehmacher's z_L is thus

$$z_{K,i} = \frac{N_i \pm 0.5 - E_i}{\sigma},$$

Küchenhoff's simulations suggest that the continuity-corrected version of Lehmacher's asymptotic test approximates the exact generalized hypergeometric distribution very well, and thus reduces the probability of nonconservative decisions considerably.

3.7 Issues of power and the selection of CFA tests

Whenever a large number of tests is available to test a statistical null hypothesis, the user needs to make a well-informed decision as to which test to select. Unfortunately, there is no one test that outperforms the others under all conditions. In addition, there exists no rounded-out body of results that can be used when making such decisions. Therefore, we piece together the knowledge that is available, and we conclude with a set of recommendations, being well aware that these recommendations are derived from incomplete information.

Three aspects are of importance in the present context of power and selecting significance tests. The first aspect is statistical power itself. Statistical tests differ in the probability with which they identify a null hypothesis as untenable. The less likely a false null hypothesis can survive, the more power is ascribed to a test. However, deviations from truth can occur in both directions. If the factual significance threshold, α, is less than the nominal α, a test is conservative, and false null hypotheses have a greater chance of surviving. If, however, the factual significance threshold is greater than the nominal α, a test is nonconservative (also called *anticonservative*), and null hypotheses will be prematurely rejected. In this section, we will encounter both variants of tests.

A second important aspect complicates this situation. This aspect is that with the exception of the binomial test, all tests discussed in the previous sections are approximative in nature. Therefore, we need to take into consideration (a) statistical power and (b) the accuracy of approximation of a sampling distribution.

The third aspect of relevance here is that the tests can be applied only under particular sampling schemes. Thus, the selection of tests will not only rest on power and nearness of approximation arguments, but also on the sampling scheme under which the data were collected.

What do we know thus far, from reading the previous sections on statistical tests in CFA? First, we reported based on results that can be found in the literature. and we illustrated that

$$\sqrt{X^2} < z_{X^2} < z_{bin} < z_K < z_L \ ,$$

where X^2 is the Pearson component test, z_{X^2} is the z-approximation of the X^2-test, z_{bin} is the z-approximation of the binomial test, z_K is Lehmacher's asymptotic hypergeometric test with Küchenhoff's continuity correction, and z_L is Lehmacher's asymptotic hypergeometric test without Küchenhoff's continuity correction. In addition, we know that $B^{Stirling} > B$, that is, that the Stirling approximation of the binomial tests suggests more conservative decisions than the exact binomial test.

However, the previous sections have not provided us with information about the Anscombe z-approximation, the power of the binomial versus the z-tests, and the power curves of any of the tests. Therefore, we present in the following sections results of power investigations for a number of the tests used in CFA (Naud, 1997, 1999), and we apply the CFA tests to real data in order to first gain insight into the relative performance of the tests, where theoretical results and simulation results are unavailable.

3.7.1 Naud's power investigations

There exists only a very small number of power investigations for the tests used in CFA. First recommendations were rules of thumb, derived from characteristics of the tests in different contexts. The context of CFA is that of tests at the cell-wise level, in the presence of other tests. Usually, application of the Pearson X^2-test is considered valid as long as $E_i > 5$. However, there exist more liberal rules, most of which come with qualifications or conditions that must be met. For example, Wise (1963) suggests considering the test still valid if $E_i \geq 2$, as long as the E_i are all about equal. Everitt (1977) references sources in which $E_i = 1$ is allowed. Larntz (1978; cf. Koehler & Larntz, 1980) presents simulation results that show that Pearson's X^2 performs better than other approximations of χ^2 even if the sample size is as small as $N = 8$ and the expected cell frequencies are as small as $E_i = 0.5$. Koehler and Larntz (1980) found that E_i can be as low as 0.25 for the null hypothesis of symmetry, that is, equal cell probabilities. Their sample size recommendation for Pearson's χ^2-test is that $N \geq 10$ and $N^2/c \geq 10$ for tables with three or more cells, where c indicates the number of cells in a table.

Although useful, rules of thumb are necessarily crude. The minimum cell expectation is not the only criterion that can be used nor is it the most useful one. Naud (1999) considers the marginal total more useful. Other factors that play a role include the number of small expectations, the size of the table, and whether the small observed cell frequencies are smaller or larger than the expected cell frequencies under the alternative hypothesis. In addition, the type of hypothesis (goodness-of-fit test, independence test, or homogeneity test), the symmetry-asymmetry of the table under study, and the nature of the expectancies as calculated versus fixed seem to play major roles.

Table 11 presents a summary of Naud's (1997) simulation results. The results on which this summary is based are by no means complete. In addition, they focus on the binomial test and Pearson's X^2-test only. Nevertheless, they shed light on the complexity of the matter, and indicate that precise rules or recommendations are hard to formulate.

The table presents results for the binomial and the Pearson tests in columns separately for symmetric and asymmetric tables. Symmetrical tables have uniform marginal totals. Asymmetric tables have marginal totals that differ from each other. The rows present three blocks that report results for three test situations. These situations are the goodness-of-fit test, the test of independence, and the homogeneity test. The goodness-of-fit test represents the case in which CFA base models are selected that can be more complex than the main effect model that is also used for the standard X^2 test. The test of independence represents the classical CFA base model of variable independence, that is, the log-linear main effect model. The homogeneity test represents the case of a two-sample CFA in which two groups of cases are compared in their frequency distributions.

Three types of sampling are considered. The first is the standard multinomial where each response is assigned a cell at random. The second is the product-multinomial sampling where each case is randomly assigned but only to an a priori specified group of cells, for example, the cells for the responses from smokers. The third is Poisson sampling, where p is very small, a case that is routine in CFA applications.

Behavior of tests is classified as either *conservative* or *liberal*, with qualifiers, where appropriate. A statistical decision is conservative if the probability of rejecting a null hypothesis is less than α. A statistical decision is liberal if the probability of rejecting a null hypothesis is greater than α. When selecting from a number of less-than-perfect tests, one opts for conservative tests, thus protecting the significance level.

Table 11: Summary of Naud's (1997) simulation results

Sampling	Binomial test		Pearson X^2 test	
	symmetric table	asymmetric table	symmetric table	asymmetric table
		Goodness-of-fit test		
multi-nomial	conservative $N = 25c$	liberal, 2 x 2, $N = 5c$; conservative 4 x 4, $N = 25c$	very liberal, 2 x 2; liberal, 4 x 4, $N = 5c$	within range
product multi-nomial	very liberal to liberal	very liberal, $N = 5c$; liberal, $N = 25c$	none detected[a], 2 x 2 very liberal, 4 x 4	liberal; only types detected
Poisson	conservative $N = 25c$	liberal, 2 x2, $N = 5c$; conservative 4 x 4, $N = 25c$	very liberal, 2 x 2; liberal, 4 x 4, $N = 25c$	conservative 4 x 4, $N = 25c$
		Test of independence		
multi-nomial	none detected, 2 x 2; very conservative 4 x 4	none detected, 2 x 2; very conservative 4 x 4	none detected, 2 x 2; very conservative, 4 x 4	conservative $N > 20$
Poisson	very few detected	none detected, 2 x 2; very conservative 4 x 4	none detected, 2 x 2; very conservative, 4 x 4	very conservative $N > 20$, only types

/ cont.

Sampling	Binomial test		Pearson X^2 test	
	symmetric table	asymmetric table	symmetric table	asymmetric table
Homogeneity test				
	very few detected	none detected, 2 x 2; very conservative 4 x 4	none detected, 2 x 2; very conservative 4 x 4	very conservative N > 20, only types

[a]*non detected* indicates that under this condition, the test identified neither types nor antitypes

Sample sizes in the simulations were kept proportional. Therefore, sample sizes are indicated as multiples of c, the number of cells in a table. The smallest sample size was $N = 20$ for 2 x 2 tables.

Naud (1997) concludes from these and other results that the binomial test is "much to be preferred over X^2" when the table is asymmetrical, because the X^2-test detects more types and fewer antitypes than the binomial test. This result confirms the conclusion drawn from the simulation that led to Figure 1, in Section 3.4. The problem of differential power for types and antitypes will be taken up again in Section 3.9.

In accordance with the complexity of the results presented in Table 11, the power curves for the binomial test and X^2 depend on the variables discussed here and on whether $N > E$ or $N < E$. Unfortunately, the current knowledge does not allow us to present a complete picture. We do know, however, that the power of the tests varies depending on sample size, sampling scheme, distribution, symmetry of table, marginal totals, type of test, and the nature of E as calculated versus fixed. We also know that the nature of the tests can swing wildly from very conservative to very liberal if just one of these parameters changes. The next section presents an application of all tests discussed here to empirical data.

3.7.2 Applications of CFA tests

In this section, we apply eight tests that have been proposed for global CFA to two empirical data sets. The first data set is *sparse*. That is, the number of cells to number of cases ratio is small. In addition, the data are very unevenly distributed. A number of cell frequencies is zero, other frequencies are relatively large[5]. The second data set contains relatively large frequencies.

3.7.2.1 CFA of a sparse table

In a study on the evaluation of job interviews, two raters, X and Y, rated 465 interviews with regard to the interviewees' ability to organize, plan, and prioritize. The ratings were given on a 7-point scale, with 7 indicating high ability. Crossed, the two raters' judgments form a 7 x 7 cross-classification. Table 12 displays this classification, along with the CFA results from eight tests: the binomial test (*bin*), the binomial test with Stirling approximation of the factorials (*bs*), the Pearson X^2-component test (X^2), the normal approximation of the binomial (*bz*), the *z*-test (*z*), Lehmacher's test (*L*), Lehmacher's test with Küchenhoff's continuity correction (*LK*), and Anscombe's *z*-approximation (*A*). Omitting the test statistics, Table 12 displays the tail probabilities. α was adjusted using Bonferroni's procedure which led to $\alpha^* = 0.00102$. Types are marked with **T**, antitypes are marked with **A**.

The results in Table 12 first confirm the results reported earlier in this chapter. The Lehmacher test is the most powerful, identifying the largest number of types and antitypes. The Lehmacher test with continuity correction is slightly less powerful, followed by the *z*-test, the about equivalent z-approximation of the binomial test, the Pearson X^2 component test, the binomial test and its approximation based on using Stirling's formula.

The *z*-approximation proposed by Anscombe did not provide any results, because the program did not complete its run. The reason for this lack of results is that Anscombe's test statistic does not have a unique solution when $E_i < 0.16667$. In the numerator of the equation in Section 3.5 we find the term $(E_i - 0.166667)^{2/3}$. If $E_i < 0.16667$, the expression inside the parentheses becomes negative and the root has two solutions. Consider, for example, the expectancy of $E_i = 0.1$. This value yields -0.066667 for the expression inside the parentheses. Raising this value to the power of 0.66667 yields the solutions -0.0822 and 0.1424. Therefore, Anscombe's *z*-approximation cannot be used when a base model yields expectancies less than 0.166667.

---

[5]We would like to thank Neal Schmitt for making this data set available to us.

The results in Table 12 also suggest that the Pearson X^2-test is less powerful in particular when $N_i < E_i$, that is, when antitypes could be detected. When compared with the binomial test, the Pearson test detects one additional type, but *six fewer antitypes*. The z-test and the binomial approximation of the z-test may also be less sensitive to antitypes than the binomial test and its Stirling approximation. However, the tail probabilities in the table suggest that this lack of sensitivity is less extreme than for the X^2-test. In addition, this result may reflect test differences in power, data characteristics, or both.

Table 12 also indicates that the Lehmacher test with Küchenhoff's continuity correction behaves in a bizarre way under certain conditions. Consider Configuration 12. The observed frequency of this configuration is zero. The expected frequency is 0.047. This value is not only that small that the accuracy of the approximation to the normal distribution can be questioned, it is also that small that the continuity correction does not achieve what it is supposed to achieve. More specifically, consider the equation for z with continuity correction (cf. Sections 3.6.1 and 3.6.2), which is for $N_i < E_i$

$$z_{LK} = \frac{N_i - E_i + k}{\sigma},$$

where k usually is set to 0.5. The effect of the continuity correction is that the discrepancy between N_i and E_i is reduced by a constant of 0.5. This can be illustrated using Configuration 15, for which we have $N_{15} = 0$ and $E_{15} = 0.637$. Inserting into the equation for z_{LK} we obtain $z_{LK} = (0 - 0.637 + 0.5)/\sigma = -0.137/\sigma$. This value is closer to zero than $z_{LK} = -0.637/\sigma$ which results without the continuity correction. However, consider again Configuration 12. For this configuration we obtain $z_{LK} = (0 - 0.047 + 0.5)/\sigma = 0.453/\sigma$. This value is *larger* than $z_{LK} = -0.047/\sigma$ without the continuity correction. In other words, as soon as k, the corrective constant in the numerator of z_{LK}, is greater than $0.5k(N_i - E_i)$, Küchenhoff's continuity correction results in an *increase of the difference between* N_i *and* E_i rather than the desired decrease.

Table 12: Results from eight CFA tests in Schmitt's interview study

Cells	Frequencies		Significance tests (tail probabilities)[a]							
XY	observed	expected	*bin*	*bs*	X^2	*bz*	*z*	*L*	*LK*	*A*
11	1	.004	.004	.005	**< α* T**	**< α* T**	**< α* T**	**< α* T**	**< α* T**	-
12	0	.047	.954	.954	.828	.414	.412	.413	.017	-
13	1	.120	.113	.123	.011	.006	.006	.004	.129	-
14	0	.327	.721	.721	.568	.284	.284	.266	.370	-
15	0	.637	.529	.529	.425	.212	.212	.167	.418	-
16	0	.692	.500	.500	.405	.202	.203	.151	.387	-
17	0	.172	.842	.842	.678	.339	.339	.332	.204	-
21	0	.017	.983	.983	.896	.448	.448	.447	**< α* A**	-
22	1	.189	.172	.186	.062	.031	.031	.029	.233	-
23	1	.482	.382	.411	.455	.227	.228	.219	.489	-
24	4	1.308	.044	.044	.019	.009	.009	.005	.017	-
25	2	2.546	.532	.559	.732	.366	.366	.338	.486	-
26	0	2.770	.062	.062	.096	.048	.048	.019	.045	-

cont./

Table 12, panel 2/4

Cells	Frequencies		Significance tests (tail probabilities)[a]							
XY	observed	expected	*bin*	*bs*	X^2	*bz*	*z*	*L*	*LK*	*A*
27	0	.688	.502	.502	.407	.203	.203	.191	.406	-
31	0	.067	.936	.936	.796	.398	.398	.395	.041	-
32	6	.733	< α* **T**	< α* **T**	< α* **T**	< α* **T**	< α* **T**	< α* **T**	< α* **T**	-
33	5	1.867	.041	.042	.022	.011	.011	.007	.020	-
34	11	5.067	.014	.015	.008	.004	.004	.001	.003	-
35	8	9.867	.346	.350	.552	.274	.276	.228	.293	-
36	1	10.733	< α* **A**	< α* **A**	.003	.001	.001	< α* **A**	< α* **A**	-
37	0	2.667	.069	.069	.102	.051	.051	.039	.076	-
41	0	0.183	.833	.833	.669	.334	.334	.318	.206	-
42	3	2.011	.326	.334	.485	.242	.243	.218	.350	-
43	16	5.118	< α* **T**	< α* **T**	< α* **T**	< α* **T**	< α* **T**	< α* **T**	< α* **T**	-
44	31	13.892	< α* **T**	< α* **T**	< α* **T**	< α* **T**	< α* **T**	< α* **T**	< α* **T**	-
45	26	27.054	.468	.470	.839	.417	.420	.393	.443	-

/ cont.

Table 12, panel 3/4

Cells	Frequencies		Significance tests (tail probabilities)[a]							
XY	observed	expected	*bin*	*bs*	X^2	*bz*	*z*	*L*	*LK*	*A*
46	9	29.430	< α* **A**	< α* **A**	< α* **A**	< α* **A**	< α* **A**	< α* **A**	< α* **A**	-
47	0	7.312	< α* **A**	< α* **A**	.007	.003	.003	< α* **A**	.002	-
51	0	.301	.740	.740	.583	.292	.292	.255	.332	-
52	0	3.312	.036	.036	.069	.034	.034	.014	.031	-
53	5	8.430	.153	.156	.237	.116	.119	.073	.107	-
54	25	22.882	.354	.355	.658	.325	.329	.281	.329	-
55	67	44.559	< α* **T**	< α* **T**	< α* **T**	< α* **T**	< α* **T**	< α* **T**	< α* **T**	-
56	41	48.473	.144	.145	.283	.128	.142	.056	.069	-
57	2	12.043	< α* **A**	< α* **A**	.004	.002	.002	< α* **A**	< α* **A**	-
61	0	.320	.726	.726	.571	.286	.286	.246	.350	-
62	1	3.525	.132	.141	.179	.089	.089	.050	.093	-
63	0	8.972	< α* **A**	< α* **A**	.003	.001	.001	< α* **A**	< α* **A**	-
64	5	24.353	< α* **A**	< α* **A**	< α* **A**	< α* **A**	< α* **A**	< α* **A**	< α* **A**	-

/ cont.

Table 12, panel 4/4

Cells	Frequencies		Significance tests (tail probabilities)[a]								
XY	observed	expected	*bin*	*bs*	X^2	*bz*	*z*	*L*	*LK*	*A*
65	41	47.424	.183	.183	.351	.162	.175	.086	.103	-
66	85	51.589	< α* **T**	< α* **T**	< α* **T**	< α* **T**	< α* **T**	< α* **T**	< α* **T**	-
67	17	12.817	.149	.149	.243	.118	.121	.069	.096	-
71	0	.108	.898	.898	.743	.371	.371	.364	.103	-
72	0	1.183	.306	.306	.277	.138	.138	.122	.251	-
73	0	3.011	.049	.049	.083	.041	.041	.029	.057	-
74	0	8.172	< α* **A**	< α* **A**	.004	.002	.002	< α* **A**	< α* **A**	-
75	4	15.914	< α* **A**	< α* **A**	.003	.001	.001	< α* **A**	< α* **A**	-
76	25	17.312	.045	.045	.065	.030	.032	.008	.012	-
77	21	4.301	< α* **T**	< α* **T**	< α* **T**	< α* **T**	< α* **T**	< α* **T**	< α* **T**	-

[a] "< α*" indicates that the tail probability is less than α* = 0.00102.

The consequence of this failure to achieve the intended decrease of the difference between N_i and E_i may in many cases be just a mis-estimation of the tail probability. However, when $N_i = 0$, otherwise unsuspiciously small discrepancies can look impressive. In extreme cases, very small discrepancies can become significant if the estimated standard errors are small. This is the case for the smallest discrepancy in Table 12, Configuration 21. Here we find $N_{21} = 0$ and $E_{21} = 0.017$. None of the other six tests identifies this discrepancy as significant or even approaching significance. The z-approximation with the Lehmacher-Küchenhoff test, however, calculates for this small discrepancy a z = 3.713 and a tail probability of $p = 0.000102$. This value is less than the Bonferroni-adjusted $\alpha^* = 0.001$. In addition, the estimated z-value for this discrepancy is positive which should occur only if $N_i > E_i$. Similar effects occur for other configurations in Table 12. We therefore recommend using Lehmacher's test with Küchenhoff's continuity correction only if $N_i - E_i > 0.5$.

3.7.2.2 CFA tests in a table with large frequencies

In this section, we present a data example with slightly larger frequencies. The small frequency-specific problems illustrated in section 3.7.2.1 will therefore not surface. We present this example with two goals in mind. The first goal is to also use Anscombe's z-approximation in comparison with the other seven tests. The second goal is to present a case in which differences in statistical power result in larger discrepancies in the numbers of types and antitypes identified.

The data are the same as in Table 1. They describe 65 students who were administered LSD 50. The three symptoms Narrowed Consciousness (C), Thought Disturbance (T), and Affective Disturbance (A) were observed. Each symptom was scaled as either present (1) or absent (2). We now apply all eight significance tests to the C x T x A cross-classification. The results appear in Table 13. As for Table 12, we employed Bonferroni adjustment which yielded $\alpha^* = 0.00625$.

The results in Table 13 confirm the earlier theoretical and empirical results concerning the power differences. In this data set, the Pearson X^2-component test is clearly the least powerful, followed by the *z*-test, the binomial test and its approximation with the Stirling formula. The Anscombe *z*-test yields an inconsistent pattern of higher and lower tail probabilities in comparison with the binomial and the *z*-tests. It seems to be more sensitive

Table 13: Results from eight CFA tests in Lienert's LSD data

Cell	Frequencies		Significance tests (tail probabilities)[a]							
CTA	observed	expected	*bin*	*bs*	X^2	*bz*	*z*	*L*	*LK*	*A*
111	20	12.51	.018	.018	.034	.009	.017	< α* **T**	.001 **T**	.023
112	1	6.85	.006 **A**	.007	.025	.009	.013	.001 **A**	.002 **A**	.003 **A**
121	4	11.40	.007	.007	.028	.008	.014	< α* **A**	.001 **A**	.006
122	12	6.24	.020	.020	.021	.008	.011	.001 **T**	.002 **T**	.017
211	3	9.46	.010	.011	.035	.012	.018	.001 **A**	.002 **A**	.008
212	10	5.18	.033	.033	.034	.014	.017	.003 **T**	.007	.026
221	15	8.63	.021	.022	.030	.010	.015	.001 **T**	.002 **T**	.022
222	0	4.73	.007	.007	.030	.012	.015	.002 **A**	.007	.001 **A**

[a] "< α*" indicates that the tail probability is less than α* = 0.00625.

to the possible existence of antitypes than the X^2-test. More detailed investigations will have to show whether this test statistic indeed approximates the normal distribution better than the z-statistic (see Section 3.9; von Eye, 2002). The most powerful test is, as can be expected, the Lehmacher test. Küchenhoff's continuity correction takes away some of this power. This loss, however, does not decrease the power to the levels of the other tests. It may, however, reduce the nonconservative characteristics of Lehmacher's asymptotic test in small and medium size samples.

When comparing the tail probabilities of the eight tests, power differences become obvious. The biggest discrepancies in Table 13 are those between the X^2-test and Lehmacher's test. The tail probability from the X^2-test are bigger than those from Lehmacher's test by a factor of up to over 90. Because of these power differences, it is possible that CFA users will look at these data and conclude that there are no types or antitypes (binomial test using Stirling formula, X^2-test, binomial approximation of the normal distribution, z-test), only a few types or antitypes (binomial test, Anscombe's test), or that (almost) all configurations constitute types or antitypes (Lehmacher tests). We therefore need guidance concerning the selection of tests. The choice of a particular test determines, as can be seen from Tables 12 and 13, to a certain degree whether types and antitypes can be found, and what the characteristics of these findings are. In the next section we present guidelines concerning the selection of significance tests for global CFA.

3.8 Selecting significance tests for global CFA

Table 14 summarizes the results available for the eight CFA significance tests discussed in the last sections for global CFA, the binomial test, the binomial test based on Stirling's approximation of factorials, the Pearson X^2-component test, the normal approximation of the binomial test, the z-test, Lehmacher's test, Lehmacher's test with Küchenhoff's continuity correction, and Anscombe's z-approximation. When selecting a significance test for global CFA based on the results presented by Naud (1997) or von Eye and Rovine (1988), it should be kept in mind that these results are incomplete. Therefore, there will be a certain degree of uncertainty in this selection until these and other possible tests have been thoroughly investigated.

The results presented by Naud (1997; see also von Eye & Rovine,

1988) and the summary in Table 14 make it difficult to select one test as always the best. Clearly, when sampling is product-multinomial and the sample is very large, Lehmacher's test is the most powerful, even with Küchenhoff's continuity correction. Researchers would therefore select this test for global CFA when the sampling is product-multinomial. However, when sampling is multinomial and a base model for higher order CFA or regional CFA was specified, one of thc other six tests must be selected.

When researchers opt for an exact test, there is currently only one pragmatic choice, the binomial test. There exist exact hypergeometric tests for product-multinomial sampling (Lehmacher, 1981; Lindner, 1984). However, none of the commonly used computer programs makes these tests available. The Sterling approximation-based test is typically selected only to reduce the time needed for computations. This benefit is minimal when state-of-the-art computers are used. The binomial test has a number of desirable characteristics. First, it is exact. Thus, there is no need to make assumptions concerning the accuracy of an approximation to some sampling distribution. Second, the test is slightly conservative (see Section 3.2). The danger of committing an α-error is close to the nominal threshold, α. Third, the test works equally well for samples of small and medium sizes. Numerical problems will occur only for very large samples. In addition, the test is nearly equally sensitive to both types and antitypes. Thus, the binomial test is a good overall choice for the search for types and antitypes.

The X^2 component test is a good choice when a numerically non-intensive test is needed. This may be the case when a pocket calculator is used to perform a CFA. The test is inconsistent in its nature as conservative versus nonconservative. As indicated in Table 11, this test can be very conservative, within range, or very liberal, depending on testing situation and size of table. For small and medium sample sizes, the test is clearly less sensitive to antitypes than to types. This was illustrated in Figure 1 which illustrates that this lack of sensitivity occurs in particular when $p > 0.5$. Thus, the Pearson X^2-component test is the test of choice when numerical simplicity is of importance and when the focus of exploration is on types.

The binomial normal approximation and the z-test are largely equivalent. These tests perform very well overall, and are not as much biased against antitypes as the X^2-test is. The approximation of the normal distribution is good when the sample is large. That is, when the expected frequency for a configuration is large, the z-tests can be trusted (Osterkorn, 1975). The test is easily calculated. Overall, this test is a good choice when samples are relatively large and when overall performance is of importance.

Table 14: Characteristics of eight tests for global CFA

Characteristic	Test							
	binomial	binomial/ Sterling	X^2	binomial/ normal	z	Lehmacher	Lehmacher/ Küchenhoff	Anscombe's z
exact (e)/ asymptotic (a)	e	e/a	a	a	a	a	a	a
sampling	any	any	any	any	any	p[c]	p	any
constraints	$E_i > 0$	$E_i > 0$	$E_i > 0$	$E_i > 0$	$E_i > 0$	$E_i > 0$	$E_i > 0$; $N_i - E_i > 0.5$	$E_i > .16667$
power rank[a]	5	6	7	4	3	1	2	≈ 2.5[b]
conservative (c)/liberal[a] (l)	c	c	c/l	c	-	l	l	-

[a] based on the existing, incomplete results; varies with type of test, size of table, p, and number of variables; depends also on whether $N_i > E_i$ or $N_i < E_i$; there exists insufficient information to classify the z-test and Anscombe's test as conservative or liberal; Rank 1 indicates most power

[b]This score is a guess based on the results presented in Table 12.

[c]p = product-multinomial

The Lehmacher (1981) test is clearly the most powerful. However, the test suffers from one major shortcoming. It requires very large samples. If samples are small, the test can be very nonconservative. Küchenhoff's (1986) continuity correction helps keep this problem under control. This approximation is meaningful only if $N_i - E_i > 0.25$. In addition, the test can be applied only when sampling is product-multinomial and when no covariates are considered. Therefore, this test is a good choice when (a) the sample is large, (b) maximum power is needed, and (c) sampling is product-multinomial.

Anscombe's (1953) *z*-approximation performed well in the empirical applications known to the authors (see Table 13). It requires that the expected cell frequencies be greater than 0.166667. It seems to have more power than the *z*-test, yet less than the Lehmacher test, even with continuity correction. It does seem to be biased in favor of antitypes. However, before the test can be generally recommended, more information is needed, in particular with regard to its small sample performance and to its performance under different testing conditions and in different table sizes.

In the examples that follow, we use most of these eight tests. The next section discusses the sensitivity of CFA tests to types and antitypes and the probability of finding types and antitypes in general.

3.9 Finding types and antitypes: Issues of differential power

Researchers seem to disagree as to the usefulness, interpretability, and even meaningfulness of types and antitypes. Von Eye, Spiel, and Wood (1996a, b) discuss CFA under the assumption that types and antitypes are equally important. Bergman (1996) considers the antitype "a central concept in CFA", and emphasizes that antitypes indicate to the researcher "what does not go together" (1996, p. 332). This can be of importance, for instance, in semantics where concepts are incompatible, a characteristic which can manifest in terms of an antitype, or in marketing, where antitypes indicate niches in the market that are not covered by any product. In contrast to the above authors, Krauth (1996b) does "not think that this concept carries much value for interpreting data" (p. 335).

Regardless of what the use of types and antitypes can possibly be, a researcher's chance of detecting types and antitypes in sparse tables is reduced for four reasons:

(1) When sample sizes are small, the number of instances in which there can be deviations from expectancy that are large enough to qualify as type or antitype is reduced.

(2) The asymptotic test statistics can be far from the theoretical sampling distribution for small sample sizes; to avoid non-conservative decisions, the statistics are typically constructed such that they err on the conservative side, a characteristic that often prevents researchers from identifying types and antitypes.

(3) CFA is typically employed in exploratory contexts where many significance tests are performed; to prevent capitalization from chance, the test-wise significance level α needs to be protected, which leads to possibly prohibitively small significance thresholds.

(4) The probability of detecting antitypes may be reduced in sparse tables even more than the probability of detecting types.

To illustrate this last issue, Indurkhya and von Eye (2000) present the following 2 x 2 x 2 table, which they analyzed using first order CFA with the z-approximation of the X^2-test. The test-wise α was adjusted to be $\alpha^* = 0.00625$. Table 15 presents the CFA results for these artificial sample data.

Table 15: CFA of Indurkhya and von Eye's (2000) sample data

Cell index	Frequencies		Statistical tests		Type/ Antitype?
	observed	expected	z	$p(z)$	
111	3	1.32	1.462	0.072	
112	0	1.08	-1.039	0.149	
121	3	1.98	0.725	0.234	
122	0	1.62	-1.273	0.102	
211	5	3.08	1.094	0.137	
212	0	2.52	-1.587	0.056	
221	0	4.62	-2.149	0.016	
222	9	3.78	2.685	0.004	T

The data in Table 15 were constructed such that they are as extreme as possible. That is, the differences between the observed frequencies are as large as possible. The only two constraints were that (a) the expected cell frequencies be no smaller than 1.0 and (b) the sample size was fixed to $N = 20$. Making the observed frequency distribution more extreme than the one presented will result in expected cell frequencies smaller than 1.0.

The results of a CFA of the data in Table 15 suggest that there is one type and no antitype. We now ask whether this result reflects a systematic pattern or is just a happenstance. Consider the case where $N_i = 0$. The z-value that must be exceeded for $\alpha^* = 0.00625$ is approximately 2.5. For $N_i = 0$, the z-value of 2.5 implies that we need an expected cell frequency of at least $E_i \geq 2.5^2$, that is $E_i = 6.25$ for an antitype to emerge. This value can be calculated from

$$\frac{N_i - E_i}{\sqrt{E_i}} \geq -z_{\alpha;\ crit} ,$$

where $-z_{\alpha;\ crit}$ is the z-value needed for an antitype to emerge. The expected cell frequencies in a 2 x 2 x 2 table can be estimated by

$$E_i = \frac{N_{i..}\ N_{.j.}\ N_{..k}}{N^2}.$$

Inserting $E_i = 6.25$ and $N = 20$, we can calculate that the product of the marginals in the numerator of this equation must be at least $400 \cdot 6.25 = 2500$ for an antitype to emerge. For the distribution in Table 15 we find that the maximum marginal product is $14 \cdot 12 \cdot 11 = 1848$. As a result, there is no possible candidate for an antitype in the distribution in Table 15, but a type did emerge.

Indurkhya and von Eye (2000) performed simulations on the performance of the Pearson X^2 test, in which they used cross-classifications of two or three variables, and critical minimal expected cell frequencies of 1.0 and 0.5. Later simulations (von Eye, 2002) also included the z-test, Anscombe's test, and Lehmacher's test. A "typical" result of these simulations appears in Figure 2 for 2 x 2 x 2 tables. Figure 2 displays the simulated sample size in the abscissa and the antitype-to-type ratio in the ordinate.

Figure 2: Type-to-Antitype Ratio for four CFA Tests

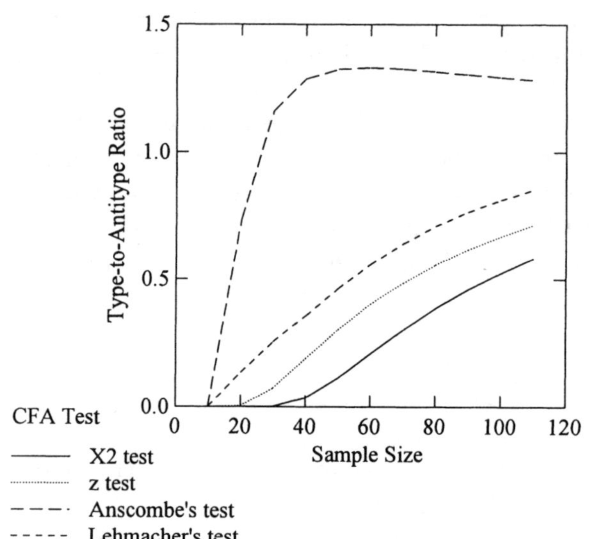

The simulation results suggest that

(1) for small and middle size samples, the X^2-test, the z-test, and Lehmacher's test identify more types than antitypes; only for very large samples, this ratio is inverted (not shown here; see von Eye, 2002);

(2) for Anscombe's test, the number of possible types is always smaller than the number of possible antitypes if $N > 20$;

(3) with the exception of the small sample sizes below 20, where Anscombe's z-approximation is the least biased, Lehmacher's test is the best; it approaches the optimal ratio of 1 the quickest; Pearson's X^2 approaches this ratio the slowest;

(4) when the sample size increases, this bias shrinks; for large samples (500 and larger for 2 x 2 x 2 tables), none of the tests shows a strong bias (not shown here; see von Eye, 2002).

Simulations for 2 x 2 and 3 x 3 tables led to very similar patterns of results. One exception was that the Lehmacher test showed no bias at all in 2 x 2 tables. Future research will have to show whether it is possible to devise methods of protecting the test-wise α-error that take this discrepancy between types and antitypes into account.

3.10 Methods of protecting α

As was hinted at repeatedly in the earlier chapters, proper application of CFA requires protection of the test-wise α. In other words, one must make sure that each the type/antitype decision made in CFA is made with an error rate as close as possible to the desired significance threshold, also called the *nominal* α. There are two major reasons why this is necessary.

The first reason is known as *mutual dependence of multiple tests*. Consider the researcher who uses one data set to test hypotheses. When this researcher performs one significance test, the probability of rejecting a false null hypothesis is α, the significance threshold. Performing only one test, however, is rather unusual. Most typically, researchers perform many tests concerning many hypotheses using the same data set. Whenever the number of tests performed on the same data set is greater than one, the possibility that these tests are dependent upon each other cannot be excluded. For example, Steiger, Shapiro, and Browne (1985) showed that, X^2-tests, when sequentially applied to the same data, can be asymptotically quite highly intercorrelated. This dependency can lead researchers to heavily underestimate the *factual* (as compared to the *nominal*) α level. Therefore, the probability of false rejections of true null hypotheses can increase.

If the same data are analyzed twice at the nominal level of $\alpha = 0.05$, this nominal α level will apply to the first test only. In extreme cases, the conditional probability for the second test to suggest a wrong statistical decision concerning the null hypothesis might be $\alpha = 1$. The null hypothesis may no longer stand a chance of surviving, regardless of whether it is true or false. Krauth and Lienert (1973) present an example of such an extreme case in which a researcher first employs Wilcoxon's Rank Sum-test. In a second step, the researcher attempts a "cross-validation" of results by employing the equivalent Mann-Whitney U-test to the same data. Both tests are nonparametric and are used for mean comparisons. If the null hypothesis is rejected by Wilcoxon's test, it is very unlikely that it will survive when the Mann-Whitney test is used. In other words, the null hypothesis does not stand a fair chance of surviving if both tests are applied to the same data one after the other. In CFA, the problem of *mutual dependence of tests* virtually always arises because usually all configurations are tested.

The problem of mutual dependence of multiple tests is both closely related to and exacerbated by the second problem, that of *multiple testing*. This problem results from the fact that each statistical test comes with an error probability of size α. When researchers perform more than one test, the likelihood of their *capitalizing on chance*, that is, making a false statistical decision concerning the null hypothesis, is high. This applies even if the tests are independent. As a result, null hypotheses are rejected and types or antitypes are said to exist even though they do not exist. Consider the following example. A researcher performs a CFA on a 3 x 3 x 3 cross-classification. That is, the researcher performs 27 tests. Let the nominal α be the usual 0.05. Then, the chance of committing three Type I errors, that is, the chance of declaring three configurations types or antitypes that in truth, do not constitute types or antitypes, is $p = 0.1505$, even if the tests are independent.

The problems of dependency of tests and multiple testing are not specific to CFA. Still, a good number of the methods available for protecting researchers from making wrong decisions concerning the null hypothesis have been devised in the context of CFA, and other methods, devised in different contexts, can be employed in CFA. All of these methods aim at controlling the significance threshold α such that the factual α is as close to the nominal α as possible. To control the factual α one can select from a number of strategies. Each of these strategies results in an adjusted significance threshold α* which can be far lower than the original, nominal threshold α. These strategies are (cf. Shaffer, 1995):

(1) *Protection of the local level α.* This method of protection guarantees that for each separate hypothesis test concerning a single configuration the factual α is not greater than the nominal α (Perli et al., 1987);

(2) *Protection of the global level α.* This method of protection guarantees that the probability of a false rejection of at least one type or antitype null hypothesis does not exceed α (Perli et al., 1987);

(3) *Protection of the multiple level α.* This method of protection guarantees that the probability of committing a Type I error when deciding about at least one null hypothesis does not exceed α, regardless of which other typal null hypotheses hold true (Perli et al., 1987). Methods of protection of multiple levels are also known

as experimentwise or familywise techniques (Cribbie, Holland, & Keselman, 1999; Dunnett & Tamhane, 1992; Keselman, Cribbie, & Holland, 1999; Williams, Jones, & Tukey, 1999)

(4) *Relaxed protection of the multiple level α.* When the number of tests is large, the adjusted significance threshold can become prohibitively low. Therefore, Benjamini and Hochberg (1995) and Cribbie et al. (1999) proposed protecting k or more decisions with $k > 1$.

Perli, Hommel, and Lehmacher (1985) proposed that researchers protect the local level α in *exploratory CFA*, that is, the routine application of CFA. For *confirmatory CFA*, that is, when testing is confined to an a priori specified selection of configurations, Perli et al. recommend controlling the multiple level α. In the following sections, we present the methods commonly used in CFA for protection against Type I errors. Dunnett and Tamhane (1992) classify procedures for adjusting significance levels into three groups: single step (SS), step down (SD), and step up (SU). For the SD and the SU procedures, hypotheses must be ordered based on their p values (or test statistics). Adjusted significance levels are calculated specifically for each individual hypothesis. The SD procedures begin with the hypothesis with the smallest p value. The SU procedures begin with the hypothesis with the largest p value. The SS procedures determine a single threshold that is applied to each individual test. There is no need to order the p values (Olejnik, Li, Supattathum, & Huberty, 1997).

3.10.1 The Bonferroni α protection (SS)

The original Bonferroni method of α adjustment is a single step procedure. Let α_i be the α error of the test for the ith configuration, for $i = 1, ..., r$, and r is the number of configurations to be examined. Let α^* be the probability that at least one test leads to a false rejection of H_0. Then, to control the local level α, the Bonferroni procedure determines each α_i such that two conditions are met. First, the sum of all α_i values does not exceed the nominal α or, more specifically,

$$\sum_i \alpha_i \leq \alpha.$$

Second, the Bonferroni procedure requires that all α_i be equal, or

$$\alpha_i = \alpha^* \qquad \textit{for all } i = 1, ..., r,$$

where α^* is the adjusted significance threshold α. The value of α^* that fulfills both conditions is $\alpha^* = \alpha/r$. Krauth and Lienert (1973) showed that adjustment according to Bonferroni renders statistical testing only slightly more conservative when all r tests are independent of each other.

To illustrate the Bonferroni adjustment, consider the following example. A 2 x 3 x 4 cross-classification is subjected to CFA. If all 24 configurations in this cross-classification are included in a CFA, and the nominal α is set to 0.05, the Bonferroni procedure yields the adjusted $\alpha^* = 0.05/24 = 0.00208333$ (a table with Bonferroni-adjusted αs appears in Appendix B).

3.10.2 Holm's procedure for α protection (SD)

The method of protecting the local level α proposed by Holm (1979) is a step down procedure. The method sets the significance level individually and sequentially. Because the significance level is determined individually for each configuration, the adjusted threshold is no longer the same for each test. The adjusted α is determined as

$$\alpha_i^* = \frac{\alpha}{r - i + 1},$$

where i is the number of the test for $i = 1, ..., r$. Before performing significance testing under the Holm procedure, the probabilities for each test must be ordered in an ascending order. That is, $i = 1$ for the smallest probability.

The Holm procedure can be illustrated as follows. The successive significance levels are for the first, second, $(r - 1)^{st}$, and rth test, respectively,

$$\alpha_1^* = \frac{\alpha}{r - 1 + 1} = \frac{\alpha}{r},$$

$$\alpha_2^* = \frac{\alpha}{r - 2 + 1} = \frac{\alpha}{r - 1},$$

$$\alpha_{r-1}^* = \frac{\alpha}{r - (r - 1) + 1} = \frac{\alpha}{2},$$

and

$$\alpha_r^* = \frac{\alpha}{r - r + 1} = \alpha.$$

These equations show that the Holm and the Bonferroni procedures start with the same adjusted α, that is, with $\alpha^* = \alpha/r$. Already at the second step, at which the Bonferroni procedure still uses $\alpha^* = \alpha/r$, the Holm procedure is less prohibitive, using $\alpha^* = \alpha/(r - 1)$. At the last possible step, the Holm procedure yields $\alpha^* = \alpha$.

The Holm procedure is computationally more intensive than the Bonferroni procedure because the p values need to be ordered and the α_i^* need to be calculated for each i. However, some of this effort is made up when fewer than r configurations constitute types or antitypes. As soon as the first null hypothesis in the sequence of tests prevails, all remaining null hypotheses in the order are retained. Please note that, ordering the p values and concluding the testing after a null hypothesis prevailed can also be done in Bonferroni's procedure.

3.10.3 Hochberg's procedure for α protection (SU)

Hochberg (1988) proposed a *step up* procedure that is based on the Simes (1986) inequality. In contrast to Holm's procedure, this approach requires the p values to be ordered in *descending* order, that is, $i = 1$ for the *largest* probability. The sequence of adjusted α values is then

$$\alpha_1^* = \alpha ,$$

$$\alpha_i^* = \frac{\alpha}{i} ,$$

and

$$\alpha_r^* = \frac{\alpha}{r} ,$$

for $i = 1, ..., r$. The testing routine is slightly different than the one for Holm's method. Null hypotheses are tested sequentially until the first of them can be rejected. Then, all remaining null hypotheses in the order are rejected.

Although Hochberg's method uses the same criterion as Holm's method, it may be slightly more powerful as was suggested by Olejnik et al.'s simulation results (1997). However, these differences in power seem to appear mostly in the third decimal. Therefore, their effect on the probability of detecting types and antitypes can be expected to be minimal.

3.10.4 Holland and Copenhaver's procedure for α protection (SD)

Holland and Copenhaver (1987) base an improvement of the original Bonferroni procedure on the Sidak (1967) inequality. For this procedure, the probabilities p_i must be arranged in an ascending order, that is, $i = 1$ for the smallest probability. Then, the adjusted threshold α* for Configuration i is calculated as

$$\alpha_i^* = 1 - (1 - \alpha)^{\frac{1}{r - i + 1}}.$$

This criterion is slightly less restrictive than the one used by Holm or Hochberg. Thus, the power of tests under this procedure can be expected to be slightly greater than the one under Holm or Hochberg. The simulation results presented by Olejnik et al. (1977) suggest however, that the power differences between these α protection procedures are minimal, and Hochberg's procedure may actually be slightly more powerful than both Holm's and Holland and Copenhaver's methods. A newly improved, adaptive method was recently proposed by Benjamini and Hochberg (2000).

3.10.5 Hommel, Lehmacher, and Perli's modifications of Holm's procedure for protection of the multiple level α (SD)

Hommel, Lehmacher, and Perli (1985; see also Hommel, 1988, 1989), proposed a further modification of Holm's procedure. These modifications can be applied to two-dimensional and three-dimensional tables (see below). This adjustment procedure uses results by Marcus, Peritz, and Gabriel (1976) which suggest that, under certain conditions hypotheses on single cells can be viewed as intersections of m other cell hypotheses. The following results hold true for two-dimensional tables with r cells:

(1) if m null hypotheses cannot be rejected, the remaining $r - m$ cannot be rejected either and therefore the global null hypothesis holds true (for $r > m \geq r - 3$);

(2) if $m = r - 5$ null hypotheses hold true, at least one additional null hypothesis does also hold true.

From these results it can be derived that in a sequence of tests, a certain number of tests for types and antitypes can be performed at the same α level. This level is less restrictive than those determined using the original Holm procedure. Specifically, consider a two-dimensional cross-classification with three or more rows and three or more columns. For this table, the sequence of significance tests is

$$\alpha_1^* = \frac{\alpha}{r},$$

$$\alpha_2^* = \ldots = \alpha_5^* = \frac{\alpha}{r - 4},$$

$$\alpha_6^* = \alpha_7^* = \frac{\alpha}{r - 6},$$

$$\alpha_8^* = \frac{\alpha}{r - 7},$$

$$\alpha_{r-1}^* = \frac{\alpha}{2},$$

and

$$\alpha_r^* = \alpha .$$

As does the original Holm procedure, this modification requires the *p* values to be arranged in an ascending order.

Hommel et al. (1985) showed that this strategy can be improved even more if one considers that the first test, $\alpha_1^* = \alpha/r$ corresponds to a Bonferroni test of the global null hypothesis that the CFA base model describes the data sufficiently. Based on this consideration, this first test can be substituted by the goodness-of-fit test of the CFA base model for the two-way table under study. If the test suggests that the base model fits, subsequent CFA tests are unnecessary. If, in contrast, the goodness-of-fit test indicates significant base model-data discrepancies, the following adjusted significance thresholds can be used:

$$\alpha_1^* = \alpha_2^* = \ldots = \alpha_5^* = \frac{\alpha}{r - 4},$$

$$\alpha_6^* = \alpha_7^* = \frac{\alpha}{r - 6},$$

$$\alpha_8^* = \frac{\alpha}{r - 7},$$

$$\alpha_{r-1}^* = \frac{\alpha}{2},$$

and

$$\alpha_r^* = \alpha .$$

Obviously, this procedure yields a less restrictive significance threshold already for the first test. Beginning with the sixth test, the two procedures proposed by Hommel et al. (1985) use the same α*.

Perli, Hommel, and Lehmacher (1985) present an extension of this procedure for use in three-way tables. One obtains the adjusted α-levels

$$\alpha_1^* = \frac{\alpha}{r}$$

$$\alpha_2^* = \ldots = \alpha_5^* = \frac{\alpha}{r - 4}$$

$$\alpha_m^* = \frac{\alpha}{r - m + 1},$$

for $m = 6, \ldots, r$. For four-way and higher-way tables, this procedure is not recommended. Perli et al. (1985; cf. Naud, 1997) suggest that the tests used in CFA tend to become nonconservative. Therefore, in higher dimensional tables, most researchers resort to using the more conservative Bonferroni and Holm procedures.

3. 10. 6 Illustrating the procedures for protecting the test-wise α

In this section, we present two illustrations of the procedures for protecting the test-wise α. First, we calculate the adjusted significance thresholds for a cross-classification of two variables with $r = 3 \times 3 = 9$ cells. Second, we apply the various procedures in the analysis of a data set.

Illustration I: Calculation of adjusted significance values. The results of the first illustration appear in Table 16.

Table 16: Adjusted significance thresholds α* for 3 x 3 cross-classification

Test number	Adjustment procedure					
	Bonferroni	Holm	Hochberg	Holland & Copenhaver	Hommel et al. I	Hommel et al. II
1	.00556[a]	.00556	.00556	.00568	.00556	.01[a]
2	.00556	.00625	.00625	.00639	.01	.01
3	.00556	.00714	.00714	.0073	.01	.01
4	.00556	.00833	.00833	.00851	.01	.01
5	.00556	.01	.01	.01021	.01	.01
6	.00556	.0125	.0125	.01274	.01667	.01667
7	.00556	.01667	.01667	.01695	.01667	.01667
8	.00556	.025	.025	.02532	.025	.025
9	.00556	.05	.05	.05	.05	.05

[a] leading and trailing zeros were omitted in the table

The adjusted significance values in Table 16 suggest that

(1) with the exception of Bonferroni's procedure, all adjustment procedures suggest that α* = α for the test with the smallest test statistic (or largest probability; see last row of table);

(2) among the methods applicable to tables of any size, Holland and Copenhaver's procedure is the most liberal one, suggesting the least prohibitive significance thresholds;

(3) the improved methods proposed by Hommel et al. are even less restrictive than Holland and Copenhaver's procedure; however, they can be used for two- and three-dimensional tables only;

(4) many researchers consider Holland and Copenhaver's procedure

largely equivalent to Holm's, because the improvements appear only in the fourth decimal or even later;

(5) Bonferroni's procedure suggests the most conservative decisions in all tests except the first, where it uses the same α^* as three of the five alternatives.

It should be noticed that although Table 16 suggests that Holm's and Hochberg's procedures are identical, they are not. As was described in the sections above, Holm's procedure starts with the smallest p and proceeds in an ascending sequence, whereas Hochberg's procedure starts with the largest p and proceeds in a descending sequence. Both procedures stop testing null hypotheses as soon as the first test suggests a different interpretation of a null hypothesis than the previous tests. The results from Table 16 are depicted in the form of a bar chart in Figure 3.

Figure 3: Comparing six methods of alpha adjustment

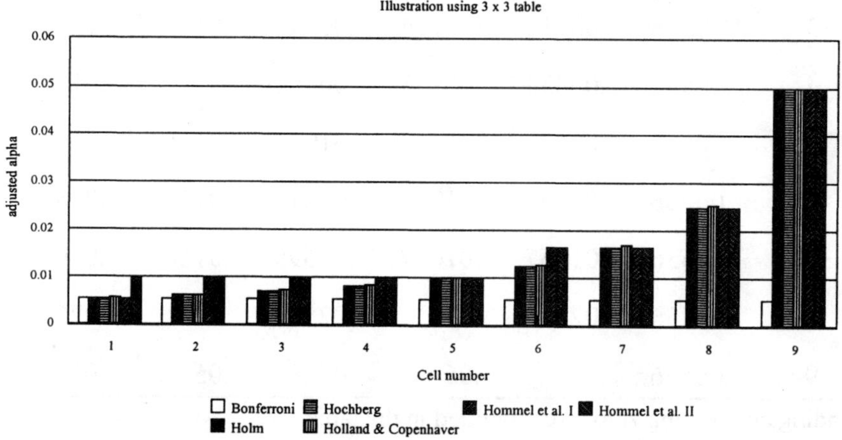

Illustration II: Analysis of empirical data. In this section, we illustrate the benefits of the alternative procedures for protecting the test-wise α by applying them in the analysis of an empirical data set. In a study on the development of aggressive behaviors in adolescents, (Finkelstein, von Eye, & Preece, 1994) assessed the Tanner stage of 83 respondents, that is, the progress of physical pubertal development. For the present purposes, we analyze the Tanner stages observed in the years 1983 and 1985. The variable indicating Tanner stage had been categorized to have four levels with Level 1 indicating prepubertal and Level 4 indicating physically mature. Level 4 did not occur in 1983, and Level 1 did not occur in 1985.

Therefore, the cross-tabulation of the two Tanner stage observations has 3 x 3 rather than 4 x 4 categories. The eliminated 7 cells would have been empty.

Table 17 displays the results of a first order CFA of the cross-classification of the two Tanner observations. Analysis was done using Lehmacher's Test with Küchenhoff's continuity correction.

The significance thresholds in Table 17 that are surpassed by empirical tail probabilities are highlighted. As a matter of course, the two types, constituted by Cells 12 and 34, and the two antitypes, constituted by Cells 14 and 22, that were identified using the conservative Bonferroni procedure, appear also under all other procedures. In addition, however, the more liberal procedures allow one to label Cell 32 as constituting an antitype. This configuration failed to make it beyond the threshold posed under Bonferroni. None of the other configurations represents a type or antitype. However, Cell 23 comes close to constituting a type when evaluated under the two procedures proposed by Hommel et al. (1985).

Substantively, the two types suggest that more adolescents than expected from the base model of independence progress from Tanner stage 1 (prepubertal) to Tanner stage 2 (beginning stage of puberty) and from Tanner stage 3 (prematurational) to Tanner stage 4 (mature body). Developments that leap two stages, that is, from Tanner stage 1 to 4, are less likely than chance and thus constitute an antitype. Also less likely than chance is lack of development when Tanner stage 2 has been reached. Configuration 22 therefore constitutes the second antitype.

It should be noted that the second of the procedures proposed by Hommel et al. (1985) can be employed only if the CFA base model fails to describe the data satisfactorily. This is the case for the present data. We calculate a Pearson $X^2 = 43.97$ ($df = 4$) that suggests significant data-model discrepancies ($p < 0.01$).

Table 17: CFA of Finkelstein et al.'s (1994) Tanner stage data

Cell	Frequencies		z	$p(z)$	rank	Adjustment procedure[a]					
	o	e				Bonferroni	Holm	Hochberg	H & C	H I	H II
12	31	20.6	4.35	< 0.01	2	0.00556	0.00625	0.00625	0.00639	0.01	0.01
13	7	10.1	-1.28	.1009	7	0.00556	0.01667	0.01667	0.01695	0.01667	0.01667
14	0	7.3	-3.79	.0001	3	0.00556	0.00714	0.00714	0.0073	0.01	0.01
22	14	21.2	-2.92	.002	4	0.00556	0.00833	0.00833	0.00851	0.01	0.01
23	15	10.3	2.06	.0196	6	0.00556	0.0125	0.0125	0.01274	0.01667	0.01667
24	10	7.5	1.09	.1360	8	0.00556	0.025	0.025	0.02532	0.025	0.025
32	0	3.3	-2.3	.01	5	0.00556	0.01	0.01	0.01021	0.01	0.01
33	0	1.6	-1.0	.1490	9	0.00556	0.05	0.05	0.05	0.05	0.05
34	6	1.2	4.64	<.001	1	0.00556	0.00556	0.00556	0.00568	0.00556	0.01

[a]H & C is Holland and Copenhaver; H I is Hommel et al.'s first procedure; and H II is Hommel et al.'s improved procedure; trailing zeros omitted

4. Descriptive Measures in Global CFA

Descriptive measures in global CFA have not been discussed until very recently, when von Eye and Gutiérrez-Peña (in preparation) proposed using two measures descriptively that have also been used in efforts of Bayesian data mining (DuMouchel, 1999). The consideration of descriptive measures can be useful for four reasons. First, the measures discussed here are most intuitive and can easily be interpreted. Second, these measures are sensitive to different data characteristics than the residual-based statistics discussed in Chapter 2 of this volume. The differences between measures are discussed below. Third, the measures do not require assumptions concerning sampling schemes or approximations of sampling distributions. Thus, they are useful under almost any condition. Fourth, because these measures are used descriptively and in an exploratory context, they are particularly useful when tables are sparse and the significance tests cannot be trusted any more, or when tables are so large that the adjusted significance levels are prohibitively small. DuMouchel (1999) illustrated the use of these measures in tables with over 1.4 million cells. The Bonferroni-adjusted α for a table of this size is $\alpha^* = 3.57E^{-8}$, a value that can be exceeded only with very large samples (or extremely small expected cell frequencies).

4.1 The relative risk ratio, *RR*

The first measure to be introduced here is the *relative risk ratio, RR*. This measure simply relates the *i*th observed cell frequency, N_i, to the corresponding expected cell frequency, E_i,

$$RR_i = \frac{N_i}{E_i},$$

where *i* goes over all cells in a cross-classification. The interpretation of *RR* is intuitive and straightforward. When $RR_i = 1$, the number of cases observed for Cell *i* is the same as the expected number. We thus conclude that for Cell *i*, the base model describes the data adequately. If $RR_i < 1$, Cell *i* contains fewer cases than expected. For example, if $N_i / E_i = 0.25$, only 25% of the expected cases did display the pattern of the *i*th configuration. If $RR_i > 1$, Cell *i* contains more cases than expected from the base model. If, for example, $N_i / E_i = 100$, 100 times as many cases were observed as had been expected.

Obviously, the relative risk ratio, *RR* is a descriptive measure. It differs from inferential measures in its use and in its characteristics. For instance, situations with, say $E_{i1} = 5$ have a different interpretation than situations with $E_{i2} = 0.0005$, even if they come with the same RR_i. In the former, all of the CFA tests discussed in this text can be applied. In the latter, none of the asymptotic tests can be applied, and many researchers even doubt whether the exact tests can be trusted - a consideration of lesser importance in the use of *RR*. $RR = 0$ if $N_i = 0$, regardless of the value assumed by E_i.

A comparison of *RR* with other measures and a sample application follow in Section 4.3.

4.2 The measure log *P*

The second descriptive measure to be introduced for use in global CFA is log *P*, defined as

$$\log P = -\log_{10}(Pr[X \geq N_i]),$$

where X ~ Possion (E_i). When log *P* becomes large, that is, $\log P > 12$, the following quantity approximates log *P*:

$$-\log_{10}\left[\sum_{n=N}^{N+3} e^{-E}\frac{E^n}{n!}\right] = \frac{E}{\log 10} - N\log_{10}E + \\ + \log_{10}N! - \log_{10}\left[1 + \frac{E}{N+1} + \right. \\ \left. + \frac{E^2}{(N+1)(N+2)} + \frac{E^3}{(N+1)(N+2)(N+3)}\right]$$

where N is the observed cell frequency and E is the expected cell frequency. If E_j is large, the normal approximation of the Poisson can be exploited and the test statistic

$$\chi_i = \frac{N_i - E_i}{\sqrt{E_i}}$$

can be used. The Pearson X^2 under $df = 1$ "would also be expected to behave similarly" (DuMouchel, 1999, p.180) to log P. Log P indicates the probability of N_i under a Poisson distribution, when E_i was expected. More specifically, log P indicates the negative value of the exponent of base 10. For example (DuMouchel, 1999, p. 180), if $N_i = 100$ and $E_i = 1$, log P = 158.4. The corresponding probability is $p = 10^{-158.4}$. DuMouchel (1999, p. 180) concludes that a probability as small as this has "no meaning except as a possible value for ranking cells" in the cross-classification under study. The corresponding X^2 component would be $X^2 = 9801$, an equally colossal and unlikely value ($p \approx 0.0$).

4.3 Comparing the X^2 component with the relative risk ratio and log *P*

In this section, we compare the two descriptive measures *RR* and log *P* with the well known Pearson X^2 component.

Comparing *RR* with the X^2 component. At first look, it seems as if *RR* and X^2 measure the same data characteristics. Indeed, simulations by von Eye and Gutiérrez-Peña (in preparation) show that the correlation between *RR* and X^2 approximates $r = 1.0$ when $(N_i - E_i) > E_i$. However, for $(N_i - E_i) < E_i$, the correlation between the two measures approximates $r = -1.0$. This somewhat surprising behavior can be explained using the two panels of Figure 4. The two panels of Figure 4 indicate that the values of both *RR* and the X^2-component increase with the difference between the observed and the expected cell frequencies, as long as $(N_i - E_i) > E_i$. This can be seen in the right-hand sides of both panels of the figure. However, when $(N_i - E_i) < E_i$, RR_i still decreases whereas the Pearson X^2-component increases, but at a lesser rate than for $(N_i - E_i) > E_i$ (cf. Figure 1 in Section 3.4).

Figure 4: **X^2-components (left panel) and *RR* (right panel) for $(N_i - E_i) > E_i$ and $(N_i - E_i) < E_i$**

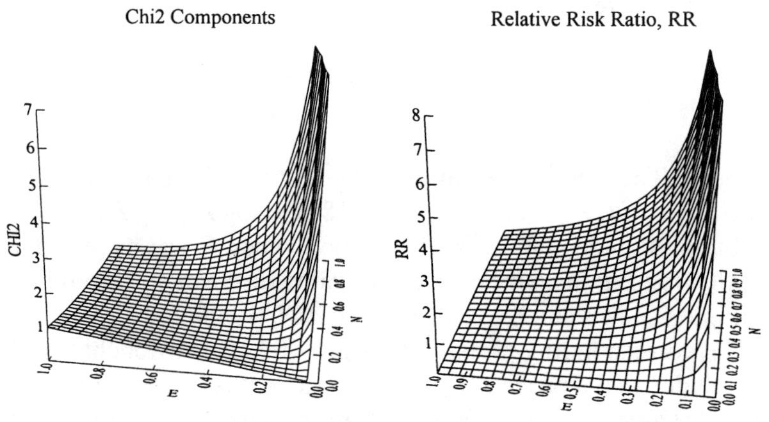

From the perspective of employing the two measures in CFA, this comparison suggests that

(1) both X^2 and *RR* indicate CFA types by large values;

(2) whereas X^2 indicates CFA antitypes also by large values (these values are smaller than the values for types; for the asymmetry in this measure see Figure 1 in Section 3.4, and Section 3.9), *RR* indicates antitypes by values that approach zero;

(3) X^2 in combination with measures of α protection allows researchers to make a statistical decision as to the status of a configuration as type or antitype. In contrast, *RR* is typically used in a descriptive context. In this context, configurations are either not labeled as constituting types or antitypes, or the most extreme cells are identified and treated as types or antitypes.

Comparing log P with RR and the X^2 component.
The simulation results by von Eye and Gutiérrez-Peña (in preparation) also suggest that the correlation between log P and the X^2 component is consistently high, and that the correlation between log P and RR changes in parallel with the correlation between X^2 and RR. However, the relationship between these three measures seems to be neither linear nor easy to interpret. More specifically, the authors show that

(1) the relationship between log P and RR is positive and approximates a straight line for very small E_i;
(2) the relationship between log P and RR becomes increasingly concave as E increases; specifically, it resembles a parabola for $E_i = 0.5N_j$ and is negative for $E_i > 0.5N_i$;
(3) the relationship between log P and X^2 is positive and convex for very small E_i;
(4) the relationship between log P and X^2 approximates a straight line as the expectancy approximates $E_i = 0.5N_i$, where it bifurcates;
(5) for $E_i > 0.5N_i$ the relationship between log P and X^2 is still positive but concave;
(6) the relationship between the Pearson X^2-component and RR is positive and convex for small values of E_i;
(7) the relationship between the Pearson X^2-component and RR can be described by a quadratic function when $E_i = 0.5N_i$;
(8) the relationship between the Pearson X^2-component and RR is convex yet negative for $E_i > 0.5N_i$.

What are the implications of these results for the user of CFA? Von Eye and Gutiérrez-Peña (in preparation) emphasize that the three measures correlate strongly when E is very small. Although the shape of the relationships is not linear, the rank order of the measures will be the same. We thus can expect that the three measures point to the same configurations as types when E_i is very small. However, when E_i increases and approximates $0.5N_i$, the three measures increasingly reflect different characteristics of the data. We thus have to expect the rank orders of the measures to differ, and patterns of types and antitypes will vary with measure. As E_i approximates N_i, the relationship among the three measures is clear again. However, the correlations between log P and RR_i, and the Pearson X^2-components and RR_i now are strongly negative. This last situation is of lesser importance in the present context, because the probability of finding types or antitypes approaches zero as the difference between N_i and E_i approaches zero.

Data example. One way to illustrate the use and usefulness of the descriptive measures of CFA is to analyze a large sparse table. The benefit would be that a CFA can be performed in such a context without using methods of statistical inference. Examples of this type have been presented by DuMouchel (1999) and von Eye and Gutiérrez-Peña (in preparation). In the present section, we present a smaller data example. However, even in this small example the rank orders of measures differ. The data for the following example were collected in a study on the interpretation of proverbs and sentences (von Eye, Jacobson, & Wills, 1990). A sample of 149 respondents described the meaning of proverbs and sentences. Two raters evaluated the written descriptions with respect to their concreteness. A description was rated as concrete (1) if the respondent interpreted the meaning of a proverb or sentence as concrete. Alternatively, interpretations were rated as abstract (3) or intermediate (2).

We now analyze the 3 x 3 cross-tabulation of the two raters' evaluations using the base model for first order CFA. We employ Pearson's X^2 test. Bonferroni-adjustment yields $\alpha^* = 0.005556$. Table 18 displays the results of standard CFA, along with the scores for log P and RR, as well as the ranks of the Pearson X^2 components and the two descriptive measures.

Standard inferential CFA reveals one type and one antitype. The type, constituted by Configuration 11, suggests that the two raters agree beyond chance in those cases in which the respondents had interpreted the sentences and proverbs as concrete. The antitype, constituted by Configuration 31, suggests that it is less likely than expected from the base model that Rater A thinks an interpretation is abstract, whereas Rater B thinks the interpretation is concrete. Not one incidence with pattern 31 was found.

We now ask whether the descriptive measures RR and log P describe different data characteristics than X^2. To obtain an overview, we correlate the rank orders of the two descriptive measures and X^2. Table 19 displays the Spearman rank correlations.

Table 18: **CFA of concreteness ratings of the interpretations of proverbs and sentences**

Cell Index	Frequencies		Statistics					Ranks		
Rater 1 2	observed	expected	X^2	$p(X^2)$	Type?	RR	log P	X^2	RR	log P
11	11	2.977	21.625	$< \alpha^*$ [a]	**T**	3.695	3.563	1	2	1
12	2	3.225	0.465	.4952		0.620	0.140	7	7	9
13	19	25.789	1.792	.1807		0.736	0.711	4	6	5
21	1	0.651	0.187	.6655		1.536	0.321	8	3	6
22	3	0.705	7.464	.0063		4.253	1.458	3	1	3
23	3	5.643	1.238	.2658		0.532	0.234	5	8	8
31	0	8.372	8.372	.0038	**A**	0.000	1.483	2	9	2
32	8	9.070	0.126	.7224		0.882	0.318	9	5	7
33	82	72.558	1.229	.2677		1.130	1.097	6	4	4

[a] $< \alpha^*$ indicates that the tail probability is smaller than can be expressed with four decimal places.

Table 19: Intercorrelations of the rank orders of X^2, RR, and log P

Correlations	X^2	RR
RR	0.07	
log P	0.78	0.40

Table 19 shows that information about the relative risk RR does not carry much information about the magnitude of X^2. Indeed, the ranks of the two measures are the same in only one instance (Rank 7 for Configuration 12). The largest discrepancy between the two rank orders can be found for Configuration 31, the antitype. The rank difference for this configuration is 7. In units of X^2, this configuration shows the second largest discrepancy between N_j and E_j. The relative risk of being in this cell, however, is zero, because the observed cell frequency is zero. The two measures X^2 and RR do not place the same configurations in the extreme ranks. Only Configuration 11 appears in the top two ranks for both measures. Therefore, the relative risk and the X^2 component may need to be considered separately for a complete interpretation of the present data. However, if $RR > 0$, the lowest RR-scores typically correspond to the extreme X^2-scores for antitypes.

The ranks of log P and X^2 are the same for three configurations (11, 22, and 31). The largest difference is three ranks (8 versus 5 for Configuration 23). Accordingly, the correlation between these two measures is relatively high, 0.78. We thus conclude that in the present data the probabilities of the discrepancies between N_j and E_j in the χ^2 and the Poisson distributions are relatively similar. Indeed, the two measures suggest the same decisions where it counts most in CFA, that is, in the extreme discrepancies: the first three ranks are the same.

The ranks of RR and log P are the same for Configurations 23 and 33. The largest difference is seven ranks. It can be found for Configuration 31 again. The correlation between these measures is 0.40. As in the comparison of RR and X^2, the comparison of RR and log P shows that the top three ranks are not occupied by the same configurations. Only Configurations 11 and 22 appear in both lists in the top three. Configuration 31 has the lowest rank for RR and the second rank for log P, and Configuration 21 has Rank 3 for RR and Rank 6 for log P.

Part II: Models and Applications of CFA

5. Global Models of CFA

In this chapter, we present sample global models of CFA. As was explained in Section 2.4, *global models of CFA* are similar in that all variables have the same status. Thus, there are no predictors or criterion variables, no dependent or independent variables, and no mediator or moderator variables. All variables have the same status. It is the goal of *exploratory global CFA* to identify configurations (profiles, cells, category patterns, etc.) that stand out because they were observed more often (*types*) or less often (*antitypes*) than expected based on chance. The chance concept is expressed using log-linear or other CFA base models that serve to (a) translate the chance concept into some numerically tractable form and (b) estimate the expected cell frequencies.

There is a *hierarchy of global CFA models* that groups models based on the assumptions made about the existence of effects. *Zero order CFA* assumes no effects whatsoever. As a result, expected frequencies are uniformly distributed. Types and antitypes can result from any kind of effect, that is, any main effect or interaction. *First order CFA* considers all main effects of variables. Therefore, types and antitypes can result only if variable interactions exist. *Second order CFA* considers all pair-wise interactions in addition to main effects. Types and antitypes can result only if second or higher order interactions exist. This applies accordingly to the *models of third and higher order CFA*.

All log-linear models that are used as base models for global CFA share the same form,

$$\log E = X\lambda,$$

where E is the array of expected cell frequencies, X is the design matrix that represents the CFA base model, and λ is a parameter vector (extended versions of this form have been proposed for CFA with covariates; see Section 10.5.2) and for CFA with structural zeros (see Section 10.1). Of particular importance in CFA applications are the design matrix X, because it represents the base model, and the estimated expected cell frequencies, because they are used in the search for types and antitypes. The interpretation of types and antitypes is always based on the specifications expressed in X.

In the following sections, we present sample global CFA models and applications. We show the design matrix for each model and interpret results with reference to the base models.

5.1 Zero order global CFA

Zero order CFA is the most basic form of CFA (Krüger, Lienert, Gebert, & von Eye, 1979; Lienert, 1980; Lienert & von Eye, 1984, 1985, 1989). In its base model, this form of CFA assumes that no effects exist at all. Therefore, the expected frequency distribution is uniform. The log-linear base model has the form

$$\log E = \mathbf{1}\lambda,$$

where **1** is a vector of constants, typically a vector of ones. λ contains only one parameter. Types and antitypes reflect the existence of main effects, interactions, or both.

To illustrate zero order CFA, we use a data set presented by Klingenspor, Marsiske, and von Eye (1993). The authors investigated the size of social networks in 258 women and 258 men over 70 years of age. For the present purposes we analyze the frequencies in the 2 x 2 x 2 cross-tabulation of the variables Marital Status (M; 1 = married, 2 = not married), Gender (G; 1 = men, 2 = women), and Size of Social Network (N; 1 = small, 2 = large). The log-linear base model for zero order CFA of this table is

$$\begin{bmatrix} \log e_{111} \\ \log e_{112} \\ \log e_{121} \\ \log e_{122} \\ \log e_{211} \\ \log e_{212} \\ \log e_{221} \\ \log e_{222} \end{bmatrix} = \begin{bmatrix} 1 \\ 1 \\ 1 \\ 1 \\ 1 \\ 1 \\ 1 \\ 1 \end{bmatrix} [\lambda].$$

Table 20 displays in its top panel the observed cell frequencies for the M x G x N cross-classification, the expected cell frequencies, the test statistics, the tail probabilities, and the type/antitype decisions. In its bottom panel, Table 20 displays descriptive statistics, and their rank orders. We use Anscombe's z-test because of its particular sensitivity to antitypes, and protect the test-wise α using the procedure proposed by Hommel et al. (1985; cf. the last columns in Tables 16 and 17). This method is the least conservative in three-dimensional tables.

Before employing the procedure proposed by Hommel et al. (1985), we have to make sure the base model does not allow us to describe the data satisfactorily. We calculate for the base model a Pearson $X^2 = 212.43$ ($df = 7$; $p < 0.01$). These values indicate significant model - data discrepancies and we reject the base model. Thus, we can use the procedure by Hommel et al. (1985).

Configurations that stand out as types or antitypes are labeled as T or A in Table 20. Obviously, each configuration constitutes a type or an antitype. Even the smallest test statistic of $z = 1.648$ for which $p(z) = 0.04968$ indicates a type, because 0.04968 is less than the Hommel et al.-adjusted $\alpha^* = 0.05$ for this cell. The result that each configuration constitutes a type or an antitype is due to the choice of the Hommel at al. procedure. Had we selected the Bonferroni procedure, Configurations 111 and 211 would not have emerged as antitype and type, respectively.

At the descriptive level, we find the highest relative risk for configuration 221. $RR_{221} = 2.02$ indicates that more than twice as many individuals than expected from the independence model were not married, female, and had small social networks, whereas 1.69 times as many

Table 20: Zero order CFA of the variables Marital Status (M), Gender (G), and Size of Social Network (N)

MGN	Frequencies		Statistics			α*
	observed	expected	*z*	*p*(*z*)		
		CFA results				
111	48	64.5	2.133	.01646	A	.025
112	87	64.5	2.681	.00375	T	.0125
121	5	64.5	9.836	< .0001	A	.00625
122	14	64.5	7.675	< .0001	A	.0125
211	78	64.5	1.648	.04968	T	.05
212	45	64.5	2.550	.00530	A	.01667
221	130	64.5	7.196	< .0001	T	.0125
222	109	64.5	5.066	< .0001	T	.0125
		Descriptives		Ranks		
MGN		*RR*	log *P*	*RR*		log *P*
111	48	0.744	1.457	5		7
112	87	1.349	2.482	3		5
121	5	0.078	18.085	8		1
122	14	0.217	11.672	7		3
211	78	1.209	1.443	4		8
212	45	0.698	1.820	6		6
221	130	1.420	4.153	2		4
222	109	1.219	1.809	4		6

respondents as expected were not married, female, and had large social networks ($RR_{222} = 1.69$).

The Poisson probability for Configuration 121 is the lowest (log $P_{121} = 18.09$). More specifically, this probability is $p = 10^{-18.085} = 0.8224 \cdot 10^{-18}$, that is, a number with 18 zeros before the first non-zero decimal. This number is only slightly larger than the probability of $p = 0.3939 \cdot 10^{-22}$ for the z-statistic for this configuration. The second most extreme Poisson probability, $p = 0.4864 \cdot 10^{-12}$, was calculated for Configuration 221. The probability for this configuration ranked third in the z-tests. Its Poisson probability was calculated to be $p = 0.3100 \cdot 10^{-12}$. The interpretation of these extreme probabilities is that these configurations are the least likely when the CFA base model is the null model. Configuration 121 is an antitype and configuration 221 is a type. We interpret the Poisson probabilities with no reference to a CFA null hypothesis because we will use the log P values even when the expected frequencies are small, for instance, in sparse tables (von Eye & Gutiérrez-Peña, in preparation), and in situations where the tail probabilities can be trusted only in the sense that they are extreme.

We now ask, why a researcher would use the base model of zero order CFA instead of any other base model. There are two answers to this question. First, it can be answered by describing the mechanism that leads to the detection of types and antitypes in zero order CFA. The log-linear model of zero order CFA results in uniformly distributed expected cell frequencies. As a result, configurations constitute types or antitypes when the distance between N_j and E_j is larger than a threshold that is determined by the number of tests, the nominal significance level, and the procedure used to protect the test-wise α. In other words, the magnitude of the distance of N_j from the constant E_j is the chief determinant of types and antitypes.

The constant E_j can be interpreted as the expected density in the space of the cross-classification spanned by the variables under study. If $N_j > E_j$, the density is greater than expected, and if $N_j < E_j$, the density is less than expected. The statistical tests in zero order CFA allow one to determine whether the $N_j - E_j$ differences are greater than could be expected based on chance alone.

The comparison of observed densities with an expected, average density can be seen as parallel to methods of cluster analysis. Using cluster analysis, researchers identify groups of cases that are closer or more similar to each other than to other cases. This is often done by minimizing the distance within a cluster and simultaneously maximizing the distance between clusters (Ward, 1963). The result of this optimization is a cluster that describes a sector of relatively high density. Zero order CFA types also describe sectors of increased density.

There are two differences between zero order CFA and clustering that need to be pointed out here. First, zero order CFA identifies not only sectors of high density (types), but also sectors of low density (antitypes). In this respect, CFA is unique. No other statistical method known to the authors allows one to identify sectors of low density. Second, CFA allows researchers to determine whether the deviations from average density are statistically significant. Standard clustering methods such as Ward's (1963) method are descriptive in the sense that significance tests are not employed.

The second answer to the question why a researcher would choose zero order CFA instead of other CFA models concerns the availability of prior knowledge. The log-linear base model given above for zero order CFA implies a design matrix with only one vector, the constant vector. No additional information is considered. Thus, the base model of zero order CFA is the model of choice if researchers do not have extensive knowledge available that can guide the search for types and antitypes, not even knowledge that concerns the representativity of their sample. If the sample is representative, the univariate marginal frequencies, that is, the frequencies of the categories of the variables, can be taken into account. This option is not available in cluster analysis.

5.2 First order global CFA

First order global CFA is the classic and original version of CFA (Lienert, 1969), and is still the most frequently used method of CFA. The base model for first order CFA is hierarchically one level higher than zero order CFA, because it considers main effects. All main effects must be considered for a base model to be global. The form of the CFA base model is

$$\log E = X\lambda,$$

where X is a design matrix that contains the constant vector and additional vectors for the main effects of all variables.

Because the base model of first order CFA takes all main effects into account, types and antitypes can emerge only because of the existence of interactions. In general, as was explained in Chapter 2, types and antitypes can emerge only because the effects *not* considered in the CFA base model do exist. In zero order CFA, these are main effects and interactions; in first order CFA, these are interactions of any order; in second order CFA, these are interactions of second or higher order, and so on. To illustrate first order CFA, we present two examples in the next two sections. In Section 5.2.1, we analyze three binary variables. in Section 5.2.2, we analyze two variables with three categories each.

5.2.1 Data example I: First order CFA of social network data

The first example uses Klingenspor et al.'s (1993) social network data again (see Section 5.1). We now analyze these data using first order CFA. The log-linear base model for this analysis is

$$\log E = \begin{bmatrix} 1 & 1 & 1 & 1 \\ 1 & 1 & 1 & -1 \\ 1 & 1 & -1 & 1 \\ 1 & 1 & -1 & -1 \\ 1 & -1 & 1 & 1 \\ 1 & -1 & 1 & -1 \\ 1 & -1 & -1 & 1 \\ 1 & -1 & -1 & -1 \end{bmatrix} \begin{bmatrix} \lambda_0 \\ \lambda_M \\ \lambda_G \\ \lambda_N \end{bmatrix} .$$

This design matrix contains four column vectors. The first is the constant vector that is also used in the base model for zero order CFA. The second vector is for the main effect of the first variable, Marital Status. This vector contrasts the two categories of this variable, married and not married, with each other. The third vector contrasts the two categories of the variable Gender, male and female, with each other. The fourth vector contrasts the two categories of the variable Network Size, small and large, with each other[1].

In its top panel, Table 21 displays the observed cell frequencies for the M x G x N cross-classification, the expected cell frequencies, the test

---

[1]In this design matrix and in all other design matrices in this volume, we use effect coding. Other methods of coding, e.g., dummy coding or corner value coding, allow one to express CFA base models equivalently. We choose effect coding for didactical reasons.

statistics, the tail probabilities, and the type/antitype decisions. In its bottom panel, Table 21 displays, the descriptive statistics, and their rank orders. To make results comparable to those in Table 20, we use Anscombe's z-test again, and protect the test-wise α using the procedure proposed by Hommel et al. (1985; cf. the last columns in Tables 16 and 17 and in the top panel of Table 20).

Before employing the procedure proposed by Hommel et al. (1985), we have to make sure the base model does not allow us to describe the data satisfactorily. We calculate for the first order CFA base model the Pearson $X^2 = 154.37$ ($df = 4$; $p < 0.01$). These values indicate significant model - data discrepancies and we reject the base model, and we can use the procedure by Hommel et al. (1985).

Table 21 shows that first order CFA identifies five configurations as constituting types or antitypes. Most interesting is the comparison with the results from zero order CFA, shown in Table 20. We compare the results in four respects. The results generated by these two methods of CFA differ in practically all respects. First, the estimated expected cell frequencies for first order CFA do not indicate a uniform frequency distribution. Rather, they reflect the marginal frequencies. These frequencies are $N_{1..} = 154$, $N_{2..} = 362$; $N_{.1.} = 258$, $N_{.2.} = 258$; and $N_{..1} = 261$, $N_{..2} = 255$. Summing the estimated expected cell frequencies over the categories of the three variables yields exactly the observed marginal frequencies. For example, the marginal $N_{..1} = 261$ results from 38.948 + 38.948 + 91.552 + 91.552 = 261. This result illustrates one of the chief characteristics of the main effect model, which is that it reproduces the marginal frequencies exactly. This is not systematically the case for the zero order CFA base model, which is also called the null model. The two models yield the same expected cell frequencies if the marginal frequencies are all the same. Second, the overall X^2 statistic for the first order CFA base model is much smaller than for the zero order CFA model (154.37 vs. 212.43). The reason for this result is that the first order base model uses much more information when estimating the expected cell frequencies than the zero order model. Specifically, the first order model uses the information that describes the univariate distribution of each of the variables under study. This information is neglected in zero order CFA. As a consequence, the model fit for first order CFA is almost always better than the fit for zero order CFA. Again, when considering the marginal frequencies leads to a uniform distribution of the expected cell frequencies, the overall goodness-of-fit X^2 values will also be the same. The degrees of freedom, however, will always differ.

Table 21: First order CFA of the variables Marital Status (M), Gender (G), and Size of Social Network (N)

MGN	Frequencies		Statistics		α*
	observed	expected	*z*	*p*(*z*)	
		CFA results			
111	48	38.95	1.426	.07691	.05
112	87	38.05	6.833	< .0001 T	.0125
121	5	38.95	6.952	< .0001 A	.00625
122	14	38.05	4.475	< .0001 A	.0125
211	78	91.55	1.436	.07544	.025
212	45	89.45	5.195	< .0001 A	.0125
221	130	91.55	3.797	< .0001 T	.0125
222	109	89.45	2.016	.02187	.01667
		Descriptives		Ranks	
MGN		*RR*	log *P*	*RR*	log *P*
111	48	1.232	1.205	3	7
112	87	2.286	11.127	1	1
121	5	0.128	8.700	8	2
122	14	0.368	3.978	7	5
211	78	0.852	1.107	5	8
212	45	0.503	5.972	6	3
221	130	1.420	4.153	2	4
222	109	1.219	1.809	4	6

Third, the number of types and antitypes is smaller in first order CFA than in zero order CFA. The reason for this result can also be found in the use of information. Zero order CFA uses only one piece of information, the sample size. The estimated expected cell frequency is calculated as the average number of cases per cell, that is, the sample size divided by the number of cells. In contrast, first order CFA uses information about the marginal distributions. It comes as no surprise that using more information typically results in better data description. For CFA, this implies a decreased probability decreases that model-data discrepancies will be large enough for types and antitypes to emerge. The current example attests to that.

Fourth, the rank orders of z-scores, *RR*- and log *P* values vary with the CFA base models. The correlations for the rank orders of these scores from Tables 20 and 21 appear in Table 22.

Table 22: Correlations among X^2-, *RR*-, and log *P*-values from Tables 19 and 20 (labels ...1 indicate results from first order CFA; labels ...0 indicate results from zero order CFA)

	RR1	LOGP1	CHI21	RR0	LOGP0	CHI20
RR1	1.00					
LOGP1	0.02	1.00				
CHI21	-0.31	0.93	1.00			
RR0	0.83	-0.17	-0.40	1.00		
LOGP0	-0.26	0.55	0.60	-0.12	1.00	
CHI20	-0.38	0.52	0.62	-0.26	0.98	1.00

The results displayed in Table 22 confirm the results from Sections 4.2 and 4.3. With only one exception, all $E_j > 0.5\ N_j$. Under this condition, the correlation between X^2 and log *P* approximates a straight line. Accordingly, the correlations between X^2 and log *P* in this example are 0.98 for zero order CFA and 0.93 for first order CFA. The correlations between *RR* and log *P* and X^2 are small or zero, thus indicating again that the relative risk is sensitive to different data characteristics than the other two measures. As was explained in Section 4.2, the relative risk continues to decrease when the difference between N_j and E_j changes from positive to negative. In contrast, X^2 will increase again and so can log *P*.

Before proceeding to the second example of global first order CFA, we ask whether in 'real life data analysis' we would have selected the zero order or the first order CFA base model for the exploration of the Klingenspor et al. (1993) data. When looking at the marginal frequencies, we notice that the samples of males and females are of exactly the same size. It is very likely that the researchers determined that the samples be of the same size (product-multinomial sampling). Therefore, CFA needs to reproduce the sample sizes. In the present example, this can be achieved by both zero order and first order CFA. The expected cell frequencies sum to $N = 258$ in both Table 20 and in Table 21. We thus conclude that both base models are appropriate, as long as the other two variables, Marital Status (M) and Network Size (N) are not fixed too. Therefore, first order CFA is the method of choice only if one wishes to exclude types and antitypes that emerge only because the marginal frequencies of the variables M and N are unequal.

5.2.2 Data example II: First order CFA of Finkelstein's Tanner data, Waves 2 and 3

We now present the second example of first order CFA. In this example, we illustrate a design matrix for a variable with more than two categories. The data we use were collected in the study by Finkelstein et al. (1994). In the third wave of data collection, the Tanner scale was employed again to assess the respondents' progress in physical pubertal development. The scale values range from 1 = prepubertal to 4 = physically mature. For the following analyses, we use the 64 adolescents that had provided data in Waves 2 and 3, and had Tanner scores of 2 or higher in both 1985 and 1987.

The cross-classification of the Tanner scores from 1985 and 1987 appears in Table 23, along with the results of first order CFA. We used Lehmacher's test with continuity correction and Holm's adjustment of the test-wise α. The a priori α was 0.05 and the first test had an adjusted $\alpha^* = 0.00556$. The goodness-of-fit of the CFA base model was assessed at an $X^2 = 24.10$ ($df = 4$; $p < 0.01$) which indicates that the base model is not tenable.

The results in Table 23 suggest that three types and two antitypes of physical pubertal development exist. The types suggest that developmental progress by one Tanner stage, that is, progress from Stage 2 to Stage 3 and from Stage 3 to Stage 4 occurs more often than under the assumption of no relationship between the developmental assessments in 1985 and 1987. In addition, stability at the maturity stage is beyond chance.

Table 23: First order CFA of Finkelstein's Tanner data, Waves 1985 and 1987

Tanner Score	Frequencies		CFA tests		α^*	Type/ Antitype ?
	observed	expected	z	$p(z)$		
22	3	1.64	1.013	0.15555	0.01667	
23	17	9.30	5.053	0.00002	0.00625	**T**
24	15	24.06	-4.602	$< \alpha^*$	0.00556	**A**
32	0	0.75	-0.339	0.36740	0.02500	
33	0	4.25	-2.432	0.00751	0.00833	**A**
34	16	11.00	2.781	0.00271	0.00714	**T**
42	0	0.61	-0.160	0.43663	0.05000	
43	0	3.45	-2.061	0.01964	0.01250	
44	13	8.94	2.369	0.00891	0.01000	**T**

[a] $< \alpha^*$ indicates that the tail probability is smaller than can be expressed with five decimal places.

Less likely than expected from the base model is a leap by two stages from Stage 2 to Stage 4. In addition, stagnation at Stage 3 is very unlikely.

Interesting is the observation that regression to earlier Tanner stages did not occur. The Configurations 32, 42, and 43 were not observed. With the means of CFA introduced thus far, we cannot test whether these three configurations jointly constitute an antitype. However, CFA of groups of cells (Section 10.3) and Bayesian CFA (Section 11.2) provide us with this option.

An issue of concern is the estimation of expected cell frequencies for those configurations that indicate developmental regression, that is, a Tanner stage 3 or 4 that is followed by a Tanner stage 2 or 3, respectively. If one assumes that developmental regression of this type is impossible, these cells must be empty by definition. Cells that must be empty by definition are termed *structural zeros*. Indeed, in Table 23 these frequencies are zero. If one proposes that any frequency in these cells that is unequal to zero must be an error, one can estimate expected frequencies based on a quasi-independence base model. This is a log-linear model that blanks out specified cells. Section 10.1 describes CFA for tables with structural zeros.

We now present the design matrix for the first order CFA base model that was used to determine the expected cell frequencies in Table 23. The base model for the design in Table 23 is

$$\log E = \begin{bmatrix} 1 & 1 & 0 & 1 & 0 \\ 1 & 1 & 0 & 0 & 1 \\ 1 & 1 & 0 & -1 & -1 \\ 1 & 0 & 1 & 1 & 0 \\ 1 & 0 & 1 & 0 & 1 \\ 1 & 0 & 1 & -1 & -1 \\ 1 & -1 & -1 & 1 & 0 \\ 1 & -1 & -1 & 0 & 1 \\ 1 & -1 & -1 & -1 & -1 \end{bmatrix} \begin{bmatrix} \lambda_0 \\ \lambda_{85,1} \\ \lambda_{85,2} \\ \lambda_{87,1} \\ \lambda_{87,2} \end{bmatrix},$$

where the subscripts for the λ parameters indicate the year of observation and the main effect parameters. As before, we use effect coding. The vectors in X are not orthogonal. However, for the purposes of CFA, this is not an issue of concern because the parameters are not interpreted.

The first column vector in the design matrix, X, is the constant vector. The next two vectors represent the main effect vectors of the first variable. This is the variable Tanner Score in 1985. The following two vectors represent the main effect of the variable Tanner Score in 1987. The strength of each effect is expressed by a parameter. The design matrix is post-multiplied by the parameter vector to calculate the logarithms of the expected cell frequencies.

Before introducing second order CFA in Section 5.3, we discuss reasons for selecting first order CFA. The difference between zero order and first order CFA is that the base model of first order CFA takes the main effects of all variables into consideration when estimating the expected cell frequencies. There are two reasons why researchers would opt for first order CFA over zero order CFA. The first reason is that they do not wish types and antitypes to emerge just because the categories of variables were observed at unequal rates. Non-uniform marginals do not necessarily result in types and antitypes. However, if all main effects are taken into account non-uniform marginals are guaranteed not to result in types and antitypes.

The second reason, going hand in hand with the first, is that first order CFA yields types and antitypes only if there exist associations between variables. These associations can be of any order. They can even be *local* (Havránek, Kohnen, & Lienert, 1986; Havránek & Lienert, 1984), that is, exist in a subtable only. If researchers want types and antitypes to result only from variable interactions, first order CFA is the base model of choice.

Zero order CFA and first order CFA coincide, that is, yield the same expected cell frequencies if the distribution of the expected cell frequencies is uniform. For example, consider an I x J x K x L cross-classification. The expected cell frequencies for this table are uniform if

$$\prod_{ijkl}^{IJKL} \pi_{ijkl} = constant,$$

where π_{ijkl} is the probability of configuration *ijkl*. An example of a case in which the expected frequency distribution is uniform is a table that (a) is spanned by variables with the same number of categories, (b) each of which occurs at the same rate.

5.3 Second order global CFA

The base model of second order CFA is one step above the base model of first order CFA in the hierarchy. When estimating expected cell frequencies, the second order CFA base model takes into account

(1) the main effects of all variables, and
(2) the first order interactions, that is, the interactions between all pairs of variables

(von Eye & Lienert, 1984; Lienert, Netter, & von Eye, 1987; von Eye, 1988).

To illustrate, consider the 2 x 2 x 2 cross-classification of the variables A, B, and C. The second order CFA base model for this classification is

$$\log E = \begin{bmatrix} 1 & 1 & 1 & 1 & 1 & 1 & 1 \\ 1 & 1 & 1 & -1 & 1 & -1 & -1 \\ 1 & 1 & -1 & 1 & -1 & 1 & -1 \\ 1 & 1 & -1 & -1 & -1 & -1 & 1 \\ 1 & -1 & 1 & 1 & -1 & -1 & 1 \\ 1 & -1 & 1 & -1 & -1 & 1 & -1 \\ 1 & -1 & -1 & 1 & 1 & -1 & -1 \\ 1 & -1 & -1 & -1 & 1 & 1 & 1 \end{bmatrix} \begin{bmatrix} \lambda_0 \\ \lambda_A \\ \lambda_B \\ \lambda_C \\ \lambda_{AB} \\ \lambda_{AC} \\ \lambda_{BC} \end{bmatrix} .$$

The first column vector in the design matrix for this model is the constant vector, present also in all other log-linear CFA base models. The following three vectors represent the main effects of the variables A, B, and C. These vectors are also present in the first order CFA base model (see the model in Section 5.2.1). The remaining three vectors represent the interactions between A and B, A and C, and B and C, respectively. These vectors result from element-wise multiplication of the main effect vectors involved in the interactions. The difference between the first order and the second order base models is constituted by these two-way interaction vectors.

To illustrate second order CFA, we use the social network data analyzed in Sections 5.1 and 5.2.1 again. The data describe the size of social networks (N; 1 = small, 2 = large) in a sample of individuals of which Martial Status (M; 1 = married, 2 = not married) and Gender (G; 1 = male, 2 = female) is known. We now analyze these data under a second order base model that is, we take all pair-wise associations into account when estimating he expected cell frequencies. These are the associations between Network and Marital Status, Network and Gender, and Marital Status and Gender. This log-linear base model has the form

$$\begin{aligned} \log E = \lambda_0 + \lambda_N + \lambda_M + \lambda_G \\ + \lambda_{NM} + \lambda_{NG} + \lambda_{MG} \end{aligned} .$$

The top line of the equation is identical to the equation for the first order base model of these data. The second line indicates the part that the base model of second order CFA adds to the base model for first order CFA.

To make the analyses comparable to the ones presented in Sections 5.1 and 5.2.1, we again employ Anscombe's z-approximation, set $\alpha = 0.05$, and try to protect the test-wise α using the procedure proposed by Hommel et al. (1985). In order to validly apply this procedure, the CFA base model must be rejected. In the present case we calculate a Pearson $X^2 = 0.011$ which, for $df = 1$ indicates no significant model-data discrepancies ($p = 0.92$). As a consequence, we cannot use the procedure by Hommel et al. (1985) and we are unable to identify types or antitypes. For illustration purposes, Table 24 presents the test results.

Table 24: Second order CFA of the social network data

Configuration MGN	Frequencies		Statistics	
	observed	expected	z	$p(z)$
111	48	47.83	.049	.48
112	87	87.17	.001	.50
121	5	5.17	.003	.50
122	14	13.83	.091	.46
211	78	78.17	.001	.50
212	45	44.83	.051	.48
221	130	129.83	.030	.49
222	109	109.17	.001	.50

Three results stand out in Table 24. First, being the source of information for the goodness-of-fit test of the base model, the discrepancies between the observed and the expected cell frequencies are that small that none of the z-scores reaches even 0.10. Accordingly, all tail probabilities are very near $p = 0.50$. The largest absolute residual is 0.17. We thus conclude that taking all two-way interactions into account when estimating the expected cell frequencies allows us to explain the variability in this cross-classification almost perfectly.

Hand in hand with the first result, we note secondly, that the discrepancies between the observed and the expected cell frequencies are not even remotely large enough to qualify configurations as constituting types and antitypes. If the second order base model is chosen for the present data, none of the observed frequencies comes as a surprise. The model reflects the observed distribution very well.

Third, in comparison with Tables 20 (zero order CFA) and 21 (first order CFA), the expected cell frequencies are much closer to the observed frequencies. This can be explained as follows. In zero order CFA, no information beyond the sample size is taken into account when estimating expected cell frequencies. Therefore, any main effect or interaction can lead to the presence of types and antitypes. In first order CFA, the main effects of all variables are taken into account. Therefore, only variable interactions can lead to types and antitypes. In second order CFA, all pair-wise interaction are taken into account in addition to the main effects. Therefore, interactions in groups of three variables must exist for types and antitypes to emerge. In other words, as one moves up the hierarchy of CFA base models, one uses more and more information when estimating the expected cell frequencies. One thus can expect the base model to provide an increasingly better rendering of the observed frequency distribution.

It should be noted that this is the standard case. It is possible that higher order models provide no improvement over lower order models. This can occur when the terms added to the base model fail to explain variability, for instance, when the variables that span a cross-classification are not intercorrelated at the level of pair-wise correlations. The most frequent case, however, reflects the pattern observed here in Tables 20, 21, and 24.

5.4 Third order global CFA

Third order global CFA represents the next step up in the hierarchy of global CFA models. The base model of third order CFA takes into account (a) the main effects of all variables, (b) the first order interactions of all variables, and (c) the second order interactions of all variables.

Therefore, types and antitypes can emerge only if interactions of third or higher order exist. The following artificial data example illustrates third order CFA. Consider the four variables A, B, C, and D with two categories each. The design matrix for third order CFA base model for these four variables is

$$X = \begin{bmatrix} I & A & B & C & D & AB & AC & AD & BC & BD & CD & ABC & ABD & ACD & BCD \\ 1 & 1 & 1 & 1 & 1 & 1 & 1 & 1 & 1 & 1 & 1 & 1 & 1 & 1 & 1 \\ 1 & 1 & 1 & 1 & -1 & 1 & 1 & -1 & 1 & -1 & -1 & 1 & -1 & -1 & -1 \\ 1 & 1 & 1 & -1 & 1 & 1 & -1 & 1 & -1 & 1 & -1 & -1 & 1 & -1 & -1 \\ 1 & 1 & 1 & -1 & -1 & 1 & -1 & -1 & -1 & -1 & 1 & -1 & -1 & 1 & 1 \\ 1 & 1 & -1 & 1 & 1 & -1 & 1 & 1 & -1 & -1 & 1 & -1 & -1 & 1 & -1 \\ 1 & 1 & -1 & 1 & -1 & -1 & 1 & -1 & -1 & 1 & -1 & -1 & 1 & -1 & 1 \\ 1 & 1 & -1 & -1 & 1 & -1 & -1 & 1 & 1 & -1 & -1 & 1 & -1 & -1 & 1 \\ 1 & 1 & -1 & -1 & -1 & -1 & -1 & -1 & 1 & 1 & 1 & 1 & 1 & 1 & -1 \\ 1 & -1 & 1 & 1 & 1 & -1 & -1 & -1 & 1 & 1 & 1 & -1 & -1 & -1 & 1 \\ 1 & -1 & 1 & 1 & -1 & -1 & -1 & 1 & 1 & -1 & -1 & -1 & 1 & 1 & -1 \\ 1 & -1 & 1 & -1 & 1 & -1 & 1 & -1 & -1 & 1 & -1 & 1 & -1 & 1 & -1 \\ 1 & -1 & 1 & -1 & -1 & -1 & 1 & 1 & -1 & -1 & 1 & 1 & 1 & -1 & 1 \\ 1 & -1 & -1 & 1 & 1 & 1 & -1 & -1 & -1 & -1 & 1 & 1 & 1 & -1 & -1 \\ 1 & -1 & -1 & 1 & -1 & 1 & -1 & 1 & -1 & 1 & -1 & 1 & -1 & 1 & 1 \\ 1 & -1 & -1 & -1 & 1 & 1 & 1 & -1 & 1 & -1 & -1 & -1 & 1 & 1 & 1 \\ 1 & -1 & -1 & -1 & -1 & 1 & 1 & 1 & 1 & 1 & 1 & -1 & -1 & -1 & -1 \end{bmatrix}$$

The fifteen columns in this design matrix are structured as follows. In the header of each column, the variable is listed for which an effect is coded[2]. I indicates the constant column. After the constant column follow the four columns for the main effects of the variables A, B, C, and D. The following six columns represent the pair-wise interactions A x B, A x C, A x D, B x C, B x D, and C x D. These columns result from multiplying the elements of the main effect vectors of the variables involved in an interactions. The last four columns contain the vectors for the four three-way interactions A x B x C, A x B x D, A x C x D, and B x C x D. A sample analysis of the artificial frequencies in the 2 x 2 x 2 x 2 cross-classification of the variables A, B, C, and D appears in Table 24. We assume multinomial sampling, use the normal approximation of the binomial test, and employ the Bonferroni-adjustment of α which yields the adjusted $\alpha^* = 0.003125$. Table 25 also displays the types and antitypes that result from first and second order CFA.

---

[2]Note that this header is included just to label the columns. They are not part of the matrix when parameters are estimated.

Table 25: Third order CFA of the variables A, B, C, and D

Cell Index ABCD	Frequencies		Statistics		Type (T)/ Antitype (A)?		
	observed	expected[b]	z^b	$p(z)^b$	1[a]	2[a]	3[a]
1111	21	11.93	2.703	.0034		T	
1112	3	12.07	-2.688	.0035	A	A	
1121	7	16.07	-2.353	.0093			
1122	28	18.93	2.184	.0145	T		
1211	2	11.07	-2.800	.0026	A	A	A
1212	29	19.93	2.134	.0164	T	T	
1221	21	11.93	2.703	.0034			
1222	5	14.07	-2.502	.0062			
2111	3	12.07	-2.688	.0036			
2112	20	10.93	2.817	.0024			T
2121	19	9.93	2.948	.0016			T
2122	8	17.07	-2.289	.0111			
2211	28	18.93	2.184	.0145	T	T	
2212	2	11.07	-2.800	.0026	A	A	A
2221	2	11.07	-2.800	.0026	A		A
2222	16	6.93	3.504	.0002			T

[a] Types and Antitypes for 1 = first order CFA; 2 = second order CFA; 3 = third order CFA
[b] Expected cell frequencies and test statistics for third order CFA

The results in Table 25 suggest that the interaction of all four variables must exist. Without it, none of the types and antitypes would have appeared. The results also suggest that only two antitypes appear in all three analyses, that is, in first, second, and third order CFA. These are

Configurations 1211 and 2212. All other types and antitypes appear only in one or two of the analyses. Most interesting is the observation that three types (Configurations 2112, 2121, and 2222) emerge as types only in third order CFA. This indicates that taking into account the three-way interactions in addition to the two-way interactions can re-shuffle the magnitudes of the estimated expected frequencies such that the corresponding observed frequencies are not well reproduced any more. In other words, in the light of the additional knowledge provided by the three-way interactions, the frequencies of these configurations appear more extreme rather than less extreme. Similar shifts in the evaluation of configurations can always result when base models are changed (see, e.g., the differences between the type/antitype pattern from first and second order CFA). They can also occur when covariates are taken into account (Glück & von Eye, 2000).

We now ask why researchers would search for types and antitypes after taking into account three-way interactions. Consider first the case of second order CFA which takes into account first order interactions. The second order CFA base model is interesting because many exploratory routine applications of statistical methods implicitly and exclusively focus on two-way interactions. Examples include factor analysis and correspondence analysis. If one can show that there is variability beyond first order interactions that is so large that types and antitypes can emerge, these routine applications may fail to depict important parts of the variability in a table. An example of such a case was given in Section 3.2 under the label of Meehl's paradox. In this example, focusing on two-way interactions would have led to the elimination of two items that allowed the psychiatrist to perfectly discriminate between two patient groups. This applies in an analogous fashion to the case where there is sizeable variability beyond second order interactions. Therefore, whenever researchers aim to explore their data beyond the routine level of bivariate interactions, second and higher order CFA provide them with the necessary tools.

6. Regional models of CFA

Regional CFA models differ from the global models discussed in Chapter 5 in one fundamental aspect. Whereas global models assign all variables the same status, regional models distinguish between groups of variables. For instance, regional CFA models allow one to explore the relationships between the groups of the motivational and the cognitive variables, or they allow one to discriminate between patterns of leisure behaviors in men and in women. The type of analysis is still exploratory. However, there is a stronger explanatory component in regional CFA than in global CFA. The researchers specify the variable groups before CFA. The method of CFA can be used to analyze groups of variables (see Sections 6.1.1 and 6.1.2). If no prior knowledge exists about the composition of variable groups, cluster analysis (Hartigan, 1975) or correspondence analysis (Greenacre, 1984) can also be used to create variable groups. The following sections introduce readers into regional CFA of existing groups of variables. We begin with Interaction Structure Analysis (ISA).

6.1 Interaction Structure Analysis (ISA)

ISA is a method for the analysis of the relationships between two groups of variables (Lienert & Krauth, 1973b; Krauth & Lienert, 1974; cf. Lienert & Bergman, 1985). In Section 6.1.1, we introduce readers to ISA of two groups of variables. In Section 6.1.2, we present ISA of three or more groups of variables.

6.1.1 ISA of two groups of variables

ISA of two groups of variables is based on a specific definition of *higher order interactions* (Krauth & Lienert, 1973a). This definition considers any two, nonempty sets of variables, A and B. The total number of variables, d_c, in these two groups must be at least three. If the two groups only contain one variable each, their 'higher' order interaction coincides with their two-way interaction. To define Krauth and Lienert's higher order interactions, let the total number of variables be $d_c \geq 3$, let the number of variables in A be $0 < d_A < d_c$ and the number of variables in B be $0 < d_B < d_c$, and the sum be $d_A + d_B = d_c$. Then, the relationship between the two groups is called *$(d_c - 1)^{st}$ order interaction*. For example, if the two groups A and B contain a total of $d_c = 5$ variables, the interaction between A and B is called a fourth order interaction.

Consider the case where the two variable groups A and B contain three variables. Then, ISA interactions among these three variables have the following characteristics (Krauth & Lienert, 1973a):

(1) If there exists no second order interaction and no first order interaction, the three variables are totally independent.
(2) If there exist no first order interactions, second order interactions can exist nevertheless.
(3) The existence of second order interactions follows from the existence of particular first order interactions (for details see Krauth & Lienert, 1974).

In most instances, there is a clear explanatory component when ISA is employed, such that variables are grouped before analysis. However, in predominantly exploratory contexts, researchers may ask questions concerning ISA interactions in any of the possible groupings. To illustrate the possible groupings, consider the four variables, 1, 2, 3, and 4. Table 26 displays all possible groupings of these four variables with groups that contain one or more variables. The table denotes groupings by variable labels that are separated by a period. The variables to the left of the period belong to one group, and the variables to the right of the period belong to the other group. Variables can belong to only one group, and no group can be empty. The order of groups is of no importance.

Table 26: ISA groupings and interactions of four variables

Number of variables in grouping	Number of groupings	Groupings
2	$\binom{4}{2} = 6$	1.2, 1.3, 1.4, 2.3, 2.4, 3.4
3	$\binom{4}{3}\binom{3}{2} = 12$	1.23, 1.24, 1.34, 2.13, 2.14, 2.34, 3.12, 3.14, 3.24, 4.12, 4.13, 4.23
4	$\binom{4}{1} + \binom{4}{2}/2 = 7$	1.234, 2.134, 3.124, 4.123, 12.34, 13.24, 14.23

The total number of groupings for d variables is

$$t = 0.5(3^d + 1) - 2^d.$$

For example, we obtain for $d = 4$ variables $t = 0.5(3^4 + 1) - 2^4 = 25$ groupings (see Table 26), and for $d = 10$ variables we obtain $t = 0.5(3^{10} + 1) - 2^{10} = 28{,}501$ groupings. Thus, researchers exploring all groupings of their variables using ISA may have to inspect mountains of output.

An ISA base model typically is specified such that

(1) it is saturated within each variable group. Thus, there may be interactions of any order in both groups.
(2) The two groups are independent of each other.

It follows from these specifications that ISA types and ISA antitypes can emerge only if relationships between the two variable groups exist. It should be emphasized again that the status of the two variable groups in ISA is the same. Thus, ISA can be viewed as a generalized version of global CFA. If one group is interpreted as containing predictor variables and the other as containing criterion variables Prediction CFA is the method of choice. ISA as a method is symmetrical. Switching the order of the two variable groups yields the exact same types and antitypes (see Section 6.2 on Prediction CFA, below).

It is important to note that the sampling schemes employed to collect data are largely irrelevant in ISA application because the base model is saturated in both groups of variables. The only constraint is that if two or more variables are jointly sampled under a multivariate, that is, cross-classified product-multinomial sampling scheme, the interactions among those variables must be taken into account for the estimation of expected cell frequencies.

To illustrate the use of ISA, we use data from an experiment on stress (Krauth & Lienert, 1973a). A sample of 159 young adults participated in an experiment. Two conditions were realized. The first group took a cognitive performance test twice under relaxed conditions. The second group took the same test first under the same, relaxed conditions, and the second time under stress. In each group, half of the participants had above-average intelligence, and the other half had below-average intelligence. Two variables were observed: change in quantitative performance (X) and change in qualitative performance (Y). X had categories 1 = more items processed in the second trial and 2 = the same number or fewer items processed in the second trial. Y had categories 1 = more items correct in the second than in the first trial, 2 = the same number of items correct in both trials, and 3 = fewer items correct in the second trial. The variable Experimental Condition, S, had categories 1 = control group and 2 = experimental group. Intelligence, I, had categories 1 = below average and 2 = above average. These four variables form the 2 (change in quantitative performance) x 3 (change in qualitative performance) x 2 (experimental condition) x 2 (intelligence) cross-tabulation.

For ISA, these four variables must be assigned to two groups. A total of 25 groups can be formed (see Table 26). Suppose we are in the situation of the researcher who has no a priori grouping of these four variables (later in this section we will use the grouping of our four variables as dependent and independent). This researcher will have to make a decision as to variable group membership. One option is to estimate all possible log-linear models to find out which grouping is the most promising for the detection of types and antitypes. In the absence of a priori knowledge, the log-linear model with the largest goodness-of-fit X^2 value is the one that has the potential of showing the most or the most extreme types and antitypes.

Table 27 contains the X^2 values for all 25 models, along with the degrees of freedom for each model and the tail probabilities.

Table 27: Log-linear models goodness-of-fit for all 25 groupings of the variables X, Y, S, and I (variable groups are separated by a period)

Number of variables in grouping	Grouping	X^2	*df*	$p(X^2)$
2	X.Y	4.10	2	0.129
	X.S	23.46	1	< 0.001
	X.I	0.23	1	0.623
	Y.S	67.63	2	< 0.001
	Y.I	0.52	2	0.772
	S.I	0.23	1	0.631
3	YS.X	31.47	5	< 0.001
	XS.Y	79.02	6	< 0.001
	XY.S	92.10	5	< 0.001
	YI.X	25.95	5	< 0.001
	XI.Y	25.41	6	< 0.001
	XY.I	22.61	5	< 0.001
	SI.X	37.94	3	< 0.001
	XI.S	37.86	3	< 0.001
	XS.I	17.16	3	< 0.001
	SI.Y	69.55	6	< 0.001
	YI.S	68.60	5	< 0.001
	YS.I	2.02	5	0.846

/ cont.

Number of variables in grouping	Grouping	X^2	df	$p(X^2)$
4	YSI.X	55.16	11	< 0.001
	XSI.Y	91.21	14	< 0.001
	XYI.S	96.10	11	< 0.001
	XYS.I	28.56	11	0.003
	XY.SI	135.01	15	< 0.001
	XS.YI	126.09	15	< 0.001
	YS.XI	57.52	15	< 0.001

The goodness-of-fit Pearson X^2 values in Table 27 indicate that the most promising model places the two response variables, quantitative and qualitative performance in one group, and experimental condition and intelligence in the second group, that is, model XY.SI. The X^2 for this model is 135.01 ($df = 15, p < 0.001$). This is also the model that we would have used as the base model, using the distinction between response variables and non-response variables. Therefore, we now perform an ISA using the base model XY.SI, that is, model

$$\log E = \lambda_0 + \lambda_j^X + \lambda_k^Y + \lambda_l^S + \lambda_i^I + \lambda_{jk}^{XY} + \lambda_{li}^{SI}.$$

The design matrix for this model is

$$X = \begin{bmatrix} 1 & 1 & 1 & 0 & 1 & 1 & 1 & 0 & 1 \\ 1 & 1 & 1 & 0 & 1 & -1 & 1 & 0 & -1 \\ 1 & 1 & 1 & 0 & -1 & 1 & 1 & 0 & -1 \\ 1 & 1 & 1 & 0 & -1 & -1 & 1 & 0 & 1 \\ 1 & 1 & 0 & 1 & 1 & 1 & 0 & 1 & 1 \\ 1 & 1 & 0 & 1 & 1 & -1 & 0 & 1 & -1 \\ 1 & 1 & 0 & 1 & -1 & 1 & 0 & 1 & -1 \\ 1 & 1 & 0 & 1 & -1 & -1 & 0 & 1 & 1 \\ 1 & 1 & -1 & -1 & 1 & 1 & -1 & -1 & 1 \\ 1 & 1 & -1 & -1 & 1 & -1 & -1 & -1 & -1 \\ 1 & 1 & -1 & -1 & -1 & 1 & -1 & -1 & -1 \\ 1 & 1 & -1 & -1 & -1 & -1 & -1 & -1 & 1 \\ 1 & -1 & 1 & 0 & 1 & 1 & -1 & 0 & 1 \\ 1 & -1 & 1 & 0 & 1 & -1 & -1 & 0 & -1 \\ 1 & -1 & 1 & 0 & -1 & 1 & -1 & 0 & -1 \\ 1 & -1 & 1 & 0 & -1 & -1 & -1 & 0 & 1 \\ 1 & -1 & 0 & 1 & 1 & 1 & 0 & -1 & 1 \\ 1 & -1 & 0 & 1 & 1 & -1 & 0 & -1 & -1 \\ 1 & -1 & 0 & 1 & -1 & 1 & 0 & -1 & -1 \\ 1 & -1 & 0 & 1 & -1 & -1 & 0 & -1 & 1 \\ 1 & -1 & -1 & -1 & 1 & 1 & 1 & 1 & 1 \\ 1 & -1 & -1 & -1 & 1 & -1 & 1 & 1 & -1 \\ 1 & -1 & -1 & -1 & -1 & 1 & 1 & 1 & -1 \\ 1 & -1 & -1 & -1 & -1 & -1 & 1 & 1 & 1 \end{bmatrix}.$$

The first column in this design matrix contains the constant. The second column represents the main effect of Variable X, a two-level variable. Columns 3 and 4 represent the main effect of variable Y, a three-level variable. Columns 5 and 6 represent the main effects of variables S and I, respectively. Columns 7 and 8 represent the interaction X x Y, and the last column represents the interaction of S with I. The remaining pair-wise interactions, X x S, X x I, Y x S, and Y x I, are not part of the model. If these interactions exist, they will manifest in the forms of types and antitypes, thus indicating relationships between the two groups of variables.

Accordingly, higher order interactions that involve variables from both groups are not part of this base model.

Table 28 displays the results of the ISA of the four variables X, Y, S, and I. For this analysis we use the binomial test and Bonferroni adjustment which yields $\alpha^* = 0.0021$.

Table 28: ISA of the experimental stress data

Configuration	Frequencies		p	T/A?
SIXY	observed	expected		
1111	13	7.25	.0307	
1112	6	5.21	.4221	
1113	2	6.12	.0534	
1121	11	3.63	.0011	Type
1122	2	5.44	.0881	
1123	0	6.35	.0015	Antitype
1211	17	7.68	.0018	Type
1212	12	5.52	.0098	
1213	3	6.48	.1080	
1221	3	3.48	.4631	
1222	0	5.76	.0028	
1223	1	6.72	.0083	
2111	1	8.96	.0010	Antitype
2112	2	6.44	.0418	
2113	19	7.56	.0002	Type
2121	1	4.48	.0595	

/ cont.

Configuration	Frequencies			
SIXY	observed	expected	p	T/A?
2122	12	6.72	.0380	
2123	7	7.84	.4723	
2211	1	8.11	.0023	
2212	3	5.83	.1619	
2213	3	6.84	.0855	
2221	1	4.05	.0848	
2222	10	6.08	.0852	
2223	20	7.09	$< \alpha$*	Type

The ISA of the data from the performance under stress experiment indicates four types and two antitypes. The first type, Configuration 11.21[1], suggests that the pattern " control group member and below- average intelligence" goes hand-in-hand with the pattern "processing the same number or fewer items in the second trial while increasing the number of correct solutions." The second type, constituted by Configuration 12.11, suggests that the pattern "control group member and above-average intelligence" goes hand-in-hand with "processing more items in the second trial and also increasing the number of correct solutions." The third type is constituted by Configuration 21.13. It describes below-average intelligent experimental group participants who process more items in the second trial but produce fewer correct solutions. These participants increase speed under stress, but at the expense of quality of work. The fourth type, constituted by Configuration 22.23 contains above average intelligent participants in the experimental group who process the same number of items or fewer as in the first trial and produce fewer correct solutions.

The first of the two antitypes is constituted by Configuration 11.23. These are below-average intelligent control group members who did not increase the speed of their work and produced fewer correct solutions

---

[1]The period separates the variables from the two groups.

nevertheless. The second antitype, constituted by Configuration 21.11, describes below-average intelligence control group members who increased both the speed of their work and the number of correct solutions.

Overall, this pattern of types and antitypes suggests that the experimental stress affected the quality of work. In contrast, absence of stress seems to improve qualitative performance. In addition, the results show that below-average intelligent participants respond to stress with increased speed. The above-average intelligent participants respond to stress with reduced speed. Absence of stress induces both groups to increase their speed. Table 29 displays the Relative Risk and the log *P* values.

Table 29: Relative Risk and log *P* values from the ISA of the experimental stress data

Configuration				
SIXY	*RR*	Rank_{RR}	log *P*	$\text{Rank}_{\log P}$
1111	1.792	6	1.483	6
1112	1.151	9	.419	15
1113	.327	16	.386	16
1121	3.033	1	2.873	3
1122	.368	15	.291	20
1123	.000	24	.911	11
1211	2.214	4	2.614	4
1212	2.174	5	1.956	5
1213	.463	13	.313	18
1221	.781	11	.192	24
1222	.000	23	.760	12

/ cont.

Configuration				
SIXY	*RR*	Rank_{RR}	log *P*	$\text{Rank}_{\log P}$
1223	.149	20	.701	13
2111	.112	22	1.250	8
2112	.311	17	.437	14
2113	2.513	3	3.483	2
2121	.223	19	.280	21
2122	1.786	7	1.397	7
2123	.893	10	.303	19
2211	.123	21	1.030	10
2212	.515	12	.249	22
2213	.439	14	.356	17
2221	.247	18	.221	23
2222	1.645	8	1.068	9
2223	2.820	2	4.297	1

The descriptive measures from this analysis largely reflect the results from inferential ISA. More specifically, the four types occupy the ranks from 1 to 4 in both the relative risk and the log *P* rank orders. The two antitypes, placed 3rd and 5th in the rank order of tail probabilities, occupy ranks 22 and 24 in the relative risk rank order, thus suggesting that they are extreme at the other end of the scale. Specifically, these are ranks 1 and 3 from the bottom.

6.1.2 ISA of three or more groups of variables

Thus far, all applications of ISA known to us have used or created two variable groups. However, ISA is also conceivable as a method for the analysis of three or more variable groups. Typically, the base models for all ISA models

(1) are saturated within each group of variables, and
(2) propose independence of the variable groups.

If ISA base models are specified this way, types and antitypes can result only from relationships among variable groups. Interactions within variable groups are taken into account, and therefore, they cannot be the cause of the existence of types and antitypes.

However, more complex ISA models can be considered. Let the ISA model of total independence among variable groups be the *first order ISA base model*. Then, higher order ISA base models can be specified in a fashion parallel to hierarchical CFA. For example, the *second order ISA base model* (1) is saturated within each group of variables and (2) takes into account all first order interactions between the variable groups. Accordingly, a *third order ISA base model* (1) is saturated in all variable groups and (2) takes into account all first and second order interactions among all variable groups.

As was illustrated in Section 6.1.1, all ISA applications involve both an explanatory and an exploratory component. The exploratory component concerns the detection of ISA types and antitypes. The explanatory component concerns the grouping of variables. Typically, researchers perform an ISA starting from a known grouping. If, however, the grouping is part of the exploration (see Table 27), the number of base models can be very large. Consider, for example, the five variables 1, 2, 3, 4, and 5. These five variables can be arranged in 90 groups of two variables. in addition, there are 80 partitions into three groups and 15 partitions into four groups. The combinations into 3 and 4 groups are listed in Table 30.

Obviously, the number of partitions is very large and increases exponentially with the number of variables in a study. Strategies have been discussed to reduce the number of analyses required for an optimal selection and grouping of variables for ISA (Fleischmann & Lienert, 1982; Havránek & Lienert, 1984; Lienert & von Eye, 1988). However, none of these strategies is efficient enough to reduce the number of models that need to be tested to an acceptably small number. Therefore, we recommend that the researchers either use their prior knowledge to create groups, or employ cluster analysis or correspondence analysis to create variable

Table 30: Arranging five variables in three and four groups

Number of variables used	Number of groups	Groups
	Partitions into three groups	
3	$\binom{5}{3} = 10$	1.2.3, 1.2.4, 1.2.5, 1.3.4, 1.3.5, 1.4.5, 2.3.4, 2.3.5, 2.4.5, 3.4.5
4	$\binom{5}{2}\binom{3}{2} = 30$	12.3.4, 12.3.5, 12.4.5, 13.2.4, 13.2.5, 13.4.5, 14.2.3, 14.2.5, 14.3.5, 15.2.3, 15.2.4, 15.3.4, 23.1.5, 23.1.4, 23.4.5, 24.1.3, 24.1.5, 24.3.5, 25.1.3, 25.1.4, 25.3.4, 34.1.2, 34.1.5, 34.2.5, 35.1.2, 35.1.4, 35.2.4, 45.1.2, 45.1.3, 45.2.3
5	$\binom{5}{3} + \binom{5}{2}\binom{3}{1} = 40$	123.4.5, 124.3.5, 125.3.4, 134.2.5, 135.2.4, 145.2.3, 234.1.5, 235.1.4, 245.1.3, 345.1.2, 12.34.5, 12.35.4, 12.45.3, 13.24.5, 13.25.4, 13.45.2, 14.23.5, 14.25.3, 14.35.2, 15.23.4, 15.24.3, 15.34.2, 23.14.5, 23.15.4, 23.45.1, 24.13.5, 24.15.3, 24.35.1, 25.13.4, 25.14.3, 25.34.1, 34.12.5, 34.15.2, 34.25.1, 35.12.4, 35.14.2, 35.24.1, 45.12.3, 45.13.2, 45.23.1

/ cont.

Number of variables used	Number of groups	Groups
	Partitions into four groups	
4	$\binom{5}{4} = 5$	1.2.3.4, 1.2.3.5, 1.2.4.5, 1.3.4.5, 2.3.4.5
5	$\binom{5}{2} = 10$	12.3.4.5, 13.2.4.5, 14.2.3.5, 15.2.3.4, 23.1.4.5, 24.1.3.5, 25.1.3.4, 34.1.2.5, 35.1.2.4, 45.1.2.3

groups. The following two criteria can be employed to create groups:

(1) The multivariate marginal frequencies of variables sampled according to a multivariate product-multinomial sampling scheme must be reproduced. When the base model is saturated in the variables that belong to a particular group (which it typically is), the multivariate marginal frequencies will automatically be reproduced. However, when variables sampled according to some multivariate product-multinomial sampling scheme belong to different groups, the order of ISA must be adjusted to account for the sampling characteristics.

(2) For the relationships between groups, the lowest possible order of relationships is specified. Typically, the ISA base model proposes independence among variable groups. However, mostly for reasons related to sampling, interactions of first or second order must occasionally be taken into account.

Types and antitypes can then result only if variable relationships among groups exist of an order above the one taken into account by the base model. If there are more than two groups of variables in an ISA, the identification of types and antitypes does not carry any information about which variable relationships exist. This is a characteristic that ISA shares in common with standard CFA of single variables. Post hoc analyses using, e.g., log-linear modeling or CFA of higher order, can then by considered to identify these relationships. It is important to note, however, that variable-level post hoc analyses are often uninformative in the explanation of person-level results (see Section 1.2).

6.2 Prediction CFA

It is one of the most interesting characteristics of ISA that the variable groups have the same status. There is no distinction between dependent and independent variables or between predictors and criteria. Inverting the order of variable groups yields the exact same types and antitypes. In many instances, however, researchers do use variable groups that differ in status, and distinguish between predictors and criteria or dependent and independent variables. For this purpose, Prediction CFA (P-CFA) was introduced (Havránek et al., 1986; Heilmann et al., 1979; Hütter, Müller, & Lienert, 1981; Lienert & Krauth, 1973a; Lienert & Rey, 1982; Lienert & Wolfrum, 1979).

The original P-CFA base model was identical to the two-group ISA base model. P-CFA thus was symmetrical, and exchanging predictors and criteria yielded the same types and antitypes. Krauth (1996a) therefore stated that in P-CFA, the difference between the two groups of variables exists only at the level of substantive interpretation[2]. To be able to distinguish between base models for ISA and base models for P-CFA, von Eye and Schuster (1998) proposed taking into account the nature of variables as fixed versus random (see Section 2.3 on sampling schemes). Based on this distinction, P-CFA models can be specified that are not ISA models.

6.2.1 Base models for Prediction CFA

Consider the two predictors, P_1 and P_2, and the three criteria, C_1, C_2, and C_3. In original P-CFA, the base model for these five variables is

$$\log E = \lambda_0 + \lambda_i^{P_1} + \lambda_j^{P_2} + \lambda_{ij}^{P_1P_2} + \lambda_k^{C_1} + \lambda_l^{C_2} + \lambda_m^{C_3} + \lambda_{kl}^{C_1C_2} + \lambda_{km}^{C_1C_3} + \lambda_{lm}^{C_2C_3} + \lambda_{klm}^{C_1C_2C_3}$$

[2]The original statement (in German) was: "Der Unterschied zwischen den beiden Teilmengen von Merkmalen besteht nur auf der inhaltlichen oder interpretativen Ebene" (Krauth, 1996, p. 138).

or, in brief, $[P_1 P_2][C_1 C_2 C_3]$[3]. The base model for an ISA of these variables would be exactly the same. This model takes into account the main effects of all five variables and all within-group interactions. Types and antitypes can emerge only if configurations in the group of predictors occur in tandem with configurations in the group of criteria more often or less often than anticipated from this base model. The interpretation of the status of variables as equal or as predictors and criteria has no effect on the harvest of types and antitypes.

The model $[P_1 P_2][C_1 C_2 C_3]$ is the routine ISA and P-CFA model. If researchers distinguish between predictors and criteria, they may entertain the assumption that (a) the interactions among the criteria and (b) the interactions between the predictors and the criteria are caused by the predictors. Von Eye and Schuster (1998) even discuss the idea that the predictors affect the marginal frequencies of the criteria. To facilitate the investigation of these two assumptions, we use a base model with the following characteristics:

(1) The marginals of the predictors are fixed. Therefore, the base model for the above example with five variables must include the term $[P_1P_2]$. This term guarantees that types and antitypes are not caused by main effects of or interactions among the predictors. It is important to realize that there are several reasons why predictor margins can be fixed. The two most important of these reasons are (a) by design and (b) because of the particular role they play in a study. The first reason applies when predictors are sampled according to a uni- or multivariate product-multinomial sampling scheme, that is, when researchers determine the number of respondents for a given predictor configuration before collecting data. The second reason applies when variables are predictors.

(2) The criterion variables are typically free in the sense that neither their univariate marginal distributions nor their interactions are set by design (multinomial sampling).

---

[3]From here on, we also use the so-called *bracket notation* (Fienberg, 1980) to denote hierarchical log-linear models. Using this notation, one places variables in a pair of brackets if a model is saturated in these variables. That is, all possible main effects and interactions of all variables in a pair of brackets are part of the model.

If researchers entertain the hypotheses that the predictor variables can cause (a) predictor-criterion interactions and (b) interactions among the criteria, base models become interesting that are different from the above one which is saturated in both the predictors and the criteria. The following are two examples of possible models for the above example with two predictors and five criteria:

$$\log E = \lambda_0 + \lambda_i^{P_1} + \lambda_j^{P_2} + \lambda_{ij}^{P_1P_2} + \lambda_k^{C_1} + \lambda_l^{C_2} + \lambda_m^{C_3},$$

and

$$\log E = \lambda_0 + \lambda_i^{P_1} + \lambda_j^{P_2} + \lambda_{ij}^{P_1P_2} + \lambda_k^{C_1} + \lambda_l^{C_2} + \lambda_m^{C_3} + \lambda_{kl}^{C_1C_2} + \lambda_{km}^{C_1C_3} + \lambda_{lm}^{C_2C_3}.$$

Neither of these two models is an ISA base model, because neither is saturated on the criterion side. The first of these two models considers interactions only on the predictor side. In fact, it is saturated in the predictors. It shares this characteristic with all P-CFA models. In addition, this model takes into account the main effects of all criterion variables. Types and antitypes can result only from interactions that involve the criterion variables. Thus, the first criterion of admissibility of base models is fulfilled (see Section 2.2). These interactions come from two groups. The first is the group of first and second order interactions among the criterion variables. The second is the group of predictor-criterion interactions. Therefore, this model is of interest if researchers entertain the assumption that the predictors can cause both local within-criteria interactions and predictor-criteria interactions.

The second of these models also takes all first order interactions among the criterion variables into account. As for the first model, types and antitypes can result here only if certain interactions exist that involve the criterion variables. These interactions involve the second order interaction among the criterion variables and all predictor-criterion interactions. This model is of interest if researchers assume that interactions beyond first order and predictor-criterion interactions can be traced back to the predictor variables.

A third base model was discussed as *CFA of directed variable relations* (von Eye, 1985). This model is also a P-CFA model but not an ISA model. It implies that any variation on the criterion side can be explained by the predictors. The model does not include any effect on the criterion side:

$$\log E = \lambda_0 + \lambda_i^{P_1} + \lambda_j^{P_2}.$$

The following paragraphs give an overview of possible P-CFA base models for cross-classifications with up to five variables (von Eye & Schuster, 1998). In this overview, we distinguish between two kinds of base models. The first comprises global CFA base models, that is, models in which all variables have the same status. The second comprises base models for two groups of variables.

In the selection of base models for P-CFA, three aspects need to be considered. The first concerns the base model for the predictors, the second concerns the base model for the criteria, and the third concerns the base model for the relationships between predictors and criteria. These relationships materialize in the form of types and antitypes.

P-CFA base models typically saturated in the predictors, mostly for two reasons. First, P-CFA, in analogy to regression analysis, models the distribution of the criterion variables, given the predictors. However, in contrast to standard regression analysis where correlations among predictors can cause multicolinearity problems[4], P-CFA yields types and antitypes regardless of the possible relationships among predictors. This is possible only if all substantial relationships among predictors are taken into account. Therefore, we retain the habit of using the saturated model on the predictor side.

In contrast, we do not always specify P-CFA base models that are also saturated on the criterion side. P-CFA models that are saturated on the criterion side coincide with ISA models for two variable groups. However, the distinction between ISA and P-CFA models is only one purpose. Another, possibly more important purpose for specifying base models that are more parsimonious than the saturated one was discussed above. If researchers assume that local relationships among criterion variables are caused by predictors, these relationships must be allowed to result in types and antitypes. This is possible only if the base model is not saturated on the criterion side.

The third aspect concerns the relationships among predictors on the one hand and criterion variables on the other. It is routine to specify base models such that no relationships are assumed, that is, such that predictors and criteria are independent. However, as was discussed in the context of

---

[4]Regression analysis using structural equations modeling does allow one to make predictor intercorrelations part of a model (Jöreskog & Sörbom, 1993).

ISA, it is conceivable that relationships up to a certain order are made part of the base model if researchers focus on higher order predictor-criterion relationships. This is an option that has not been treated in the literature but may be worth discussing.

Table 31 presents an overview of possible base models for global CFA and P-CFA for a total of up to five variables (from von Eye & Schuster, 1998). Possible models are marked with a "✔", and models that are excluded are marked with a "-". Models are excluded either because of the sampling scheme used for data collection or because of the researchers' interests in relationships among criterion variables. The first column in Table 31 lists the number of variables, out of five, with fixed margins. If two or more variables have fixed margins, it is assumed that their sampling scheme is multivariate product multinomial. The center panel of Table 31 displays the order of possible base models for global CFA, as it depends on the number of variables with fixed margins. The right hand panel displays the order of possible base models for the criterion variables in a P-CFA. (Remember that the P-CFA base model is always saturated in the predictor variables.) The last row in the right hand panel in Table 31 displays n/a for each number of criterion variables. The reason for this is that there can be no global CFA nor P-CFA if all variables have fixed margins.

The checkmarks in Table 31 suggest that the base models for global CFA must be of increasingly higher order when the number of variables with fixed margins increases. The last row in the panel for global CFA models shows that CFA cannot be performed at all if multivariate product multinomial sampling involves all variables, because a saturated model would result which explains 100% of the variability in a cross-classification.

To illustrate the use of the overview in Table 30, we present sample base models in Table 32 (from von Eye & Schuster, 1998). The top half of the table presents sample base models for global CFA. The bottom half presents sample base models for the criterion variable side of P-CFA.

In the following paragraphs, we present two data examples. In the first example, we illustrate two variants of P-CFA and compare them to first order CFA. In the second example, we compare P-CFA with ISA.

Data example I. In the first data example, we re-analyze a data set that has repeatedly been used for illustration of P-CFA (Lienert, 1978; Mellenbergh, 1996; von Eye et al., 1996b). The data describe suicide attempts in a sample of 482 individuals who had attempted suicide. Three variables are used: Gender (G; 1 = male, 2 = female); Motive for Suicide (M; 1 = illness, 2 =

Table 31: Possible CFA and P-CFA base models for up to five variables

# of Variables with fixed margins	Order of Global CFA Base Model					Order of Base Models for Criterion Variables in P-CFA-type Models			
	0	1	2	3	4	0	1	2	3
0	✔	✔	✔	✔	✔	✔	✔	✔	✔
1	-	✔	✔	✔	✔	-	✔	✔	✔
2	-	-	✔	✔	✔	-	-	✔	✔
3	-	-	-	✔	✔	-	-	-	✔
4	-	-	-	-	✔	-	-	-	-
5	n/a[a]	n/a	n/a	n/a	n/a	n/a[b]	n/a	n/a	n/a

[a]n/a indicates that these models are not conceivable because multivariate product multinomial sampling involves all variables and the base model would be saturated·
[b]n/a indicates that a model is not conceivable because there are no criterion variables left.

Table 32: Sample base models for Global CFA (top panel) and for the criterion side of P-CFA (bottom panel) with varying numbers of fixed variables (multivariate product multinomial sampling)

Global CFA Base Model	Model Specification in Bracket Notation
Zero order Global Model	no effects, no fixed margins
First Order Global Model	[P1][P2][C1][C2][C3]
Second Order Global Model	[P1, P2][C1, C2][P1, C1][P1, C2][P2, C1][P2, C2]

/ cont.

Global CFA Base Model	Model Specification in Bracket Notation
Third Order Global Model	[P1, P2][C1, C2, C3]
Fourth Order Global Model	[P1][C1, C2, C3, C4]
Fifth Order Global Model	[C1, C2, C3, C4, C5] (Model is saturated)

Base Model for P-CFA Criterion Variables	Model Specification in Bracket Notation
Zero Order P-CFA Base Model	[P1][P2]
First Order P-CFA Base Model	[P1][P2][C1][C2][C3]
Second Order P-CFA Base Model	[P1, P2][C1, C2][C1, C3][C2, C3]
Third Order P-CFA Base Model	[P1, P2][C1, C2, C3]
Fourth Order P-CFA Base Model	[P1][C1, C2, C3, C4]

psychiatric disorder; 3 = alcoholism); and Outcome of Suicide (O; 1 = survived, 2 = dead). The observed frequency distribution of the 2 x 3 x 2, G x M x O cross-classification appears in Table 33.

The frequency distribution in Table 33 will now be analyzed under the assumption that Gender and Motive are the predictors and Outcome is the criterion. We consider three base models (von Eye & Schuster, 1998):

1. First order CFA [G][M][O]. This model is included here only for comparison with results from classical first order CFA. A comparison with results from P-CFA is not possible because the model [G][M][O] does not discriminate between predictors and criteria or even the two groups of variables. This model of variable independence is of interest when researchers are interested in typesand antitypes that can be traced back to local interactions of variables but not to main effects. Only single variables can be fixed by design or sampling characteristics. As soon as sampling is multivariate product multinomial, first order CFA cannot be employed any longer.

Table 33: Observed cell frequencies of suicide attempt data

Cell index	Observed frequencies for outcome of suicide attempt	
GM	Survived	Died
11	64	18
12	76	47
13	7	8
21	86	16
22	61	25
23	47	27

2. Mixed-sampling P-CFA [GM]. This model considers the margins of the predictor variables, Gender and Motive as fixed. The criterion variable, Outcome of Suicide Attempt, is considered random. The predictors may interact because it seems plausible that males and females differ in motive for suicide (Chipuer & von Eye, 1989). This model is, on the criterion side, a zero order CFA (von Eye, 1985). Types and antitypes can result from (a) the main effect of the criterion, and (b) interactions among the predictors and the criterion.

3. First order P-CFA [GM][O]. In the present example, this is the standard P-CFA (and ISA) base model. It considers both the margins of the predictor variables and the margins of the criterion as fixed. Types and antitypes can only result from interactions between the predictors and the criterion.

Table 34 displays results from these three models. For each model we adjusted $\alpha = 0.05$ using the Bonferroni method which resulted in $\alpha^* = 0.0042$. The table presents the deviance residuals, z_{ijk}, and their one-sided tail probabilities.

Table 34: **z-values and their tail probabilities for three CFA models for suicide attempt data**

Cell Index GMO	n_{ijk}	[G][M][O] z_{ijk}	[G][M][O] $p(z_{aijk})$	[GM] z_{ijk}	[GM] $p(z_{aijk})$	[GM][O] z_{ijk}	[GM][O] $p(z_{aijk})$
111	64	.59	.28	3.15	<α* **T**	.77	.22
112	18	-1.39	.08	-4.18	<α* **A**	-1.28	.10
121	76	1.02	.16	3.77	<α* **T**	-1.21	.11
122	47	3.29	<α* **T**	-.10	.46	1.75	.04
131	7	-4.87	<α* **A**	-3.42	<α* **A**	-1.18	.12
132	8	-1.19	.12	-3.12	<α* **A**	1.54	.06
211	86	1.75	.04	4.61	<α* **T**	1.58	.06
212	16	-2.86	<α* **A**	-5.62	<α* **A**	-2.78	<α* **A**
221	61	-2.26	.01	.55	.29	.02	.49
222	25	-1.49	.07	-4.75	< α*	-.03	.49
231	47	2.07	.02	4.10	< α*	-.75	.22
232	27	3.03	<α***T**	.56	.29	1.11	.13

[a]Tail probabilities with three or more leading zeros are displayed as "< α*;" types are indicated by **T**; antitypes are indicated by **A**.

Table 34 suggests that first order CFA identifies two types and two antitypes. However, if researchers are interested in making the distinction between predictors and criteria a part of their analyses, interpretation of this result can be a problem, because in first order CFA all variables have the same status. In contrast, the mixed-sampling P-CFA base model [GM] does support the distinction between predictors and criteria. This model assumes that Gender and Motive for suicide attempt allow one to predict the outcome of a suicide attempt. In addition, the model [GM] is based on a mixed sampling scheme where the uni- and bivariate marginals of G and M are fixed and the marginals of O are random. As was indicated above, not considering the main effect of the criterion implies the assumption of a uniform marginal distribution for the criterion. If this assumption is of no substantive interest, criterion marginals must be made part of the base model (see below).

The model [GM] identifies three types and four antitypes. These types and antitypes overlap with the ones identified by first order standard CFA only in part. Specifically, the two types found by first order CFA are not found here, and there are three types and three antitypes found in the [GM] model but not in the first order CFA model [G][M][O].

When interpreting the results from the model [GM], one has to take into account the nature of the base model which treats G and M as predictors and O as the outcome variable. For instance, the pattern "male, suffering from some illness" allows one to predict that a suicide attempt is survived (Configuration 111). The first antitype suggests the prediction that men who suffer from some illness succeed less often in committing suicide[5] than one might expect from the [GM] base model.

Rather than interpreting the three types and four antitypes in detail, we now inspect the results from the third model, [GM][O]. This model takes more information into account than the other two models when estimating the expected cell frequencies. Therefore, there are fewer ways to deviate from the expected cell frequencies and it does not come as a surprise that the number of types and antitypes is smaller than for the other two models. The model identifies no type and only one antitype, that is, Configuration 212, which had emerged as an antitype in the other two analyses also. It suggests the prediction that fewer females that attempt a suicide for illness-related reasons succeed in this attempt than was expected from the base model.

The models [GM] and [GM][O] differ only in the term [O]. Therefore, one can conclude that the many types and antitypes identified by [GM] reflect the main effect of the outcome variable.

In sum, this example illustrates again that different models of CFA can lead to dramatically different patterns of types and antitypes (Mellenbergh, 1996).

---

[5]One might suspect that when the outcome variable is dichotomous, types and antitypes always go hand in hand. This is obviously not the case. For instance, Pattern 121 constitutes a type, but Pattern 122 is far from constituting an antitype. Another example is the pattern pair 131 and 132 which both constitute antitypes.

Data example II. In the second data example, we compare P-CFA with ISA. To do this, we use a data set published by (Mahoney, 2000). Mahoney used cluster-analytic methods to create four clusters of adolescents. The first cluster describes boys and girls who are highly competent in academic matters, less physically mature, younger, low in aggression, above average in popularity, and high in SES. The second cluster is similar to the first except that these adolescents are physically rather normal and below average in SES. Cluster III describes adolescents moderately low in academic competence, popularity and SES, and with moderately high levels of aggression. Cluster IV is described by a multiple risk profile. These individuals are older than their classmates, high in aggression, and below average in academic competence, popularity and SES. For the following analyses we follow Mahoney who fused the first two clusters because of their similarity. Thus, the clusters form the first variable, Pattern (P) with the three categories 1 = Profiles I+II, 2 = Profile III, and 3 = Profile IV. The second variable is Gender (G) with categories 1 = males and 2 = females. The third variable is School Dropout (S) with categories 1 = no and 2 = yes, and the last variable is Criminal Arrest (A) with categories 1 = no and 2 = yes.

For P-CFA we now form predictors and criteria. We attempt to predict the outcome variables School Dropout and Criminal Arrest from the predictors Pattern and Gender. We analyze these data under the following two P-CFA base models:

1. [P, G][S, A]: this is the original P-CFA base model that is identical with the two-group ISA model.
2. [P, G][S][A]: this is a P-CFA base model that allows types and antitypes to emerge from predictor-criterion relationships and from the association between the criterion variables, S and A. It should be noted that this P-CFA base model cannot be interpreted as a standard two-group ISA base model.

For both analyses we use the z-test and the Bonferroni-adjusted $\alpha^* = 0.0020833$. The results of these analyses are summarized in Table 35.

The results in Table 35 suggest that the original P-CFA model [P, G][S, A] yields fewer types and the same number of antitypes than the more parsimonious model [P, S][S][A]. The reason for this difference lies in the strong association between the variable School Dropout and Criminal Arrests. The strength of this association can be assessed by comparing the goodness-of-fit X^2 values of the two base models. We calculate for the model [P, G][S, A] the likelihood ratio $X^2 = 162.43$ ($df = 15$; $p < 0.01$), and for the model [P, G][S][A] the likelihood ratio $X^2 = 212.74$ ($df = 16$; $p <$

0.01). The improvement from the more parsimonious model to the less parsimonious one is solely due to the association between S and A which is taken into account in the less parsimonious model. The improvement is significant ($\Delta X^2 = 50.31$; $\Delta df = 1$; $p < 0.01$).

Table 35: Results from two P-CFA analyses of Mahoney's (2000) adolescent adjustment variables

Cell index GPSA	Observed frequencies	P-CFA model [P, G][S, A]		P-CFA model [P, G][S][A]	
		expected	$p(z)$	expected	$p(z)$
1111	155	134.13	.0358	127.79	.0080
1112	9	10.67	.3046	17.01	.0261
1121	6	18.55	.0018 **A**	24.89	.0001 **A**
1122	3	9.65	.0161	3.31	.4318
1211	63	71.33	.1620	67.96	.2738
1212	10	5.67	.0347	9.05	.3756
1221	11	9.86	.3585	13.23	.2696
1222	8	5.13	.1029	1.76	<α* **T**
1311	26	46.52	.0013 **A**	44.32	.0030
1312	8	3.70	.0127	5.90	.1936
1321	13	6.43	.0048	8.63	.0685
1322	13	3.35	< α* **T**	1.15	< α* **T**
2111	188	160.49	.0150	152.91	.0023
2112	7	12.77	.0533	20.35	.0015 **A**
2121	12	22.19	.0152	29.78	.0005 **A**
2122	0	11.55	.0003 **A**	3.96	.0232

/ cont.

Cell index GPSA	Observed frequencies	P-CFA model [P, G][S, A]		P-CFA model [P, G][S][A]	
		expected	$p(z)$	expected	$p(z)$
2211	76	74.43	.4279	70.91	.2729
2212	8	5.92	.1964	9.44	.3197
2221	6	10.29	.0905	13.81	.0178
2222	6	5.36	.3905	1.84	.0010 **T**
2311	20	42.09	.0005 **A**	39.15	.0011 **A**
2312	0	3.27	.0353	5.21	.0112
2321	25	5.68	< α* **T**	7.62	< α* **T**
2322	8	2.96	.0017 **T**	1.02	< α* **T**

[a] < α* indicates that the tail probability is smaller than can be expressed with four decimal places.

In the following paragraphs, we give a brief description of the types and antitypes that resulted from the more parsimonious P-CFA base model (last two columns in Table 35). Starting from the top of the table, the first antitype is constituted by Configuration 1121. This antitype suggests that one can predict that male adolescents from Clusters I or II are unlikely to drop out of school and have no criminal arrests. Almost 25 respondents were expected to display this pattern, but only 6 were found. The first type is constituted by Configuration 1222. This configuration suggests the prediction that more male adolescents from Cluster III than expected both drop out of school and have criminal arrests. Fewer than 2 respondents were expected for this configuration, but 8 were counted. The second type is constituted by Configuration 1322. These are male respondents from Cluster IV who more frequently than expected dropped out of school and have been arrested for criminal offenses. Only about one individual was expected for this profile, but 13 were found. The second antitype is constituted by Configuration 2112. This antitype suggests that fewer female adolescents from Clusters I and II than expected can be predicted to stay in school but have criminal arrests. In a similar fashion, based on the antitype designation of Configuration 2121, we can state that fewer female adolescents from Clusters I and II than expected can be predicted to drop out of school without criminal arrests. The third type, constituted by Configuration 2222, suggests that more female adolescents from Cluster III than expected can be predicted to both drop out of school and have criminal arrests. The fourth antitype is noted for Configuration 2311. These are female adolescents from Cluster IV. Fewer respondents with this profile than expected can be predicted to stay in school and have no criminal arrests. The fourth type is constituted by Configuration 2321. We can predict that more female adolescents from Cluster IV than expected drop out of school but have no criminal arrests. Accordingly, more female adolescents from Cluster IV than expected both drop out of school and have criminal arrests (Configuration 2322).

What are the criteria that researchers can use when they decide whether to use the original P-CFA base model or one of the more parsimonious ones? The most compelling argument in favor of the original base model which (a) is saturated in both the predictors and the criteria and (b) proposes independence of the predictors from the criteria is that types and antitypes can emerge only if there exist predictor-criteria relationships. If researchers focus on these relationships, the original base model is the model of choice. If, however, the researchers consider it interesting to keep the door open for local within-criterion relationships that may (or may not) be caused by the predictors, then the more parsimonious models exemplified in Table 31 must be considered.

6.2.2 More P-CFA models and approaches

In this section we discuss special issues in the context of P-CFA. Specifically, we discuss conditional P-CFA, biprediction P-CFA, and prediction coefficients.

6.2.2.1 Conditional P-CFA: Stratifying on a variable

Aggregating, that is, *collapsing* over the categories of a variable is defensible only if this variable does not interact with any other variable (Bishop, Fienberg, & Holland, 1975). The *collapsibility theorem* states that if one divides a d-dimensional array into three mutually exclusive groups, one group is *collapsible* with respect to the parameters that involve a second group "if and only if the first two groups are independent of each other." That is, the parameters that link the first two groups are zero

(Bishop et al., 1975, p. 47). This definition has two important implications for log-linear modeling:

1. If all two-way interactions exist, aggregating over any variable changes all parameters.
2. If any variable is independent of all other variables, one can remove this variable by aggregating over its categories. No parameter will change, not even when this variable is "condensed," that is, when some if its categories are combined.

The implications for configural analysis are analogous:

(1) If all two-way interactions exist, aggregating over any variable can change the emerging pattern of types and antitypes.
(2) If any variable is independent of all other variables, one can remove this variable by aggregating over its categories. In addition, one can condense this variable. The emerging pattern of types and antitypes will not change.

All this applies accordingly if two or more completely independent variables exist.

P-CFA has thus far been applied mostly in the context of differential, sociological, and psychological research. Sample applications concern the prediction of performance in elementary school from gender, SES, and test scores (Lienert & Klauer, 1983); the prediction of success of psychotherapy techniques (Lienert & Wolfrum, 1979); the prediction of psychiatric diagnoses from the variables hospital, SES, and family status (Hütter et al., 1981); the prediction of dyslexia from performance in reading, vocabulary, and reading tests (Krauth & Lienert, 1973b); and the prediction of temperamental types among preschoolers (Aksan et al., 1999). All these and a number of other applications included stratification variables such as SES and gender.

Wermuth (1976) discussed the possibility that types and antitypes can emerge even if variables are independent at certain levels of the stratification variables. In these cases, collapsibility may not be given for the stratification variables. To prevent this from happening, Krauth and Lienert (1982) proposed *conditional CFA* (cf. Krauth, 1980; Lienert, 1978). This variant of CFA searches for types and antitypes without aggregating over categories of stratification variables. More specifically, conditional CFA estimates the expected frequencies separately for each category of the stratification variables. This approach can be viewed parallel to the methods of conditional log-linear models (Fienberg, 1980).

Krauth and Lienert (1982) recommend using conditional CFA if

(1) one of the variables is a stratification variable, and
(2) the association between other predictors and the criteria varies with the levels of the stratification variable.

A sample application of conditional CFA can be found in Chipuer and von Eye (1989). The authors searched for types and antitypes of suicidal behavior separately in female and male samples from Canada and Germany. The results justified the application: Patterns of types and antitypes differed considerably across the strata.

In the context of P-CFA, conditional CFA allows one to answer the question whether prediction patterns differ across the levels of one or more stratification variables.

In the following paragraphs, we present two data examples. The first example presents artificial data (Schuster, 1997). It illustrates the dangers of falsely aggregating across the categories of a stratification variable. The second example presents real data (Krauth & Lienert, 1973a). It illustrates gender-specific suicide patterns. Both examples illustrate conditional CFA.

Data example I. A sample of 100 female and 100 male alcohol consumers were diagnosed as to alcoholism at two points in time. At the first point in time, 50% of the entire sample was classified as alcoholics. Using P-CFA, we now ask whether the alcohol diagnosis at Time 2 can be predicted from the alcohol diagnosis at Time 1. Table 36 displays the P-CFA results for the entire sample, that is, aggregated over the two gender categories. We use the binomial test and Bonferroni-adjustment of α which led to $\alpha^* = 0.0125$.

The results in Table 36 suggest that more respondents than expected were diagnosed to be alcoholic at the first observation point and can be predicted to still be alcoholic at the second observation point (Configuration 11). Fewer individuals than expected can be predicted to change their diagnostic status from alcoholic to nonalcoholic (Configuration 12) or from nonalcoholic to alcoholic (Configuration 21). More individuals than expected can be predicted to stay in the nonalcoholic category (Configuration 22).

Table 36: CFA of repeated alcoholism diagnoses

Configuration[1]	Frequencies		Binomial p	Type/ Antitype?
T1 T2	observed	expected		
11	65	40.5	.00003	Type
12	25	49.5	.00001	Antitype
21	25	49.5	.00001	Antitype
22	85	60.5	.00016	Type

[1] 1 = alcoholic; 2 = not alcoholic

While these results seem to speak in favor of the general assumption that both alcoholism and lack of alcoholism are stable over time, we have to keep in mind that bias can occur when variables are eliminated that are related to the remaining variables. In the present example, the categories of the gender variable were summed across. To assess the bias created by aggregating, we perform a conditional CFA. This method involves calculating a P-CFA separately for each gender. The results for the female respondents appear in Table 37. The results for the male respondents appear in Table 38. To make the results of the gender-specific analyses comparable to the ones from the aggregated sample in Table 36, we use the binomial test and Bonferroni-adjustment for each analysis.

The results in Tables 37 and 38 suggest a different appraisal of the diagnostic status of alcoholism in two observations. Specifically, there emerge no types or antitypes in the female sample nor in the male sample. We thus conclude, quite in contrast to the conclusions that might falsely be drawn from Table 36, that in both the female and the male samples the two diagnoses are independent of each other. It thus seems unpredictable from Time 1 diagnoses whether respondents change or stay stable.

From a variable perspective, this result can be confirmed. The log-linear base model [G][T1][T2] clearly must be rejected ($LR\text{-}X^2 = 220.32$; $df = 4$; $p < 0.01$). In contrast, the model of all two-way interactions describes the data perfectly ($LR\text{-}X^2 = 0.0$; $df = 1$; $p = 1.0$). In addition, all parameters are significant except the one for the T1 x T2 interaction. Therefore, because Gender interacts significantly with both T1 and T2, one must not collapse over the gender categories.

Table 37: CFA of repeated alcoholism diagnoses in the female sample

Configuration[1]	Frequencies		Binomial p	Type/ Antitype?
T1 T2	observed	expected		
11	1	1.0	.74	-
12	9	9.0	.59	-
21	9	9.0	.59	-
22	81	81.0	.56	-

[1] 1 = alcoholic; 2 = not alcoholic

Table 38: CFA of repeated alcoholism diagnoses in the male sample

Configuration[1]	Frequencies		Binomial p	Type/ Antitype?
T1 T2	observed	expected		
11	64	64.0	.54	-
12	16	16.0	.57	-
21	16	16.0	.57	-
22	4	4.0	.63	-

[1] 1 = alcoholic; 2 = not alcoholic

Data example II. In this data example, we perform conditional P-CFA on data that describe the year of occurrence and method of suicide in a sample of males and a sample of females. The variable categories are 1952 (y = 1) and 1944 (y = 2) for year of occurrence; and gassing (m = 1), hanging (m = 2), soporifics (m = 3), drowning (m = 4), cutting veins (m = 5), shooting (m = 6), and jumping (m = 7) for method of suicide. The results for the male sample appears in Table 39a, and the results for the female sample appear in Table 39b. For both analyses, we use the *z*-test and Bonferroni-adjustment which led to $\alpha^* = 0.00357$.

Table 39a: Conditional P-CFA of the variables Gender, Year of Occurrence, and Method of Suicide (Male Sample)

Index	Frequencies		Test statistics		Type/ Antitype?
YM	observed	expected	*z*	*p*(*z*)	
11	52	33.71	3.15	.0008	**T**
12	31	53.04	-3.03	.0012	**A**
13	44	25.28	3.72	.0001	**T**
14	20	19.33	0.15	.4397	
15	22	18.34	0.85	.1965	
16	3	18.84	-3.65	.0001	**A**
17	2	5.45	-1.48	.0696	
21	16	34.29	-3.12	.0009	**A**
22	76	53.96	3.00	.0013	**T**
23	7	25.72	-3.69	.0001	**A**
24	19	19.67	-0.15	.4403	
25	15	18.66	-0.85	.1985	
26	35	19.16	3.62	.0001	**T**
27	9	5.55	1.47	.0713	

The results in Tables 39b and 39b suggest quite discrepant patterns of types and antitypes for male and female suicides. In the *male* sample we find four types suggesting that

(1) in 1952, males committed suicide typically by gassing themselves or using soporifics (Configurations 11 and 13);

(2) in 1944, males committed suicide typically by hanging or shooting themselves (Configurations 22 and 26).

Table 39b: Conditional P-CFA of the variables Gender, Year of Occurrence, and Method of Suicide (Female Sample)

Indices	Frequencies		Test statistics		Type/ Antitype?
YM	observed	expected	z	$p(z)$	
11	47	53.85	-0.93	.1754	
12	14	24.43	-2.11	.0174	
13	97	52.85	6.07	< α*	**T**
14	10	31.91	-3.88	.0001	**A**
15	5	4.49	0.24	.4044	
16	0	5.48	-2.34	.0096	
17	2	1.99	0.00	.4984	
21	61	54.15	0.93	.1761	
22	35	24.57	2.10	.0177	
23	9	52.15	-6.06	< α*	**A**
24	54	32.09	3.87	.0001	**T**
25	4	4.51	-0.24	.4046	
26	11	5.52	2.34	.0098	
27	2	2.01	-0.00	.4984	

[a] < α* indicates that the tail probability is smaller than can be expressed with four decimal places.

We also find four antitypes in the male sample suggesting that

(1) in 1952, hanging and shooting were unexpectedly infrequent means of committing suicide (Configurations 12 and 16);

(2) in 1944, gas and soporifics were unexpectedly infrequent means of committing suicide (Configurations 21 and 23).

In the *female* sample we find two types that indicate that in 1952, women used soporifics more often than expected based on chance, and in 1944, drowning occurred more often than expected. The antitypes indicate that in 1952 drowning and in 1944 the use of soporifics occurred less often than expected.

These are the results of *conditional P-CFA*. Readers are invited to perform a (Gender x Year) x Means of Suicide P-CFA and a P-CFA after aggregating over Year and Gender, and then to compare results.

6.2.2.2 Biprediction CFA

Biprediction CFA (BCFA), introduced by Lienert and Netter (1987; see also Fleischmann & Lienert, 1992; Lienert & von Eye, 1987; Netter, 1996), allows one to test two predictions simultaneously:

(1) Predictor Configuration A leads to Criterion Configuration a
(2) Predictor Configuration B leads to Criterion Configuration b

(with A ≠ B and a ≠b). BCFA was derived from a method proposed by Havránek and Lienert (1984) for the analysis of parts of contingency tables, and from two-cell outlier analysis as discussed by Kotze and Hawkins (1984). To illustrate, consider the following example. In a drug effect study, a drug is applied in the three doses 1, 2, and 3. The response is measured as positive (+) neutral (0), or negative (-). The two variables, Drug Dose and Response can be crossed to form the 3 x 3 tabulation shown in the upper panel of Table 40.

To test these BCFA predictions simultaneously, one extracts a subtable that allows one to depict these predictions. In the present example suppose

(a) Dose 1 leads to positive responses, and
(b) Dose 3 leads to negative responses.

To test these two predictions, a 2 x 2 subtable is examined that includes only Cells 11, 13, 31, and 33 from the original tabulation. This subtable appears in the bottom part of Table 40. Variable levels not included in the predictions are labeled with an x.

To test the bipredictive hypotheses, the frequencies *a, b, c, d, A, B, C, D*, and *N* can be inserted into Kimball's (1954) equation for the exact partitioning of χ^2 in two-way tables,

$$X^2 = \frac{[A(Ba - Cb) - B(Dc - Cd)]^2}{ABCD(A + B)(C + D)/N}.$$

The test statistic X^2 is distributed approximately as χ^2 with 1 degree of freedom. For 2 x 2 tables, the test statistic is

$$X^2 = \frac{N(ad - bc)^2}{ABCD}.$$

Table 40: 3 x 3 tabulation and sample subtable for BCFA

Predictor configurations	Criterion configurations 1	2	3	Marginal totals
	Complete tabulation			
1	f_{11}	f_{12}	f_{13}	
2	f_{21}	f_{22}	f_{23}	
3	f_{31}	f_{32}	f_{33}	
	BCFA subtable			
1	$f_{11} = a$	f_{12}	$f_{13} = b$	$a + b = A$
2	f_{21}	f_{22}	f_{23}	x
3	$f_{31} = c$	f_{32}	$f_{33} = d$	$c + d = B$
Totals	$a + f_{21} + c = C$	x	$b + f_{23} + d = D$	$A + x + B = N$

Bipredictions are particularly useful for the comparison of two treatments that are expected to have divergent outcomes.

Data example. The following data example is taken from Lienert and Netter (1987). The data describe the effects of nicotine in a sample of 48 male (m) and female (f) young adults. The participants took in balanced order for one week a placebo (P), 0.5 mg (H), and 1 mg (F) of nicotine. The responses to each dose were measured as either an increase (1) or a decrease (2) in finger pulse volume. Thus, each of the variables P, H, and F had categories 1 and 2. Participants were either smokers (S) or nonsmokers (N). Sampling was univariate product-multinomial for P, H, and F, and multinomial for Gender (G) and Smoker status (C).

We now analyze the P x H x F x G x C cross-classification in two steps. First, we employ standard P-CFA, and second we use biprediction CFA. For P-CFA we use the binomial test and Bonferroni-adjustment of α which yields α* = 0.0015625. We consider Gender and Smoking Status as predictors and the responses to the three experimental conditions as criterion variables; thus, the P-CFA base model is [G, S][P][H][F]. Table 41 displays the results of P-CFA.

The overall Pearson X^2 for the P-CFA base model [G, C][P][H][F] is 53.79 ($df = 25$; $p = 0.0007$), suggesting that there must be predictor-criteria relationships that P-CFA may be able to detect. P-CFA indicates that the configuration [f, N][1, 2, 2] constitutes a type. We conclude that there are more female nonsmokers than expected from the base model for whom it can be predicted that the placebo leads to an increase, and the two nicotine conditions to a decrease in finger pulse volume. Maybe because of the small sample size, no other types of antitypes surfaced.

Table 41: P-CFA of the predictors Gender (G) and Smoking Status (C) and the criteria Placebo (P), 0.5 mg (H), and 1 mg (F) of nicotine

Cell index	Frequencies		Binomial	Type/
PHFGC	observed	expected	*p*	Antitype?
111mS	4	1.833	.1103	
111mN	3	1.031	.0841	
111fS	0	1.604	.1956	
111fN	0	1.031	.3526	
112mS	3	2.167	.3693	
112mN	1	1.219	.6549	
112fS	1	1.896	.4298	
112fN	0	1.219	.2910	
121mS	1	1.833	.4482	

/ cont.

Cell index	Frequencies		Binomial	Type/
PHFGC	observed	expected	p	Antitype?
121mN	0	1.031	.3526	
121fS	2	1.604	.4792	
121fN	0	1.031	.3526	
122mS	1	2.167	.3561	
122mN	1	1.219	.6549	
122fS	0	1.896	.1445	
122fN	7	1.219	.0002	**T**
211mS	3	1.833	.2774	
211mN	1	1.031	.7242	
211fS	3	1.604	.2159	
211fN	0	1.031	.3526	
212mS	2	2.167	.6307	
212mN	0	1.219	.2910	
212fS	3	1.896	.2945	
212fN	0	1.219	.2910	
221mS	1	1.833	.4482	
221mN	2	1.031	.2758	
221fS	2	1.604	.4797	
221fN	0	1.031	.3526	
222mS	1	2.167	.3561	

/ cont.

Cell Index	Frequencies		Binomial	Type/
PHFGC	observed	expected	p	Antitype?
222mN	1	1.219	.6549	
222fS	3	1.896	.2945	
222fN	2	1.219	.3451	

We now ask the bipredictive question whether males and females differ in their nicotine response. Specifically, we ask whether

(1) males respond to nicotine with an increase in finger pulse volume, whereas
(2) females respond with a decrease.

This prediction does not involve the variable Smoking Status. Therefore, the table for the analysis of the biprediction pools over the categories of Smoking Status. Readers are invited to perform the necessary analyses to determine whether this can be done without creating bias[6]. Table 42 displays the resulting cross-classification.

Inserting the frequencies from Table 42 into Kimball's X^2 yields

$$X^2 = \frac{[7(9\cdot 7 - 25\cdot 0) - 9(23\cdot 2 - 25\cdot 7)]^2}{7\cdot 9\cdot 25\cdot 23(7 + 9)(25 + 23)/48} = 4.4279.$$

The tail probability for this value is $p = 0.035$ ($df = 1$). We thus reject the null hypothesis of no bipredictive relationship, and retain the bipredictive type 111|m versus 122|f. This type suggests that males respond with an increase in finger pulse volume, and females respond with a decrease.

[6]To do this, one first needs to determine whether Smoking Status is unrelated to all other variables.

Table 42: Biprediction CFA of the predictor Gender (G) and the criteria placebo (P), 0.5 mg (H), and 1 mg (F) of nicotine

Configurations	Frequencies for strata		Sums
PHF	males	females	
111	$a = 7$	$b = 0$	$A = 7$
112	4	1	
121	1	2	
122	$c = 2$	$d = 7$	$B = 9$
211	4	3	
212	2	3	
221	3	2	
222	2	5	
Sums	$C = 25$	$D = 23$	$N = 48$

6.2.2.3 Prediction coefficients

When a configuration is evaluated using CFA null hypotheses tests, the result is a statement concerning the statistical significance of this configuration. The fact, however, that the difference between an observed and expected cell frequency is significant, does not imply that substantial portions of variability are explained. It is well known from analysis of variance or regression analysis that effects can be significant without explaining large portions of variance.

In this section, we present three estimators of portions of variability that are accounted for by a type or antitype (Funke, Funke, & Lienert, 1984). These measures are applicable to P-CFA (and to two-sample CFA; see Chapter 7). Each of them is derived from Pearson's Phi-coefficient for 2 x 2 tables.

$$\Phi = \frac{ad - bc}{\sqrt{(a + b)(c + d)(a + c)(b + d)}} ,$$

where a, b, c, and d denote the cell frequencies in a 2 x 2 table, read clockwise, starting from the upper left. The null hypothesis for a test of the magnitude Φ is $\Phi = 0$. Under this null hypothesis we can use the test statistic

$$z = \sqrt{N}\, \Phi ,$$

or

$$\chi^2 = N\Phi^2,$$

with $df = 1$.

The coefficients to be introduced in the following paragraphs are related to the significance tests used in CFA. Three coefficients are introduced. The first can be used for single types or antitypes. The second is useful when the criterion variable is dichotomous and when one configuration is contrasted with all others. The third coefficient measures the portion of variability that is accounted for by a biprediction hypothesis.

Practical significance of single configurations. If the Pearson X^2 component is used to test the CFA null hypothesis for an individual configuration, the measure

$$\varphi = \frac{n - e}{\sqrt{Ne}}$$

can be calculated. If the binomial test is used, the measure is

$$\varphi = \frac{n - e}{\sqrt{Ne - e^2}}.$$

These two coefficients are called *φ prediction coefficients*. For both coefficients the null hypothesis $\varphi = 0$ can be tested. This test alone does not carry information beyond the P-CFA significance test. More important is that the square of the coefficients, φ^2 can be interpreted as the portion of variability of the criterion that is accounted for by the predictors.

Consider the following example. The only type in Table 41 was found for Configuration fN.122 (the period separates the predictors from the criterion variables). Inserting N, the observed and the expected cell frequencies into the φ-formula for the binomial test, one obtains

$$\varphi = \frac{7 - 1.219}{\sqrt{48 \cdot 1.219 - 1.219^2}} = 0.766.$$

Based on the square of this φ-value, 0.586, we can conclude that Gender and Smoking Status allow one to predict 58.6% of the variability of the outcome (change) pattern 122.

Contrasting one configuration with all others. The second coefficient, also proposed by Funke et al. (1984), allows one to contrast one configuration with all others combined. Consider, for example, the hypothesis that a particular configuration allows one to discriminate between psychotics and nonpsychotics. To test this hypothesis, one can create a 2 x 2 table in which one row represents the configuration of the hypothesis, and the second row is the sum of all other configurations. The two columns represent the two comparison groups of psychotics and nonpsychotics. If the configuration under study indicates a type, one can ask what portion of the variation is covered by the relationship between the configurations and the comparison groups. The φ-coefficient to answer this question is

$$\varphi = \frac{ad - bc}{\sqrt{ABCD}},$$

where a, b, c, and d are the cell frequencies of the 2 x 2 table, clockwise, starting from the upper left, and A and B are the row sums, and C and D are the column sums. Consider the following numerical example: $a = 112$, $b = 0$, $c = 125$, and $d = 215$. For these numbers we obtain

$$\varphi = \frac{112 \cdot 215 - 0 \cdot 125}{\sqrt{(112 + 0)(125 + 215)(112 + 125)(0 + 215)}} = 0.5467.$$

Squaring this φ-value yields $\varphi^2 = 0.2988$ which suggests that about 30% of the variability in this 2 x 2 table is explainable using φ.

Obviously, this φ-coefficient is applicable only to 2 x 2 tables. That is, the criterion has only two categories. If the criterion has more than two categories, one uses

$$\varphi = \frac{ad - bc}{\sqrt{ABCD(C + D)/N}},$$

thus taking into account the total sample size.

Variability accounted for by biprediction hypotheses. The first of the three φ-coefficients presented in this section focused on individual configurations. The second contrasted one configuration with all the other configurations combined. The coefficient to be introduced here contrasts one configuration with one other configuration (biprediction), given all other configurations. The coefficient is

$$\varphi = \frac{B(Da - Cb) - A(Dc - Cd)}{\sqrt{ABCD(A + B)(C + D)}} ,$$

where the symbols are as introduced in Table 42. This coefficient is best applied if there exists a biprediction type, that is, if one configuration allows one to predict a particular outcome, and another configuration allows one to predict the opposite outcome.

For the following numerical example we use data published by Funke et al. (1984). The authors report a study in which 692 alcoholics were presented with the Alcohol Use Inventory (Wanberg, Horn, & Foster, 1977). The four subscales of the instrument are:

Scale A: positive effects of drinking,
Scale B: compulsory, permanent drinking,
Scale C: anxiety as reason for and effect of drinking, and
Scale D: impairment and damage done by drinking.

In addition, the authors estimated for each respondent a *General Alcoholism Score* (G). The four scales were dichotomized with 1 indicating scores above the median and 2 indicating scores below the median. The General Alcoholism Score was split into three groups at the 33 and 67 percentile points with 1 indicating high alcoholism scores, 2 indicating scores in the middle range of G, and 3 indicating low alcoholism scores.

For the following analyses, we consider the responses in the four subscales the predictors of the General Alcoholism Score, G. Thus, the P-CFA base model is [A, B, C, D][G]. Table 43 presents the results of P-CFA. We used Lehmacher's test with Küchenhoff's continuity correction (1986) and Bonferroni-adjustment, which led to $\alpha^* = 0.00104$.

The results in Table 43 suggest that the General Alcoholism Score is strongly related to a number of patterns of the Alcohol Use Inventory. Specifically, the P-CFA indicates the existence of 11 prediction types and 12 prediction antitypes. The *prediction types* suggest that

- high scores on the General Alcoholism scale can be predicted from the Patterns 1111, 1211, 2111, and 2211,
- medium scores on the General Alcoholism scale can be predicted from the Patterns 1112, 1212, 2112, and 2121, and
- low scores on the General Alcoholism scale can be predicted from the Patterns 1222, 2122, and 2222.

Table 43: Prediction CFA of Wanberg, Horn, and Foster's (1977) alcohol data

Cell index	Frequencies		Test statistics		Type/ Antitype
ABCDG	observed	expected	z	$p(z)$	?
11111	112	42.13	14.53	< α*	**T**
11112	11	42.84	-6.54	< α*	**A**
11113	0	38.04	-8.07	< α*	**A**
11121	13	14.04	-0.18	.4271	
11122	24	14.28	3.11	.0009	**T**
11123	4	12.68	-2.85	.0022	
11211	8	9.59	-0.44	.3290	
11212	17	9.75	2.73	.0032	
11213	3	8.66	-2.15	.0157	
11221	0	8.56	-3.46	.0003	**A**
11222	15	8.71	2.48	.0067	
11223	10	7.73	0.78	.2180	
12111	33	17.47	4.61	< α*	**T**
12112	16	17.76	-0.39	.3501	
12113	2	15.77	-4.18	< α*	**A**
12121	6	11.65	-1.91	.0284	
12122	25	11.84	4.67	< α*	**T**
12123	3	10.51	-2.67	.0038	

/ cont.

Table 43, Panel 2/3

Cell index	Frequencies		Test statistics		Type/ Antitype ?
ABCDG	observed	expected	z	$p(z)$	
12211	8	9.93	-0.57	.2836	
12212	17	10.10	2.55	.0054	
12213	4	8.97	-1.83	.0334	
12221	0	15.75	-4.90	< α*	**A**
12222	15	16.02	-0.17	.4339	
12223	31	14.23	5.37	< α*	**T**
21111	27	10.62	6.15	< α*	**T**
21112	4	10.80	-2.43	.0076	
21113	0	9.59	-3.61	.0002	**A**
21121	6	10.96	-1.70	.0446	
21122	25	11.15	5.07	< α*	**T**
21123	1	9.90	-3.29	.0005	**A**
21211	2	5.82	-1.72	.0429	
21212	14	5.92	3.90	< α*	**T**
21213	1	5.26	-2.00	.0230	
21221	1	14.04	-4.25	< α*	**A**
21222	14	14.28	0.08	.4702	
21223	26	12.68	4.46	< α*	**T**
22111	16	8.22	3.19	.0007	**T**

/ cont.

Table 43, Panel 3/3

Cell index	Frequencies		Test statistics		Type/ Antitype
ABCDG	observed	expected	z	$p(z)$	?
22112	8	8.36	0.06	.4754	
22113	0	7.42	-3.11	.0009	A
22121	3	9.59	-2.47	.0067	
22122	17	9.75	2.73	.0032	
22123	8	8.66	-0.07	.4736	
22211	1	9.59	-3.29	.0005	A
22212	12	9.75	0.71	.2396	
22213	15	8.66	2.44	.0074	
22221	1	39.04	-8.10	< α*	A
22222	7	39.70	-6.92	< α*	A
22223	106	35.25	15.56	< α*	T

[a] < α* indicates that the tail probability is smaller than can be expressed with four decimal places.

The *prediction antitypes* suggest that

- high scores on the General Alcoholism scale can be predicted to be less likely than chance if the Patterns 1122, 1222, 2122, 2221, or 2222 were observed,
- medium scores on the General Alcoholism scale can be predicted to be less likely than chance if the Patterns 1111 or 2222 were observed, and
- low scores on the General Alcoholism scale can be predicted to be less likely than chance if the Patterns 1111, 1211, 2111, 2112, or 2211 were observed.

Readers are invited to discuss the substantive interpretation of these types and antitypes. In the present context, we ask what the portion of variability is that the Predictor Configuration 1111 contributes to the explanation of the two extreme values, 1 and 3, of the General Alcoholism scale. To answer this question we create a cross-classification that contains the information needed for the φ coefficient that contrasts one configuration with all others, when the criterion variable has more than two categories. This cross-classification appears in Table 44.

Inserting the frequencies from Table 44 into the formula for φ yields

$$\varphi = \frac{112 \cdot 214 - 0 \cdot 125}{\sqrt{123 \cdot 569 \cdot 237 \cdot 214(237 + 214)/692}} = 0.498.$$

Squaring yields $\varphi^2 = 0.248$. We thus conclude that the predictor pattern 1111 allows one to predict the criterion level 1, and this relationship accounts for 24.8% of the variability in Table 43.

Table 44: Contrasting the Predictor Configuration 1111 with all others with respect to the extremes, 1 and 3, of the criterion variable

Predictor configuration	General Alcoholism Score			Row totals
ABCD	1	2	3	
1111	$a = 112$	11	$b = 0$	$A = 123$
all others	$c = 125$	230	$d = 214$	$B = 569$
Column totals	$C = 237$	241	$D = 214$	$N = 692$

We now ask whether the two Predictor Patterns 1112 and 2221 allow us to establish the biprediction that respondents with pattern 1112 score high on the General Alcoholism scale and respondents with Pattern 2221 score low. Table 45 displays the cross-classification with the frequencies needed for the biprediction φ coefficient when the criterion has more than two categories.

Table 45: Cross-tabulation to assess the biprediction contrasting the Predictor Patterns 1112 and 2221

Predictor configuration	General Alcoholism Score			Row totals
ABCD	1	2	3	
1112	$a = 13$	24	$b = 4$	$A = 41$
2221	$c = 1$	12	$d = 15$	$B = 28$
all others	223	205	195	
Column totals	$C = 237$	241	$D = 214$	$N = 692$

Inserting the frequencies from Table 45 into the formula for φ results in

$$\varphi = \frac{28(214 \cdot 13 - 237 \cdot 4) - 41(214 \cdot 1 - 237 \cdot 15)}{\sqrt{41 \cdot 28 \cdot 237 \cdot 214(41 + 28)(237 + 214)}} = 0.1399.$$

Squaring yields $\varphi^2 = 0.0196$. We thus conclude that the bipredictive relationship that involves the two Predictor Patterns 1112 and 2221 explains less than 2% of the variability in Table 45.

7. Comparing k Samples

Comparative statements are in the heart of both differential (Anastasi, 1994) and person-oriented (Bergman, Magnusson, & ElKhouri, 2000) research. Not surprisingly, CFA methods of group comparison have experienced considerable attention and development. Methods for the comparison of two groups have been the farthest developed. In addition, there exist methods for the comparison of three and more groups. The present chapter reflects these developments. The most space will be devoted to the description of methods for the comparison of two groups.

The null hypothesis behind all CFA methods of group comparisons is that all groups were drawn from the same population. Thus, their parameters and frequency distributions should differ only randomly. Using CFA, researchers identify those sectors in the data space where this assumption is violated and types and antitypes can be established.

In Section 7.1, we introduce two-sample CFA as proposed by Lienert (1971). In Section 7.2, we discuss alternative approaches to two-sample CFA (von Eye, Rovine, & Spiel, 1995).

7.1 Two-sample CFA I: The original approach

Two-sample CFA is a method for comparison of two independent groups with respect to the configurations of a number of categorical variables (Lienert, 1971b; Krauth & Lienert, 1973a; for the comparison of paired groups see Lienert & Barth, 1987). The method allows one to answer the question whether the two groups differ in the distribution of the configurations. The null hypothesis is that of no differences. This null hypothesis is tested locally, that is, for each configuration. The base model for two-sample CFA has the following characteristics:

(1) it is saturated in the variables used to compare the two groups;
(2) it proposes independence between the grouping variable(s) and the comparison (also called discriminant) variables; and
(3) if two or more variables are used to specify the groups, the model is also satisfied in these variables (see below in Section 7.3).

This base model implies homogeneity of frequency distributions in the two groups. Types can emerge only if a relationship between the discriminant variables and the grouping variable exists. This characteristic is shared by two-sample CFA and prediction CFA (see Section 6.2). A type in two-sample CFA suggests that in one of the two samples a particular configuration was observed more often than expected based on the above base model. It is important to note that two-sample CFA does not distinguish between types and antitypes. If one group was observed exhibiting a particular configuration more often than the other group, then this configuration always constitutes both a type and an antitype. The test for the antitype is redundant if a type was observed and vice versa. Therefore, the type/antitype-pairs in two-sample CFA are called *discrimination types*.

To illustrate two-sample CFA, consider a 2 x 2 x 2 design where the last variable indicates group membership. Table 46 displays the typical arrangement of the cross-classification of this design.

Two-sample CFA compares the two groups in each configuration. In most cases, this comparison requires forming a 2 x 2 cross-classification that contains the target configuration in its first row and the combined remaining configurations in its second row. A scheme for this cross-classification is given in Table 47.

To test the null hypothesis of no association between discriminant configuration ij and the grouping variable, a number of tests has been proposed. Five of these tests are presented here. An additional four will be presented in Section 7.2. A sign test for use in two-sample CFA was proposed by Müller, Netter, and von Eye (1997).

Table 46: 2 x 2 x 2 cross-classification for two-sample CFA

Configuration	Groups		Row totals
P_1P_2 [a]	A	B	
11	N_{11A}	N_{11B}	$N_{11.}$
12	N_{12A}	N_{12B}	$N_{12.}$
21	N_{21A}	N_{21B}	$N_{21.}$
22	N_{22A}	N_{21B}	$N_{22.}$
Column totals	$N_{..A}$	$N_{..B}$	N

[a] P_1 and P_2 are the discriminant variables, A and B are categories of the grouping variable.

Table 47: 2 x 2 cross-classification for two-sample CFA testing

Configuration	Groups		Row totals
P_1P_2	A	B	
ij	$a = N_{ijA}$	$b = N_{ijB}$	$A = N_{ij.}$
all others combined	$c = N_{..A} - N_{ijA}$	$d = N_{..B} - N_{ijB}$	$B = N - N_{ij.}$
Column totals	$C = N_{..A}$	$D = N_{..B}$	N

A first test for two-sample CFA is Fisher's exact test which gives the probability of cell frequency a as

$$p(a) = \frac{\binom{a+c}{a}\binom{b+d}{b}}{\binom{N}{a+b}}$$

where *a, b, c,* and *d* are defined as in Table 47, and *N* is the total sample size. An equivalent formulations of Fisher's exact test is

$$p(a) = \frac{A!B!C!D!}{N!a!b!c!d!},$$

where *a, b, c,* and *d,* and *A, B, C, D* and *N* are defined as in Table 47. Fisher's test has the virtue of being exact. Thus, no assumptions need to be made concerning the characteristics of an approximation of some test statistic to some sampling distribution. However, the test is tedious to calculate. Therefore, it appears only rarely as an option in CFA programs.

The following three tests are approximative tests. They are best applied to large samples. The first of these tests is the X^2-test,

$$X^2 = \frac{N(a \cdot d - b \cdot c)^2}{ABCD}.$$

The X^2-test works best when the sample sizes are large. When a sample size is not quite large, it is recommended to employ a continuity correction which leads to

$$X^2 = \frac{N(|a \cdot d - b \cdot c| - 0.5 \cdot N)^2}{ABCD}.$$

Both X^2 statistics are approximately distributed as χ^2 with $df = 1$.

The second approximative test involves the approximation of the binomial test. To perform this approximation, one first estimates the relative frequencies $p_1 = a/N_A$ and $p_2 = b/N_B$, where N_A is the sample size of group *A* and N_B is the sample size of group *B*. The approximately normally distributed test statistic is then

$$z = \frac{p_1 - p_2}{\sqrt{h(1 - h)(\frac{1}{N_A} + \frac{1}{N_B})}},$$

where $h = \frac{a + b}{N}$ (Krause & Metzler, 1984). The Krause-Metzler approximation is recommended if $\frac{a + b}{N} > 0.1$ and $N \geq 60$. If, however, $\frac{(a + b)N_A}{N} \geq 4$, the following approximately normally distributed test statistic may be preferable:

$$z = \frac{a - \frac{AC}{N}}{\sqrt{\frac{ABCD}{N^3 - N}}} .$$

For a quick comparison of these five tests consider the cell frequencies $a = 9$, $b = 46$, $c = 152$, and $d = 155$ (from Krauth & Lienert, 1973a, p. 91). Inserting these frequencies yields the results given in Table 48.

Table 48: Comparison of five tests for two-sample CFA

Test	Test statistic	p
Fisher's exact test	-	0.000002
X^2	20.754	0.000005
X^2 with continuity correction	19.433	0.00001
binomial approximation	-3.291	0.00050
z	-4.549	0.000033

The results in Table 48 suggest that Fisher's exact test and the z-test are the most powerful ones, and that the binomial approximation is the least powerful one. However, more detailed investigations are needed before general statements can be made.

There are several ways to estimate the expected cell frequencies for cross-classifications as the one given in Table 46. One way involves, as indicated in the formulas, estimating as part of the calculation of a test statistic. Another way involves specifying a log-linear base model that yields the expected cell frequencies. Consider again the example in Table 46. The example involves the two dichotomous discriminant variables, P_1 and P_2, and one dichotomous criterion, the grouping variable G.

The base model for two-sample CFA of these variables is $[P_1, P_2][G]$. The design matrix for this model is

$$X = \begin{bmatrix} 1 & 1 & 1 & 1 & 1 \\ 1 & 1 & 1 & -1 & 1 \\ 1 & 1 & -1 & 1 & -1 \\ 1 & 1 & -1 & -1 & -1 \\ 1 & -1 & 1 & 1 & -1 \\ 1 & -1 & 1 & -1 & -1 \\ 1 & -1 & -1 & 1 & 1 \\ 1 & -1 & -1 & -1 & 1 \end{bmatrix}.$$

The first column in this matrix is the constant vector. The second, third, and fourth columns represent the main effects of the variables P_1, P_2, and G, in that order. The last column represents the interaction between P_1 and P_2. Data example. For the following data example we use Mahoney's (2000) data again (see the second data example in Section 6.2). From the four variables in the earlier analysis we now use Gender (G) with categories 1 = male and 2 = female, School Dropout (S) with categories 1 = no and 2 = yes, and Criminal Arrest (A) with categories 1 = no and 2 = yes. These three variables can be crossed to form the (S x A) x G tabulation, where the parentheses indicate the discriminant variables. In this example, we ask whether the two gender groups differ from each other in the two discriminant variables S and A. To answer this question we perform two-sample CFA. We use the normal approximation of the binomial test and Bonferroni adjustment which gives us $\alpha^* = 0.05/4 = 0.0125$. Please notice that we *divide α in two-sample CFA not by the total number of cells, t, but by t/2, because each test compares two cells*. Table 49 displays the results of two-sample CFA of Mahoney's (2000) data.

The results in Table 48 suggest that male and female adolescents differ in only one configuration. More boys than girls who do not drop out of school have been arrested for criminal offenses.

7.2 Two-sample CFA II: Alternative methods

In this section, we introduce six new measures for two-sample CFA. The first has mostly been used descriptively; the other five have been used both descriptively and for significance testing. The measures are π^*, λ, $\tilde{\lambda}$, ρ, δ, and θ. (The measures are explained in the following sections.)

Table 49: Gender comparison based on School Dropout and Criminal Arrest information

Cell index SAG	Observed frequencies	Statistical tests		Discrimination Type?
		z	$p(z)$	
111	234	-1.69	.0454	
112	284			
121	27	2.33	.0100	✔
122	15			
211	30	-1.06	.1444	
212	43			
221	24	2.06	.0196	
222	14			

7.2.1 Gonzáles-Debén's π*

Based on a particular definition of goodness of fit originally proposed by Rudas, Clogg, and Lindsay (1994), Gonzáles-Debén (1998; cf. Gonzáles-Debén & Méndez Ramírez, 2000) introduced the coefficient π^*. This measure is an indicator of the discrepancy between the observed and the expected frequencies, estimated under the CFA base model for two-sample CFA. If this discrepancy is large, the two groups differ in the relative frequency with which the configuration under study is observed. Therefore, π^* can be interpreted as an indicator of strength of association in a 2 x 2 cross-classification. Consider again Table 47. For data arranged as in this table, the coefficient is

$$\pi^* = \frac{ad - bc}{aN} .$$

The range of possible values for π^* is $0 \leq \pi^* < 1$. If $bc > ad$, π^* can theoretically become negative. However, because of the nominal level nature of the variables involved, a sign for the association is not defined (the order of categories is arbitrary). Therefore, the terms in the numerator and the denominator of π^* are exchanged if $bc > ad$, and one obtains

$$\pi^{*\prime} = \frac{bc - ad}{bN} .$$

To illustrate π^*, consider again the above example in which $a = 9$, $b = 46$, $c = 152$, and $d = 155$. Inserting into the equation for π^* yields

$$\pi^* = \frac{46 \cdot 152 - 9 \cdot 155}{46 \cdot 362} = 0.3361.$$

This value indicates an association of medium strength. Thus far, π^* has mostly been used as a descriptive measure. The use of π^* is exemplified in the following sections and in Section 8.2.3.

7.2.2 Goodman's three elementary views of non-independence

In this section, we discuss Goodman's (1991) three elementary views of deviation from independence from the perspective of two-sample CFA (von Eye, Rovine, & Spiel, 1995). In I x J cross-classifications, the model of statistical independence states that p_{ij}, the probability that a case falls in row i and column j (with $i = 1, ..., I$ and $j = 1, ..., J$) equals

$$p_{ij} = p_{i.}\, p_{.j} ,$$

where $p_{i.} = \sum_j p_{ij}$ and $p_{.j} = \sum_i p_{ij}$ are the row and the column marginals, respectively. Standard indicators of the *degree of non-independence*, include the *odds ratio* and the *correlation coefficient*. For the 2 x 2 tables that we study in two-sample CFA, the odds ratio is

$$\theta = \frac{p_{11}/p_{21}}{p_{12}/p_{22}} .$$

The correlation coefficient is

$$\rho = \frac{p_{11}\, p_{22} - p_{12}\, p_{21}}{\sqrt{p_{1.}\, p_{2.}\, p_{.1}\, p_{.2}}} .$$

The correlation coefficient ρ is identical to Pearson's φ which can be estimated using

$$\varphi = \frac{N_{11}N_{22} - N_{12}N_{21}}{\sqrt{(N_{11} + N_{12})(N_{21} + N_{22})(N_{11} + N_{21})(N_{12} + N_{22})}} ,$$

which is equivalent to the $\varphi = \dfrac{ad - bc}{ABCD}$, given in Section 6.2.2.3. If statistical independence as given above holds, one obtains $\theta = 1$ and $\rho = \varphi = 0$. The following measures of non-independence are extensions of the odds ratio θ and the correlation ρ.

Three measures of independence. Goodman (1991) discusses the three measures of non-independence, the *non-weighted interaction* λ, the *relative difference* Δ, and the *weighted interaction* $\tilde{\lambda}$. Within a log-linear framework, these three measures can be introduced as follows. Let G_{ij} be the natural logarithm of the probability p_{ij}, that is, $G_{ij} = \ln p_{ij}$. Then, the *non-weighted interaction* is

$$\lambda_{ij} = G_{ij} - G_{i.} - G_{.j} + G_{..} ,$$

where $G_{i.} = \frac{1}{J}\sum_j G_{ij}$, $G_{.j} = \frac{1}{I}\sum_i G_{ij}$, and $G_{..} = \frac{1}{IJ}\sum_i \sum_j G_{ij}$. The non-weighted interaction is a relative of the odds ratio. The usual conditions apply, that is, both the row sum and the column sum of the λ_{ij} are equal to zero.

The *relative difference* Δ_{ij} , a relative of the correlation coefficient ρ, is

$$\Delta_{ij} = \frac{p_{ij} - p_{i.}\, p_{.j}}{p_{i.}\, p_{.j}} .$$

The row sum and the column sum for Δ_{ij} are zero too, that is, $\sum_i \Delta_{ij}\, p_{i.} = 0$ and $\sum_j \Delta_{ij}\, p_{.j} = 0$.

The third measure, the *weighted log-linear interaction* $\tilde{\lambda}_{ij}$, is defined as

$$\tilde{\lambda}_{ij} = G_{ij} - \tilde{G}_{i.} - \tilde{G}_{.j} + \tilde{G}_{..} ,$$

where $\tilde{G}_{i.} = \sum_j G_{ij}\, p_{.j}$, $\tilde{G}_{.j} = \sum_i G_{ij}\, p_{i.}$, and $\tilde{G}_{..} = \sum_i \sum_j G_{ij}\, p_{i.}\, p_{.j}$. As for Δ_{ij}, the row and column sums are $\sum_i \tilde{\lambda}_{ij}\, p_{i.} = 0$ and $\sum_j \tilde{\lambda}_{ij}\, p_{.j} = 0$.

Thus far, we have defined the cell-wise measures λ_{ij}, Δ_{ij}, and $\tilde{\lambda}_{ij}$. Using these definitions, we can specify the three measures of non-independence for I x J tables as

$$\lambda = \sqrt{\sum_i \sum_j \frac{\lambda_{ij}^2}{IJ}} ,$$

$$\Delta = \sqrt{\sum_i \sum_j \Delta_{ij}^2 \, p_{i.} \, p_{.j}} ,$$

and

$$\tilde{\lambda} = \sqrt{\sum_i \sum_j \tilde{\lambda}_{ij}^2 \, p_{i.} \, p_{.j}} .$$

Characteristics of the measures of non-independence, λ, Δ, and $\tilde{\lambda}_{ij}$. In this section, we illustrate characteristics of ρ, θ, λ, Δ, and $\tilde{\lambda}$. Specifically, we focus on the characteristics *marginal-free* and *marginal-dependent*. Marginal-free measures of non-independence assess strength of non-independence without weighting by the relative frequencies of the marginals. Consider the expression for λ, above. Under the square root of this expression, there are no margin-specific weights. In contrast, the expressions for Δ and $\tilde{\lambda}$ do contain weights that reflect the relative frequencies of the row- and column categories. This implies in the present context of two-sample CFA, that whereas λ is not affected by differences in marginal probabilities, Δ and $\tilde{\lambda}$ are.

In the following paragraphs, we give a summary of characteristics of λ, Δ, and $\tilde{\lambda}$ (for more detail and results from simulation studies, see von Eye et al., 1995). The characteristics of the odds ratio are well known and do not need to be repeated here (see Rudas, 1998). ρ and Δ differ only in the sign (ρ can become negative; however, this sign can only be interpreted if the order of categories is defined). In general, we find that $\lambda_{ij} = \Delta_{ij} = \tilde{\lambda}_{ij} = 0$ if the usual log-linear (and CFA) model of independence holds. For the individual coefficients we find

(1) Δ is a marginal-free quantity. It is related to ρ and to Pearson's X^2.
(2) λ is a marginal-free quantity. It is related to the odds ratio θ (both

the rwo and column sums of the λ_{ij} are zero). λ is the log-linear interaction.

(3) $\tilde{\lambda}$ is a marginal-dependent quantity. It is also related to the odds ratio θ. It can be interpreted as a weighted log-linear interaction with the row- and column probabilities as weights.

(4) $\tilde{\lambda}$ is similar to λ in terms of the G_{ij}.

(5) $\tilde{\lambda}$ is similar to Δ in its use of the marginal probabilities.

Based on these characteristics, von Eye et al. (1995) proposed searching for discrimination types in two-sample CFA using the above three concepts of non-independence by Goodman (1991). Specifically, the authors suggested three forms of discrimination types:

- *interaction types* (coefficient λ)
- *correlation types* (coefficient ρ), and
- *weighted interaction types* (coefficient $\tilde{\lambda}$).

To illustrate the possible differences between standard two-sample CFA which only uses marginal-dependent measures, we now re-analyze Mahoney's (2000) data as given in Table 49, above. Table 49 suggests that configuration 121/122 discriminates between female and male adolescents. Table 50 displays the results for λ, $\tilde{\lambda}$, ρ, and θ. We omit Δ because, as we said before, it is identical to ρ except for the sign. As in Table 49, the Bonferroni-adjusted α^* is 0.0125. Please note that, with the exception of θ, there exist no significance tests for these measures. Therefore, jack-knife methods were used to determine the standard errors, se (for details on this procedure see von Eye et al., 1995). The significance test for the null hypothesis that $\theta = 1$ is $z = \dfrac{\theta}{se_\theta}$, with

$$se_\theta = \sqrt{\frac{1}{a} + \frac{1}{b} + \frac{1}{c} + \frac{1}{d}}.$$

where *a, b, c*, and *d* are defined as in Table 47.

The comparison of the results in Tables 49 and 50 suggests that results from the four measures can be quite different. Configuration 12 appears as discrimination type again when θ is used. When ρ is used, this configuration is far from constituting a discrimination type. In contrast, based on ρ, Configuration 11 constitutes a discrimination type, whereas the

Table 50: Re-analysis of Mahoney's (2000) data (from Table 49, above), using four measures of deviation from independence

Cell index	Frequencies		Measures[a]				
SA	males	females	Estimate	λ	$\tilde{\lambda}$	ρ	$\log \theta$
11	234	284	parameter	.08	.07	-.07	-.31
			se	.05	.04	.03	.18
			z	1.68	1.68	-2.54	-1.69
			p(*z*)	.05	.047	**.01**	.05
12	27	15	parameter	.19	.09	.09	.76
			se	.09	.04	.14	.33
			z	2.22	2.22	.66	2.28
			p(z)	.01	.013	.25	**.01**
21	30	43	parameter	.07	.04	-.04	-.27
			se	.06	.04	.10	.25
			z	1.04	1.05	-.42	-1.06
			p(*z*)	.15	.15	.34	.14
22	24	14	parameter	.18	.08	.08	.70
			se	.09	.04	.14	.35
			z	1.97	1.97	.56	2.03
			p(*z*)	.02	.02	.29	.02

[a] Discrimination types are printed in **bold**.

other measures agree with the z-test in Table 49 that this configuration is unsuspicious. Configurations 21 and 22 fail to constitute discrimination types in both Tables 49 and 50. However, the tail probabilities show that the evaluation of group differences can be quite different, depending on which concept of deviation from independence is selected.

One question of importance is whether these differences reflect differences in statistical power rather than in the concepts of non-independence. The answer to this question is that the jack-knife methods should give these tests relatively similar power. Possible differences in power to the z-approximation, the χ^2-approximations, the binomial test, and the test for θ still need to be investigated. However, the results in Table 50 suggest that not one test is consistently more powerful than the others. More detailed investigations of the characteristics of the tests are not available at the moment of this writing.

Which of these tests shall be selected in an empirical study? To make a decision, we use two types of information. First, we look at the nature of a coefficient. Second, we consider the weighting. As was indicated before, we select from

- the correlation coefficient, ρ,
- the log-linear interaction, λ,
- the relative difference, Δ,
- the weighted log-linear interaction, $\tilde{\lambda}$, and
- the odds ratio.

Of these coefficients, Δ and ρ differ only in the sign. Therefore, only one of the two needs to be discussed. Correlation coefficients in 2 x 2 tables can be interpreted as standard correlations. Strong correlations suggest that high frequencies in a category of the row variable go hand-in-hand with high frequencies in a category of the column variable. If the variables are nominal level, the order of the categories is arbitrary. Thus, the sign of the correlation is arbitrary and researchers focus on the magnitude of the correlation. If the variables are at the ordinal or a higher scale level, the order is no longer arbitrary and the sign can be interpreted.

The log-linear interaction indicates the degree to which the observed frequencies deviate from the model of variable independence, that is, the main effect model for the 2 x 2 table. In many instances, strong deviations come in the form of correlational frequency patterns. However, there are other instances, where such deviations come with almost constant correlations.

Δ is in the definition of the Δ_{ij} identical to Pearson's X^2 components. However, unlike the X^2 components which are summed without weighting, the Δ_{ij} are summed with the row and column probabilities as weights. In an analogous fashion, the components of $\tilde{\lambda}$ have a similar form as the components of λ, but they are weighted by the row and column probabilities.

The odds ratio θ is a ratio of two ratios. It is independent of the marginals.

In general, one selects θ, ρ, X^2 or the exact Fisher test if one focuses on that part of the variability in a 2 x 2 table that is not already contained in the marginal distribution. In contrast, one selects Δ or $\tilde{\lambda}$, if the marginal distribution is deemed so important that it may affect the appraisal of the deviation from independence (see Goodman, 1991; Rudas, 1998; von Eye et al., 1995).

7.2.3 Measuring effect strength in two-sample CFA

It is well known that the sample size plays a major role in statistical significance testing. More specifically, null hypotheses have, all other things being equal, a reduced chance of prevailing when the sample size increases. For example, for a sample of size 10, a Pearson correlation coefficient of $r = 0.4$ comes with a $t = 1.23$ ($df = 8$) and $p = 0.126$, thus indicating a non-significant relationship. For a sample of size 30, the correlation of 0.4 comes with a $t = 2.31$ ($df = 28$) and $p = 0.014$. This indicates a significant relationship.

This applies to CFA in an analogous fashion. Consider the X^2 component $X^2 = \frac{(N - E)^2}{E}$. For I x J tables, this component can equivalently be expressed as $X^2 = \frac{(Np_{ij} - Np_{i.}\,p_{.j})^2}{Np_{i.}\,p_{.j}}$ and $X^2 = N\frac{(p_{ij} - p_{i.}\,p_{.j})^2}{p_{i.}\,p_{.j}}$. This last expression shows that the magnitude of the Pearson X^2 component is directly dependent upon the sample size. Thus, keeping the ratio of the probabilities constant, one can manipulate whether a configuration constitutes a type or an antitype by manipulating the sample size.

Therefore, measures of effect strength are important. These measures tell researchers what portions of variability are explained by a particular configuration. In addition to the above φ measures, the *binomial effect size* (BES; Rosenthal & Rubin, 1982) has been proposed as a measure of effect size in two-sample CFA (von Eye, Spiel, & Rovine, in press). The BES is a measure of effect size in 2 x 2 tables. Its range is $-1 \leq \text{BES} \leq +1$. The measure is defined as

$$BES = \frac{N_{11}}{N_{11} + N_{12}} - \frac{N_{21}}{N_{21} + N_{22}}$$

for $N_{i.}$ and $N_{.j} > 0$. If the 2 x 2 table is symmetric, the BES is identical to the correlation between the row and the column variable (Cohen, 1988; Rovine & von Eye, 1997). In addition, and this is the reason why we use it in two-sample CFA and in Prediction CFA, the BES allows statements about the size of the effect. In two-sample CFA, this is the discrimination effect, and in Prediction CFA, this is the prediction effect. The BES indicates the proportionate surplus of cases in one group over the other (assuming that the comparison groups are listed in the first row of the 2 x 2 table).

Data examples. The following two data examples, taken from Krauth and Lienert (1973a; see von Eye et al., in press), describes 162 inpatients with symptoms of aphasia. Three variables were observed: A = pointing at objects ("please point at the boat"), D = creating alliterations ("please list as many words as you can that begin with an M"), and E = number of verbal and phonemic mistakes made during the assessment procedure. Each of these variables was dichotomized with + indicating pathological and - indicating normal behavior. In the following two examples, we cross the variables D and E separately for the patients in the A+ and the A- categories. The two-sample CFA hypothesis is that the number of alliterations allows one to discriminate the patients with a pathologically high number of verbal and phonemic mistakes from those with a normal number of mistakes. Table 51 displays the 2 x 2 table for the patients in the A+ category.

The Pearson X^2 for this table is 43.52 ($df = 1$, $p < 0.01$). Performance in the Alliterations task does allow one to discriminate patients with pathologically high numbers of mistakes[1]. Considering the sample size dependence of X^2 measures, we now ask what the effect size of this discrimination statement is. We calculate

---

[1]Readers are invited to test a biprediction hypothesis for this table.

Table 51: BES display for the E x D table for patients with pathological scores in Pointing

Number of mistakes (E)	Performance in Alliterations (D)		Totals
	+	-	
+	5	28	33
-	37	2	39
Totals	42	30	72

$BES = \frac{5}{33} - \frac{28}{30} = -0.81$. This value suggests that those patients who make pathologically many mistakes in Pointing, perform in the Alliteration task rather like normals. The proportion of cases predicted correctly is over 80%. Please note that the sign is defined here because of the ordinal nature of the +/- classification.

The second sample table is for the patients with normal scores in Pointing. Table 52 displays the frequencies.

Table 52: BES display for the E x D table for patients with normal scores in Pointing

Number of mistakes (E)	Performance in Alliterations (D)		Totals
	+	-	
+	8	37	45
-	18	27	45
Totals	26	64	90

The Pearson X^2 for this table is 5.41 ($df = 1, p = 0.02$). This value, while small compared to the X^2 for Table 51, still indicates a significant association between the rows and columns of this table. The $BES = -0.27$, however. indicates that the discrimination of the number of mistakes-groups is far weaker than in Table 51[2]. Less than 30% of the variability in Table 52 is explained. As in the A+ group, the direction of the relationship is negative. Positive BES scores are interpreted accordingly.

Discussion of the BES (see von Eye et al., in press). The φ measures introduced for prediction CFA and the BES presented in this section add two important facets to CFA:

(1) When searching for types and antitypes, researchers have the choice from a large number of statistical tests (see Chapter 3, and Sections 7.1, 7.2.1, and 7.2.2). Differences between these tests often reflect differences in power rather than differences in the data characteristics these tests can depict (compare the description of test characteristics in Sections 3.7 - 3.9 with the discussion of alternative tests in the present section). The BES allows one to describe the strength of effects regardless of the statistical test used for the identification of types and antitypes, regardless of the concept of deviation from independence employed, and independent of sample size. This applies accordingly to the φ^2 measures.

(2) The BES has been criticized because it may encourage misinterpretations in particular when the marginal frequencies are uneven (Thompson & Schumacker, 1997). In addition, the BES does not carry different information than ρ when the marginals are uniform. Unfortunately, the alternative measures proposed by Thompson and Schumacker fail to have the easy interpretational appeal of the BES. Thus, we recommend the BES for careful evaluation of effect sizes in two-sample CFA and in Prediction CFA.

(3) A number of alternatives to the BES have been proposed. Examples include Lienert's *coefficient of conciseness*, Q, (Krauth & Lienert, 1973) and Vogel's (1997) *deviation-from-independence coefficient*, *V* (see Lautsch, 2000).

---

[2]Here again, readers are invited to entertain a biprediction hypothesis.

7.3 Comparing three or more samples

The comparison of three or more samples with CFA follows the same scheme as the comparison of two groups (Lienert, 1971b). Computationally, two equivalent procedures have been used for the comparison of three or more groups. The first procedure involves creating a 2 x k cross-tabulation. In the first row, we place the frequencies of the configuration in which the k groups are compared with each other. The second row contains the frequencies of the combined remaining configurations. This table can be analyzed by comparing any two pairs of groups. Table 53 displays the scheme of such a 2 x k cross-tabulation in analogy to the scheme in Table 47. The configuration in which the *k* groups are compared, is labeled ij.

Table 53: 2 x k cross-classification for the CFA comparison of k groups in Configuration *ij*

Configuration	Groups				Totals
P_1P_2	A	B	...	K	
ij	M_{ijA}	M_{ijB}	...	M_{ijK}	$M_{ij.}$
All others combined	$N_{..A} - M_{ijA}$	$N_{..B} - M_{ijB}$	...	$N_{..K} - M_{ijK}$	$N - M_{ij.}$
Sample sizes	N_A	N_B	...	N_K	N

For each configuration, a cross-classification can be created as shown in Table 53. Then, in each of these tables, a total of $\binom{K}{2}$ comparisons can be made. Thus, the number of tests per t x K table, where t is the total number of configurations is $t\binom{K}{2}$, and the Bonferroni-adjusted significance threshold is $\alpha^* = \alpha / (t\binom{K}{2})$. This approach implies that all pairs of configurations are subjected to a two-sample CFA. This way, the significance level will be prohibitively extreme.

Therefore, we propose analyzing the table as a whole and

performing just one P-CFA-type CFA. The base model for this approach would be [P][G], where P indicates all variables used to discriminate among the K groups, and G is the grouping variable. This model is saturated in the P-variables and proposes independence between the P-variables and the grouping variable. Thus, types and antitypes can emerge only if there exists a relationship between the P-variables and the grouping variable. Please note that when three or more groups are discriminated using CFA, both types and antitypes can emerge. The existence of a type does not necessitate the existence of a particular antitype, as is the case in two-sample CFA.

Data example. The following data example uses the data published by Aksan et al. (1999). The data describe Control (C), Negative Affect (A), and Approach (H) in a sample of 488 children aged 3 years and 6 months. Each of the three variables was classified in three levels, with C = 1 indicating low control, C = 2 averge control, C = 3 high control; A = 1 low score in negative affect, A = 2 average, A =3 high; and H = 1 high score in approach, H = 2 average, and H = 3 low. We now analyze these data in two steps. First, we perform a first order CFA. Results of this analysis will tell us whether configurations stand out and constitute types and antitypes. In the second step, we perform three-group and two-group comparisons. The results from this analysis tell us where the group differences are significant. To save space, we perform only one three-group comparison and a small selection of the total of $27\cdot\binom{3}{2} = 81$ two-group comparisons.

For the first order CFA, we use Lehmacher's test with Küchenhoff's continuity correction and Bonferroni-adjust the significance threshold. For the 3 (Control; C) x 3 (Approach; H) x 3 (Affect; A) cross-classification, the adjusted α is $\alpha^* = 0.0018519$. Table 54 displays the results from first order CFA of the temperament data.

The results in Table 54 suggest that there exist three types and three antitypes. Rather than interpreting each of these in detail (see Aksan et al., 1999), we now ask whether the three groups of control differ in Configuration 31, which is high approach and high negative affect. To perform this comparison, we first create a 2 x 3 cross-classification of the format given in Table 53, in which the first row displays the frequencies of the configuration under study, and the second row contains the combined frequencies of all other configurations. This cross-classification appears in Table 55.

Table 54: First order CFA of children's temperament data (Aksan et al., 1999)

Cell index	Frequencies		Test statistics		Type/ Antitype ?
AHC	observed	expected	z	$p(z)$[a]	
111	3	12.97	-2.965	.0015	**A**
112	2	17.32	-4.129	< α*	**A**
113	4	12.19	-2.471	.0067	
121	23	15.55	2.020	.0217	
122	23	20.77	0.450	.3265	
123	6	14.61	-2.418	.0078	
131	39	14.84	7.006	< α*	**T**
132	33	19.82	3.348	.0004	**T**
133	9	13.94	-1.350	.0886	
211	11	17.54	-1.672	.0472	
212	29	23.42	1.256	.1045	
213	13	16.48	-0.846	.1987	
221	19	21.02	-0.392	.3474	
222	36	28.08	1.714	.0433	
223	19	19.75	-0.067	.4734	
231	21	20.06	0.115	.4542	
232	26	26.79	-0.069	.4725	
233	18	18.85	-0.094	.4625	

/ cont.

Cell index	Frequencies		Test statistics		Type/ Antitype ?
AHC	observed	expected	z	$p(z)$[a]	
311	13	14.07	-0.172	.4318	
312	30	18.79	2.887	.0019	
313	41	13.22	8.473	$< \alpha^*$	**T**
321	12	16.86	-1.226	.1101	
322	14	22.52	-2.013	.0221	
323	23	15.84	1.919	.0275	
331	8	16.09	-2.174	.0149	
332	6	21.49	-3.830	.0001	**A**
333	7	15.12	-2.239	.0126	

[a] $< \alpha^*$ indicates that the probability has more than four zero decimals after the decimal point.

Table 55: Three-group comparison of the children at the three control levels (data from Aksan et al., 1999)

Cell index	Control levels			Totals
HA	Low control	average control	high control	
31	39	21	8	68
all others combined	103	171	146	420
Totals	142	192	154	488

The base model for the analysis of Table 55 is [HA][C], where HA has no more than 2 categories. The Pearson X^2 for this model is 32.92 ($df = 2$; $p < 0.01$), thus suggesting significant group differences. We now ask where these differences are. We form the 2 x 2 cross-classifications for all three two-group comparisons. These cross-classifications appear in Table 56. The Bonferroni-adjusted α for these three comparisons is α* = 0.05/4 = 0.0125. We set $t = 4$ in the denominator, because we also count the test for Table 55.

Table 56: Pair-wise comparisons of the three control level groups in Configuration 31

Cell index	Comparison groups (levels of control)					
HA	low	average	low	high	average	high
31	39	21	39	8	21	8
all others	103	171	103	146	171	146
z and $p(z)$	3.890 $p(z) = 0.0001$ **discrimination type**		5.237 $p(z) < \alpha^*$ **discrimination type**		1.916 $p(z) = 0.0277$	

[a] < α* indicates that the tail probability is smaller than can be expressed with 4 decimal places.

The results in Table 56 suggest that the low control group differs from both the high and the average control groups. The latter two groups do not differ from each other. The frequency distributions in the two middle panels of Table 56 show that in both cases there are more children displaying high scores both in Approach and in Negative Affect (Configuration 31) in the low control group than in the average and the high control groups. Readers are invited to estimate the ρ statistic for the non-significant comparison and to determine if it is significant for α* = 0.0125.

We now ask what the effect sizes of these discrimination types are. We calculate the following BES values (see Section 7.2.3): $BES_{low\text{-}average}$ = 0.165; $BES_{low\text{-}high}$ = 0.223; and $BES_{average\text{-}high}$ = 0.057. These values suggest that the effect sizes for the discrimination types are in the medium to small range. The effect size for the comparison of the average with the high control group is practically zero.

7.4 Three groups of variables: ISA plus k-sample CFA

Multivariate research often involves not only multiple variables, but also multiple groups of variables. Often, factor models are used to establish groups of variables (Bartholomew & Knott, 1998; von Eye & Clogg, 1994), in particular when variables are continuous. At the level of manifest categorical variables, CFA provides the option to analyze groups of variables using ISA (see Section 6.1) or P-CFA (Section 6.2). In this section, we illustrate how ISA of two groups can be performed in the context of a k-sample comparison.

The method of *k-sample ISA* allows one to examine the relationships among two or more groups of variables differentially in *k* groups (Kohnen & Rudolf, 1981; Lienert & Netter, 1984). Consider the case in which researchers investigate the five dichotomous variables A, B, C, D, and E. A and B form one group of variables and C and D the other. E is the grouping variable. Then, the base model for *2-sample ISA* is [A, B][C, D][E], or

$$\log f = \lambda_0 + \lambda_i^A + \lambda_j^B + \lambda_k^C + \lambda_l^D + \lambda_m^E + \lambda_{ij}^{AB} + \lambda_{kl}^{CD}.$$

This model

(1) is saturated in the variable group that contains A and B;
(2) is saturated in the variable group that contains C and D;
(3) considers the main effect of the grouping variable, E; and
(4) assumes independence of the three variable groups that contain A and B, C and D, and E, respectivley.

As in standard ISA, types and antitypes can result from this design if relationships between the three variable groups exist. Two-sample ISA is employed to identify types that allow one to discriminate between the two groups of respondents. The design matrix for the base model of a two-sample ISA of the variables A, B, C, D, and E is

$$X = \begin{bmatrix}
1 & 1 & 1 & 1 & 1 & 1 & 1 & 1 \\
1 & 1 & 1 & 1 & 1 & -1 & 1 & 1 \\
1 & 1 & 1 & 1 & -1 & 1 & 1 & -1 \\
1 & 1 & 1 & 1 & -1 & -1 & 1 & -1 \\
1 & 1 & 1 & -1 & 1 & 1 & 1 & -1 \\
1 & 1 & 1 & -1 & 1 & -1 & 1 & -1 \\
1 & 1 & 1 & -1 & -1 & 1 & 1 & 1 \\
1 & 1 & 1 & -1 & -1 & -1 & 1 & 1 \\
1 & 1 & -1 & 1 & 1 & 1 & -1 & 1 \\
1 & 1 & -1 & 1 & 1 & -1 & -1 & 1 \\
1 & 1 & -1 & 1 & -1 & 1 & -1 & -1 \\
1 & 1 & -1 & 1 & -1 & -1 & -1 & -1 \\
1 & 1 & -1 & -1 & 1 & 1 & -1 & -1 \\
1 & 1 & -1 & -1 & 1 & -1 & -1 & -1 \\
1 & 1 & -1 & -1 & -1 & 1 & -1 & 1 \\
1 & 1 & -1 & -1 & -1 & -1 & -1 & 1 \\
1 & -1 & 1 & 1 & 1 & 1 & -1 & 1 \\
1 & -1 & 1 & 1 & 1 & -1 & -1 & 1 \\
1 & -1 & 1 & 1 & -1 & 1 & -1 & -1 \\
1 & -1 & 1 & 1 & -1 & -1 & -1 & -1 \\
1 & -1 & 1 & -1 & 1 & 1 & -1 & -1 \\
1 & -1 & 1 & -1 & 1 & -1 & -1 & -1 \\
1 & -1 & 1 & -1 & -1 & 1 & -1 & 1 \\
1 & -1 & 1 & -1 & -1 & -1 & -1 & 1 \\
1 & -1 & -1 & 1 & 1 & 1 & 1 & 1 \\
1 & -1 & -1 & 1 & 1 & -1 & 1 & 1 \\
1 & -1 & -1 & 1 & -1 & 1 & 1 & -1 \\
1 & -1 & -1 & 1 & -1 & -1 & 1 & -1 \\
1 & -1 & -1 & -1 & 1 & 1 & 1 & -1 \\
1 & -1 & -1 & -1 & 1 & -1 & 1 & -1 \\
1 & -1 & -1 & -1 & -1 & 1 & 1 & 1 \\
1 & -1 & -1 & -1 & -1 & -1 & 1 & 1
\end{bmatrix}.$$

The design matrix for the two-sample ISA base model contains eight columns. The first column is the constant. The main effects of the five variables follow in the next five columns: variable A in column 2, variable B in column 3, variable C in column 4, variable D in column 5, and variable E in column 6. In addition, the base model contains the interactions between variables A and B as well as variables C and D. These interactions appear in columns 7 and 8.

Data example. For the following example, we use data published by Netter (1982; Netter & Lienert, 1984). In an experiment on stress responses, a sample of 162 participants worked under two stress conditions. The first condition was a response time task, and the second condition was a verbal fluency task. The order of participation was balanced. Under each condition, plasma samples were taken to measure the levels of adrenaline and noradrenaline. The four resulting dependent variables were adrenaline (A_1 and A_2), measured under the two experimental conditions, and noradrenaline (N_1 and N_2), also measured under the two experimental conditions. Each of these four variables was scaled as either 1 = not elevated or 2 = elevated. The participants were classified as either hypertonics or controls.

We now analyze the 2 x 2 x 2 x 2 x 2, or A_1 x A_2 x N_1 x N_2 x P cross-classification in two steps. First, we perform an ISA. The base model for this analysis considers the two variables from the first experimental condition in one group, the two variables from the second experimental condition in a second group, and participant classification in a third group. That is, the base model is [A_1, N_1][A_2, N_2][P], where P is the participant classification with categories h= hypertonic and n = normal. The log-linear base model can also be represented as

$$\log F = \lambda_0 + \lambda_i^{A_1} + \lambda_j^{N_1} + \lambda_k^{A_2} + \lambda_l^{N_2} + \lambda_m^{P} + \lambda_{ij}^{A_1 N_1} + \lambda_{kl}^{A_2 N_2} .$$

The design matrix for this model is the same as the one given in the five-variable example above.

For this analysis, we use Anscombe's test and Bonferroni-adjustment which led to $\alpha^* = 0.05/32 = 0.0015625$. The results appear in Table 57. In the second step, we perform a two-sample comparison.
In spite of the low critical α^* and the small sample size, the Pearson X^2 for the two-sample ISA base model is a large 113.71 ($df = 24$; $p < 0.01$), so that we can expect types and antitypes to emerge. Indeed, three types and two antitypes were detected. The first type is constituted by configuration 1111n. These are the normals who do not respond strongly to either stress condition. Less than four individuals were expected to display this pattern, but 11 were observed to show it. The second type is constituted by configuration 1212n. This type suggests that more normals than expected respond to both stress conditions with increased noradrenaline levels without increasing adrenaline output. The third type, constituted by configuration 2222h, contains those hypertonics who respond to both stress

conditions with increases in both adrenaline and noradrenaline.

Table 57: ISA of the adrenaline/noradrenaline data from Netter's (1982) stress experiment

Cell index	Frequencies		Test statistics		Type/ Antitype?
$A_1N_1A_2N_2P$	observed	expected	z	$p(z)$	
1111h	6	3.468	1.323	.0930	
1111n	11	3.643	3.206	.0007	**T**
1112h	3	4.118	0.497	.3097	
1112n	2	4.326	1.174	.1202	
1121h	4	5.960	0.786	.2158	
1121n	8	6.262	0.733	.2319	
1122h	1	4.010	1.730	.0418	
1122n	1	4.213	1.817	.0346	
1211h	4	4.335	0.082	.4672	
1211n	1	4.554	1.957	.0251	
1212h	9	5.147	1.610	.0537	
1212n	14	5.408	3.160	.0008	**T**
1221h	4	7.450	1.328	.0920	
1221n	1	7.828	3.072	.0011	**A**
1222h	5	5.012	0.070	.4723	
1222n	7	5.266	0.792	.2141	
2111h	4	4.431	0.129	.4489	
2111n	5	4.655	0.236	.4070	

/ cont.

Cell index	Frequencies		Test statistics		Type/ Antitype?
$A_1N_1A_2N_2P$	observed	expected	z	$p(z)$	
2112h	2	5.262	1.562	.0591	
2112n	0	5.528	3.456	.0003	A
2121h	14	7.616	2.133	.0165	
2121n	17	8.001	2.828	.0023	
2122h	1	5.123	2.179	.0147	
2122n	3	5.383	1.051	.1466	
2211h	0	3.371	2.663	.0039	
2211n	1	3.542	1.519	.0644	
2212h	7	4.004	1.438	.0751	
2212n	1	4.206	1.814	.0348	
2221h	2	5.795	1.765	.0388	
2221n	5	6.088	0.387	.3492	
2222h	13	3.898	3.734	.0001	T
2222n	6	4.096	0.963	.1678	

The first of the two antitypes is constituted by Configuration 1221n. These are participants who respond to the first stress condition with an increased adrenaline level and to the second stress condition with an increased noradrenaline level. Only one individual from the normal group displayed this pattern, but almost eight had been expected to display it. The second antitype, constituted by configuration 2112n, is the complement to the first antitype. Not a single individual from the normal group responded to the first stress condition by only increasing the noradrenaline level and to the second stress condition by only increasing the adrenaline level. Yet, more than five had been expected.

These interpretations of the findings of two-sample ISA do not allow us to directly compare the two groups of normals and hypertonics. Therefore, we now perform the second step of analysis which involves two-sample comparisons. Table 58 displays the results of two-sample ISA. Because of the relatively small sample size we used the X^2-test with continuity correction (see Section 7.1). The significance level was protected using the Bonferroni procedure which led to $\alpha^* = 0.003125$.

Table 58: Two-sample ISA of the adrenaline/noradrenaline data from Netter's (1982) stress experiment (cf. ISA in Table 57)

Comparison pair	observed Frequencies	Test statistics		π^*
$A_1N_1A_2N_2P$		X^2	$p(X^2)$	
1111h	6	0.843	.3586	.208
1111n	11			
1112h	3	0.003	.9553	.187
1112n	2			
1121h	4	0.658	.4172	.231
1121n	8			
1122h	1	0.458	.4987	.025
1122n	1			
1211h	4	0.931	.3346	.390
1211n	1			
1212h	9	0.597	.4396	.158
1212n	14			
1221h	4	0.931	.3346	.390
1221n	1			

/ cont.

Comparison pair	observed Frequencies	Test statistics		π^*
$A_1N_1A_2N_2P$		X^2	$p(X^2)$	
1222h	5	0.045	.8328	.122
1222n	7			
2111h	4	0.006	.9392	.078
2111n	5			
2112h	2	0.558	.4551	.512
2112n	0			
2121h	14	0.061	.8052	.066
2121n	17			
2122h	1	0.208	.6481	.317
2122n	3			
2211h	0	0.001	.9802	.488
2211n	1			
2212h	7	3.554	.0594	.443
2212n	1			
2221h	2	0.499	.4800	.283
2221n	5			
2222h	13	2.497	.1141	.287
2222n	6			

The combination of lack of power due to the small cell frequencies and small adjusted α prevented us from identifying any discrimination types. Using the exact Fisher test, the z-test or any of the four measures λ, $\tilde{\lambda}$, ρ, or θ would have lowered the tail probability for configuration 2222h/2222n to values around 0.03. However, even this value is larger than the adjusted significance threshold of 0.003125. We thus conclude that although employing ISA allows one to identify types and antitypes, none of the two-group comparisons indicates differences between normotonics and hypertonics that are large enough to establish discrimination types.

Part III: Methods of Longitudinal CFA

Longitudinal data are pervasive in the social sciences. Whenever intervention effects, change, or development are of concern, longitudinal data are being collected. There is a number of reasons why this is the case. The most important reason is that change within the individual can be assessed only if the individual is observed more than once. Thus, one of the chief characteristics of person-oriented research, the focus on the individual (Bergman et al., 2000), can be achieved only using longitudinal data, when development and change are of interest.

Longitudinal data provide an interesting blend of features, sharing characteristics with multivariate data and with time series data. Multivariate data describe an individual in a string of data. Each data point represents a different variable, and the number of data points is often relatively small, that is, less than 20. In cross-sectional research, samples of individuals are observed, and for each individual, such a string of data is created. An example of cross-sectional data is the investigation of performance of adolescents in school, in which motivational variables are observed together with performance variables, ability variables, work-behavior variables, and parental variables.

In contrast, time series typically describe just one individual, data carrier, or phenomenon. Times series also consist of strings of data. However, these strings are typically much longer, often comprising hundreds or thousands of observations. These observations typically represent just one, or a small number of variables. An example of a time series is the alcohol consumption behavior of an alcoholic, observed daily over three years.

Longitudinal data are similar to time series data in that the same variables are observed repeatedly over time. Thus, longitudinal data share many of the structured characteristics of time series data that may not be present in cross-sectional data. However, the number of observation points is small, sometimes as small as two, and the number of variables is typically greater than one. In addition, longitudinal data are typically observed in samples of individuals, some times in different samples for comparison purposes.

The present chapter introduces readers to CFA methods of analysis of longitudinal data (cf. von Eye & Niedermeier, 1999). A large number of characteristics of longitudinal data can be considered. Examples of such characteristics include the mean, and changes in the mean, the variability, patterns of change, slope, and acceleration/deceleration processes. Even the change in correlational structures and multivariate distances can be analyzed using CFA. Some of the challenges faced by the analyst of longitudinal data are unique, that is, new to those otherwise concerned only with cross-sectional data. Problems unique to longitudinal data analysis will be discussed in detail. Issues covered already in the first two parts of this volume will only be briefly reviewed.

8. CFA of Differences

At the nominal scale level, a difference tells us whether two scores, that is, two labels, are the same. The magnitude of a difference cannot be interpreted. The decision is only between *equal* or *different*. At the ordinal level, a difference tells us how many ranks apart two objects are. The magnitude of the difference can be interpreted only in the sense that it counts the number of ranks that separate two objects. The same number of ranks between pairs of objects can correspond to different distances (if these are defined). At the interval level, a difference tells us how many units apart two objects are, and the units are the same across the entire range of admissible scores. The same holds true for the ratio scale level.

It is well known that the distinction of four scale levels, proposed by Stevens (1946, 1951) is being hotly disputed (Velleman & Wilkinson, 1993), and other distinctions have been proposed and used (e.g., Clogg, 1995). Still, for the present purpose of introducing the methods of differences and their use in CFA, the four scale levels of Stevens are sufficient for the description of data characteristics used.

The methods of differences have been used as substitutes for derivatives, to determine the shape of a series of measures, to identify errors in the data, and to estimate polynomial parameters (for an overview, see Abramowitz & Stegun, 1972). Depending on the scale level at which variables are observed, differences between scores come with specific meaning. For instance, estimating polynomial parameters requires interval or ratio scale level. Determining whether scores in a series increase in a linear, some accelerated, or some decelerated fashion, only requires ordinal level information, as long as no interpolation or smoothing of values is attempted. In the following section, we review methods of differences. In the subsequent sections, we embed methods of differences in the context of CFA.

8.1 A review of methods of differences

For the following description of methods of differences consider a series of measures, $x_0, x_1, ..., x_n$ with corresponding measures $y_0, y_1, ..., y_n$. The difference between two measures y_i and y_j is termed $\Delta^v y_k$, where v indicates the order of differences and k counts the y-scores ($i \neq j$; $k = 0, 1, ..., n$). For example, $\Delta^1 y_2$ indicates that the second ($k = 2$) raw score is subtracted from the first. Differences $\Delta^1 y_i$ are termed *first differences*. The term $\Delta^2 y_3$ indicates that the third difference of raw scores ($v = 2$; $k = 3$) is subtracted from the second difference of raw scores. Differences $\Delta^2 y_i$ are termed *second differences*. In general, we obtain

$$\Delta y_0 = y_1 - y_0, \qquad \Delta y_1 = y_2 - y_1 , \quad ...$$
$$\Delta^2 y_0 = \Delta y_1 - \Delta y_0, \quad \Delta^2 Y_1 = \Delta y_2 - \Delta y_1 , \ ...$$
$$...... ,$$

or, in short,

$$\Delta^v y_k = \Delta^{v-1} y_{k+1} - \Delta^{v-1} y_k .$$

Table 59 displays this scheme for a series of five x-scores and five y-scores. All possible differences between directly adjacent raw scores or differences are presented.

The scheme displayed in Table 59 is termed *scheme of descending differences*, because in each descending row in the scheme the y-differences have the same subscript. This scheme is also called *scheme of forward differences*, because the previous score is subtracted from the following one.

Occasionally, it may be advantageous to change the reference in the subtraction, that is, to subtract y_{k-1} from y_k (instead of subtracting y_k from y_{k+1}). The general term in this procedure is

$$\Delta^v y_k = \Delta^{v-1} y_k - \Delta^{v-1} y_{k-1} .$$

The result of this procedure is that the subscripts in the descending rows increase rather than remain the same. The calculated difference values in the scheme are the same. This scheme is termed *ascending differences* or *backward differences*. Other schemes have been proposed, for example, the scheme of central differences.

Table 59: Scheme of descending differences

Raw scores		Differences			
X	Y	ΔY first differences	$\Delta^2 Y$ second differences	$\Delta^3 Y$ third differences	$\Delta^4 Y$ fourth differences
x_0	y_0				
		Δy_0			
x_1	y_1		$\Delta^2 y_0$		
		Δy_1		$\Delta^3 y_0$	
x_2	y_2		$\Delta^2 y_1$		$\Delta^4 y_0$
		Δy_2		$\Delta^3 y_1$	
x_3	y_3		$\Delta^2 y_2$		
		Δy_3			
x_4	y_4				

Two conditions must be fulfilled for the method of differences to be properly employed. First, the data points subjected to the differences procedures must be *equidistant*. That is, the distances between data points on X must be the same. This is the case, for instance, when observations are made every day or every hour. If equidistance cannot be realized, for instance because there are no data collections over weekends, the distances must be the same for each case. The second condition that must be met is that the scores that are subtracted from each other are at the interval level. It may be defensible to use ordinal level variables, if differences meaningfully describe the number of ranks apart two observations are. However, second and higher order differences may not easily be interpretable anymore.

In the following paragraphs, we present two illustrations of the usefulness of differences in the analysis of longitudinal data. The first is the description of the curvature of a series. The second example shows the effects of errors in the data.

Illustration 1: Scheme of ascending differences for $Y = X^3$. Table 60 displays the scheme of differences for cubic numbers for X = {1, 2, 3, 4, 5}, $Y = X^3$.

Table 60: Scheme of differences for $Y = X^3$

Scores		Differences			
X	Y	ΔY, 1st differences	$\Delta^2 Y$, 2nd differences	$\Delta^3 Y$, 3rd differences	$\Delta^4 Y$, 4th differences
0	0				
		1			
1	1		6		
		7		6	
2	8		12		0
		19		6	
3	27		18		
		37		...	
4	64		...		

The illustration in Table 60 shows that

(1) The nth differences of a polynomial of order n are constant. Specifically,

$$\Delta^n y_k = h^n n!,$$

where h indicates the difference $x_1 - x_0$ (the x-scores are equidistant). In the example in Table 60 we have $h = 1$ and obtain $\Delta^3 y_k = 1^3 \cdot 3! = 3 \cdot 2 \cdot 1 = 6$. This is the value for the 3rd differences found in Table 60.

(2) The $(n + 1)$st differences and all higher order differences disappear. This is indicated in the last column of Table 60.

Thus, the method of differences can be used to determine the order of the polynomial that describes a series of measures. In the example in Table 60, this polynomial is of third order, and the third differences are perfectly constant. In the analysis of real data, there typically are errors in both X and Y, and one would expect slight variations in the third differences. The second illustration shows how the method of differences can be used to identify errors of size ε in Y.

Illustration 2: The effects of errors in Y. The method of differences can also be used to detect inconsistencies in series of measures. For this illustration consider a *y*-score afflicted with an error of size ε. This error carries on in the scheme of differences in a way that can easily be understood. Specifically, the error causes the higher differences to display increasing rather than decreasing variances. Table 61 displays a scheme with eight *y*-scores, one of which has the error of size ε.

Table 61: Using the method of differences to detect an error of size ε

Scores		Differences			
X	Y	ΔY	$\Delta^2 Y$	$\Delta^3 Y$	$\Delta^4 Y$
0	0				
		0			
1	0		0		
		0		0	
2	0		0		ε
		0		ε	
3	0		ε		-4ε
		ε		-3ε	
4	ε		-2ε		6ε
		-ε		3ε	
5	0		ε		-4ε
		0		-ε	
6	0		0		
		0			
7	0				

Table 61 shows that the error ε is located exactly where the variation in the higher differences reaches a maximum. Please notice that the variation increases with the order of the differences. To illustrate this numerical effect consider the equation Y = cos X. Table 62 displays the scheme of differences for this function (in radian units). One *y*-value contains the very small error ε = 0.0005.

Table 62: Scheme of differences and the detection of an error in the trigonometric function Y = cos X

Scores		Differences		
X	Y	ΔY first differences	$\Delta^2 Y$ second differences	$\Delta^3 Y$ third differences
1.50	0.07074			
		-0.09994		
1.60	-0.02920		0.00030	
		-0.09964		**0.00048**
1.70	-0.12884		**0.00078**	
		-0.09886		**0.00249**
1.80	**-0.22770**		**0.00327**	
		-0.09559		**-0.00054**
1.90	-0.32329		**0.00273**	
		-0.09286		**0.00143**
2.00	-0.41615		0.00416	
		-0.08870		
2.10	-0.50485			

The *y*-value for *x* = 1.80 is incorrect. As a consequence, all differences printed in bold face are also wrong. The correct Y-value is -0.22720. The error appears in the fourth decimal. Still, its effect is large. Table 63 displays the scheme of differences based on the correct y-score.

Table 63: Correct scheme of differences for the function Y = cos X

Scores		Differences		
X	Y	ΔY first differences	Δ^2Y second differences	Δ^3Y third differences
1.50	0.07074			
		-0.09994		
1.60	-0.02920		0.00030	
		-0.09964		*0.00098*
1.70	-0.12884		*0.00128*	
		-0.09836		*0.00099*
1.80	*-0.22720*		*0.00227*	
		-0.09609		*0.00096*
1.90	-0.32329		*0.00323*	
		-0.09286		*0.00093*
2.00	-0.41615		0.00416	
		-0.08870		
2.10	-0.50485			

Obviously, the third differences in Table 63 show far less variation than their counterparts in Table 62. In addition, they are, on average, smaller. The differences printed in italics are correct (compare with the differences printed in bold face in Table 62).

8.2 The method of differences in CFA

In CFA, the method of differences plays a particular role. It is not used to estimate polynomial parameters or to smooth series of measures. Rather, the method of differences is used to describe characteristics of time series. Specifically, first differences are used to describe the ups and downs, that is, the linear trend in data. Second differences are used to describe the accelerations in the ups and downs, that is, the quadratic trend in data, and so on.

In many instances, the data available for analysis are at the ordinal level rather than the interval or ratio scale levels. Therefore, the differences themselves are often categorized. Thus, sometimes only particular characteristics of a time series are reflected in a pattern of first, second, or higher order differences. The following sections first present an example of the method of differences in the description of curves (Section 8.2.1). Then, methods of categorization are introduced (Section 8.2.2). The following sections discuss CFA for data at the ordinal scale level and after categorization of differences (cf. Lienert & Krauth, 1973c; Lienert & Netter, 1985).

Figure 5: Illustration of the method of differences

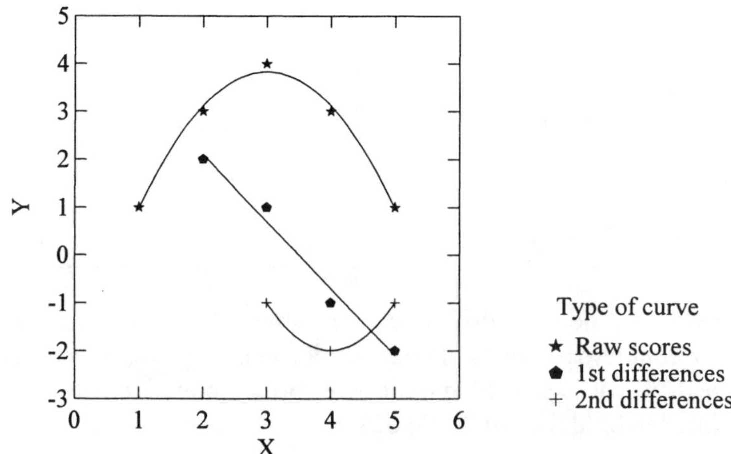

8.2.1 Depicting the shape of curves by differences: An example

In this section, we illustrate how the shape of curves can be depicted using differences. Consider the series of Y-values 1, 3, 4, 3, 1. This series can be described well by a quadratic polynomial (see Figure 5). As can be seen in the figure, this polynomial fails to describe the data points perfectly. Still, a quadratic polynomial provides a reasonably good description. We now illustrate that the method of differences allows one to capture characteristics of the series of five points. In addition, we illustrate the effects of categorization. Table 64 displays the scheme of differences for the series of five data points.

Table 64: Scheme of differences for a series of five data points

Scores		Differences		
X	Y	ΔY first differences	$\Delta^2 Y$ second differences	$\Delta^3 Y$ third differences
1	1			
		2		
2	3		-1	
		1		-1
3	4		-2	
		-1		1
4	3		-1	
		-2		
5	1			

The raw scores, the first differences ΔY, and the second differences $\Delta^2 Y$ are displayed in Figure 5. The third differences appear only in the scheme of differences. This scheme shows that

(1) Up to the level of third differences, the difference scores are not constant. Thus, a fourth order polynomial is needed to perfectly capture the five data points. Alternatively, there may be error in the *y*-scores.

(2) The signs of the ascending first differences indicate that the second raw score is larger than the first, the third raw score is larger than the second, the fourth raw score is smaller than the third, and the fifth raw score is smaller than the fourth. This pattern is typical of one-peaked curves which is also suggested by the values of the first differences (for a classification of curves based on the pattern of the signs of their first differences see Lienert & Krauth, 1973d; Stemmler, 1998). We see from this pattern that the signs of first differences can be used to describe the shape of a curve.

(3) The signs of the second differences indicate that the ascending differences between the first differences are all negative, thus indicating that all k-1st first differences are bigger than the *k*th first differences. The values of the second differences indicate in addition that this negative acceleration is strongest between the second and the third first differences.

(4) The signs of the third differences suggest that the second differences are positively accelerated (Figure 5 suggests that the curve of the second differences has one minimum). The values of the third differences show that the curve is symmetric.

In sum, this example shows that a scheme of differences provides a description of the shape of a curve. In addition, this example indicates that the signs of differences also contain valid information concerning the shape of a curve.

8.2.2 Transformations and the size of the table under study

Focusing on the information carried by the signs of differences, CFA researchers have discussed and used these signs in the analysis of short time series of measures. The following two methods of categorizing differences have been proposed for the creation of signs (Lienert & Krauth, 1973b, c; cf. Lienert & von Eye, 1984b; Lienert & zur Oeveste, 1985). Consider the variable Y, observed T times at the ordinal or some higher level. Then, the sign Δ of the first differences of Y is

$$\Delta_Y = \begin{cases} + & if \ y_2 - y_1 < 0 \\ 0 & if \ y_2 - y_1 = 0 \\ - & if \ y_2 - y_1 > 0. \end{cases}$$

For each variable that is transformed from Y to Δ_Y, one obtains T - 1 first differences that can be analyzed using CFA. Each of thc differences has the three categories +, =, and -. In an analogous fashion, second, third, and higher order differences can be transformed this way.

If variables are observed with very high reliability and at very high resolution levels, identical scores do not exist and even the smallest differences are interpretable. Therefore, the following alternative transformation has been discussed (Lienert & Krauth, 1973b, c; Lienert & Rudolph, 1983; Lienert & Straube, 1980):

$$\Delta_Y^{/} = \begin{cases} + & if \ y_2 - y_1 < 0 \\ - & if \ y_2 - y_1 > 0. \end{cases}$$

This transformation from Y to $\Delta_Y^{/}$ also yields T - 1 differences. However, each of these has only the two categories + and -. As before, second, third, and higher order differences can be transformed this way.

It is important to mention that the decision that leads to the assignment of +, -, and = to the difference between two adjacent scores can be based on various criteria. Most frequently, researchers calculate the numerical difference between two scores and assign categories solely based on the result of this subtraction. This is defensible if the reliability of the scale that is used to describe behavior is high enough so that differences between single scale units can be trusted. If, however, the reliability is lower than that, a statistical decision as to whether two scores differ from each other may be preferable. In the following examples, we simply assume that the researchers who made the decisions involved in the categorization process can defend these decisions.

Obviously, CFA of differences reduces the size of a data analysis problem. The cross-classifications analyzed by CFA of differences can be much smaller than the cross-classifications that result from using the raw scores, even if the raw scores have the same number of categories (ranks) as the differences. For example, consider a study in which researchers observe d variables. The ith variable has c_i categories. Then, the cross-classification of these d variables has

$$c = \prod_{i=1}^{d} c_i$$

cells. If each variable has $c_i = 3$ categories, the number of cells is 3^d, and if each variable has $c_i = 2$ categories, the number of cells is 2^d.

If first differences are analyzed instead of raw scores, the number of variables is reduced from d to $d - 1$. If each difference has 3 categories, the number of cells is 3^{d-1}. This represents a savings of $3^d - 3^{d-1}$ cells, that is, a savings of one third. In other words, the cross-classification from the raw scores has three times as many cells as the cross-classification from the first differences, if both the raw scores and the differences have three categories. If the raw scores have more than three categories, the savings can be even bigger.

If both the raw scores and the first differences have two categories, the cross-classification from the raw scores has 2^d cells. Because of the reduction in the number of variables from d to $d - 1$, the cross-classification of the differences has only 2^{d-1} cells. This represents a savings of $2^d - 2^{d-1}$ or a savings of one half. In other words, the cross-classification from the raw scores has twice as many cells as the cross-classification from the first differences, if both the raw scores and the first differences have two categories. If the raw scores have more than two categories, the savings can be much bigger.

8.2.3 Estimating expected cell frequencies for CFA of differences

There is a certain temptation to analyze cross-classifications of difference variables just as any other cross-classification. Indeed, most textbooks illustrate CFA of differences by subjecting the difference variables to standard CFA (Krauth & Lienert, 1973a; Krauth, 1993). However, this procedure can be problematic because patterns of differences have different a priori probabilities. Therefore, estimating expected cell frequencies can lead to biased, sample-specific expected cell frequencies and thus to the identification of types and antitypes that may not exist. In the following section, we present examples of the calculation of a priori probabilities for difference variables and data examples.

8.2.3.1 Calculating a priori probabilities: Three examples

In this section we consider three examples. The first two examples concern first differences, the third example concerns second differences. The first example discusses a priori probabilities for the case in which scores in a series of measures are always different. The second example includes no-change scores.

Example I: First differences, all scores are different. A researcher observes the performance of a singer at the ordinal scale level three times. Suppose the researcher uses the three ranks 1, 2, and 3 to evaluate the performance. For these ranks, the following sequences are possible: 123, 132, 213, 231, 312, and 321. The signs of the first differences for these sequences are ++, +-, -+, +-, -+, and - -. The frequencies of occurrence of these sign patterns of differences are: ++ appears once, +- appears twice, -+ appears twice, and - - appears once. In other words, the a priori probabilities of these difference sign patterns are $p(++) = 0.167$, $p(+-) = 0.333$, $p(-+) = 0.333$, and $p(- -) = 0.167$. Thus, estimating the expected cell frequencies from the marginals can lead to different decisions about the existence of types and antitypes than using the a priori probabilities. This applies accordingly when time series are longer, when more than one variable is investigated, when zero-differences are considered, and when second or higher order differences are analyzed.

Example II: First differences, scores can be equal (von Eye & Niedermeier, 1999). A researcher observes a child playing piano three times. Each time the child's progress is rated on a three-point scale. There is no constraint on the ratings. All three ratings can be different, and two ratings or all three can be the same. Thus, three series of scores can be created. The first contains those patterns in which all three scores are different: 123, 132, 213, 231, 312, 321. The second series contains those patterns in which two values are the same: 112, 113, 121, 131, 211, 311, 122, 322, 212, 232, 221, 223, 133, 233, 313, 323, 331, 332. The third series contains those patterns in which all three scores are the same: 111, 222, 333. From these 27 patterns, nine first difference sign patterns are possible: ++, +0, +-, 0+, 00, 0-, -+, -0, and - -. The a priori probabilities of these patterns appear in Table 65.

The a priori probabilities in Table 65 differ from each other by a factor of up to five. This suggests again that estimating expected frequencies from the marginal frequencies may cause bias.

Table 65: A priori probabilities of first differences (three points in time; three-point scale)

First differences pattern	Frequency of occurrence	Probability
+ +	1	0.037
+ 0	3	0.111
+ -	5	0.185
0 +	3	0.111
0 0	3	0.111
0 -	3	0.111
- +	5	0.185
- 0	3	0.111
- -	1	0.037

Example III: A priori probabilities of second differences. Consider a therapist who observes a client's change in behavior over four weeks. At each observation, the therapist uses a four-point scale to assess the behavior. Focusing only on changes in behavior leads to the 24 sequences 1234, 1243, 1324, 1342, 1423, 1432, 2134, 2143, 2314, 2341, 2413, 2431, 3124, 3142, 3214, 3241, 3412, 3421, 4123, 4132, 4213, 4231, 4312, and 4321. Readers are invited to calculate the first differences for these 24 patterns and to determine their a priori probabilities. From these 24 first differences, one can calculate 24 second differences. For example, the signs of the second differences from the signs of the first differences - + - are + -. The number of different change patterns is nine, if the no-change patterns are included. These second differences patterns, their frequencies, and their a priori probabilities appear in Table 66.

The a priori probabilities in Table 66 differ from each other by a factor of up to five.

Table 66: A priori probabilities of second differences (four points in time; four point scale; no-change patterns disregarded at the level of first differences)

Second differences	Frequency	Probability
+ +	2	0.083
+ 0	1	0.042
+ -	7	0.292
0 +	1	0.042
0 0	2	0.083
0 -	1	0.042
- +	7	0.292
- 0	1	0.042
- -	2	0.083

Two methods have been proposed to deal with the problem of a priori probabilities in CFA. The first method (von Eye, 1990) estimates expected cell frequencies based on the a priori probabilities. The second method (Görtelmeyer, 2000, 2001) calculates a priori probabilities for patterns of raw scores, difference patterns, and sign patterns. These two methods share the same virtues and problems which are described in the following sections.

Determining expected cell frequencies based on a priori probabilities. Consider a situation in which t configurations c_i, for i = 1, ..., t, are subjected to a CFA. The a priori probabilities of these configurations are known to be p_{c_i}. Then, the expected cell frequency for configuration c_i can be calculated as

$$\hat{E}_i = p_{c_i} N,$$

where N is the sample size. If configurations represent more than one differences variable or first, second, and higher differences of the same variable, expected cell frequencies are estimated using the same formula. However, the p_{c_i} are created to conform with the propositions of CFA base models. For instance, if the base model is the model of variable independence and two differences variables are used, the p_{c_i} are calculated from $p_{c_i} = p_{c_{1i}} \, p_{c_{2j}}$, where the 1 and the 2 in the subscripts index the two differences variables, and the i and the j in the subscripts index the categories of these variables.

In the following section, we present three data examples. The first two examples involve the use of a priori probabilities for the estimation of expected cell frequencies. The first example illustrates that the significance tests can result in very different tail probabilities in the CFA tests. The second example shows that only using a priori probabilities poses the danger of labeling configurations discrimination types that are inconspicuous. It is also shown how this danger can be averted. The third data example illustrates multivariate CFA of first differences. Three variables will be analyzed. A priori probabilities in this example suggest a uniform frequency distribution.

8.2.3.2 Three data examples

Data example 1: First order CFA of differences. In the following data example, we compare the results from first order CFA of differences for two methods of estimating expected cell frequencies. The first method is standard CFA, the second method estimates expected cell frequencies based on the a priori probabilities of the configurations. The base model for both approaches is the same. The data stem from a study published by Bartoszyk and Lienert (1978). A sample of 30 participants took an intelligence test. The pulse rates of the subjects were measured before, during, and after administration of the test. The changes in pulse rate were scored as either increasing (+) or decreasing (-). The four possible change patterns for the three observations are ++, + -, - +, and - -. The pattern ++ suggests that a participant's pulse rate increases from before the test to during the test, and again from during the test to after the test. The pattern + - indicates that a pulse rate increased from before to during the test administration, and decreased from during administration to after administration. The other patterns can be interpreted accordingly. The a priori probabilities of these four configurations are $p(++) = 1/6$, $p(+\,-) =$

2/6, $p(- +) = 2/6$, and $p(- -) = 1/6$.

Table 67 summarizes the results of the two CFA analyses. For both analyses we used the binomial test and Bonferroni adjustment, that is, $\alpha^* = 0.0125$.

Table 67: CFA of first differences: Estimation of expected cell frequencies from marginals (left panel) and from a priori probabilities (right panel)

Cell index	Observed frequencies	Standard CFA		Estimation based on a priori probabilities	
		E	p	E	p
+ +	5	7.367	.2185	5	1.0000
+ -	8	5.633	.1878	10	.5271
- +	12	9.633	.2296	10	.5271
- -	5	7.367	.2185	5	1.0000

The X^2-value for the base model of independence of the two change scores is 3.0964 ($df = 1$; $p = .078$). Thus, there is no chance for types or antitypes to emerge. Accordingly, none of the discrepancies between observed and expected cell frequencies is significant. This applies to both panels of the table. However, the right-hand panel still differs considerably from the left-hand panel. The expected cell frequencies, estimated based on the a priori probabilities differ clearly from those estimated based on the log-linear model $\log m = \lambda_0 + \lambda_i^{\Delta_1} + \lambda_j^{\Delta_2}$, where Δ_1 and Δ_2 are the two difference variables. Accordingly, the tail probabilities of the binomial tests differ from those from standard CFA.

Data example 2: Two-sample CFA of differences. The following artificial data example describes the concentration of two groups of respondents, observed three times in hourly intervals. Difference scores were calculated such that only change is reflected. Table 68 displays the results from two approaches to two-sample CFA. The first approach (top panel) is standard two-sample CFA as introduced in Section 7.1. The second approach (bottom panel) estimates expected cell frequencies based on a priori probabilities. For both analyses we used the Pearson X^2-test without continuity correction, and we Bonferroni-adjusted α to be α* = 0.0125.

Table 68: Two approaches to two-sample CFA of differences: Standard CFA (top panel) and CFA based on a priori probabilities (bottom panel)

Cell index $\Delta_1\Delta_2G$	Observed (expected) frequencies	Statistical tests X^2	$p(X^2)$	Discrimination Type?
standard two-sample CFA				
+ +1	22 (12)	25.00	< α*	✔
+ + 2	2 (12)			
+ - 1	3 (11)	16.76	< α*	✔
+ - 2	19 (11)			
- + 1	10 (12)	1.00	.317	
- + 2	14 (12)			
- - 1	1 (1)	0.00	1.000	
- - 2	1 (1)			
two-sample CFA based on a priori probabilities				
+ +1	22 (6)	54.40	< α*	✔
+ + 2	2 (6)			
+ - 1	3 (12)	16.25	< α*	✔
+ - 2	19 (12)			
- + 1	10 (12)	1.00	.317	
- + 2	14 (12)			
- - 1	1 (6)	10.00	0.002	✔
- - 2	1 (6)			

[a] < α*: tail probability is smaller than can be expressed with three decimal places.

Standard two-sample CFA identifies two discrimination types. Two-sample CFA based on a priori probabilities identifies the same two discrimination types. However, there seems to be a third discrimination type for the last pair of configurations. The significance test results suggest that configuration - - allows one to discriminate between the two groups of respondents. This appears to be a mistake, because the two samples are equal in size ($N_1 = N_2 = 36$) and the observed frequencies are 1 for both cells.

How then can the test suggest that there exists a discrimination type? The truth is, the test does not make this suggestion. Here is the explanation. In standard two-sample CFA, to test the hypothesis that a configuration allows us to discriminate between two groups, we create a 2 x 2 cross-classification, and employ the Pearson X^2-test. This test is equivalent to a log-linear main effect model. This model posits that there is no association between the two variables Group and Configuration. If this hypothesis can be rejected, there exists an association and, in the context of two-sample CFA, this configuration constitutes a discrimination type.

When the expected cell frequencies are estimated based on a priori probabilities rather than a log-linear main effect model, the Pearson X^2-test examines a different hypothesis. The hypothesis is that the observed frequency distribution does not deviate significantly from the expected frequency distribution which reflects the a priori probabilities rather than the log-linear main effect model. As the present example shows, there can be significant discrepancies even if there exists no discrimination type.

How then can the researcher determine whether a significant test result indicates the existence of a discrimination type or just the discrepancy between two distributions without the existence of a discrimination type? We recommend looking at the X^2 from standard two-sample CFA, estimating the odds ratio or Gonzáles-Debén's π^*[1]. As was given in Section 7.2.1, the odds ratio is

$$\theta = \frac{p_{11}/p_{21}}{p_{12}/p_{22}},$$

---

[1]The following considerations apply only if the observed frequencies for a configuration are both greater than zero. If both are zero, none of the measures and statistics discussed here can be estimated, and researchers may consider invoking the Delta option, that is, adding a small constant Δ to each cell, e.g., $\Delta = 0.5$.

and Gonzáles-Debén's π* is

$$\pi^* = \frac{p_{11}\,p_{22} - p_{12}\,p_{21}}{p_{11}} .$$

The θ and π* values for the data example in Table 67 appear in Table 69.

Table 69: Odds ratios and π-values for the data in Table 68

Cell index $\Delta_1\Delta_2G$	Observed (expected) frequencies	Statistical tests			Discrimination Type?
		π*	log θ	$p(\theta)$[a]	
+ +1	22 (6)	0.455	3.285	< α*	✔
+ + 2	2 (6)				
+ - 1	3 (12)	0.421	-2.509	< α*	✔
+ - 2	19 (12)				
- + 1	10 (12)	0.143	-.504	0.160	
- + 2	14 (12)				
- - 1	1 (6)	0.000	0.000	0.500	
- - 2	1 (6)				

[a] < α* indicates a probability that small that it cannot be expressed using three decimal places

Table 69 indicates that the π*-estimates for the two configurations that constitute discrimination types in both analyses are relatively large and the log-odds also support the identification of these configurations as discrimination types. For Configuration - -, π* = 0 and log θ = 0. The value of π* = 0 suggests that there exists no association whatsoever in the 2 x 2 table that was created to examine configuration - -. This table has the cell frequencies $N_{11} = 1$, $N_{12} = 1$, $N_{21} = 35$, and $N_{22} = 35$. Similarly, the odds ratio suggests that the odds of being in Group 1 as compared to Group 2 are the same. Thus, the significance test suggests that there is no significant discrepancy from the null hypothesis of even odds. This applies accordingly if one uses the a priori probabilities.

We conclude from this example:

(1) The use of a priori probabilities allows researchers to consider information that otherwise remains unused.

(2) Solely relying on a priori information can lead to artifacts such that the existence of discrimination types is suggested that do not exist. These artifacts surface only if the a priori information is used. The Pearson X^2-test and the z- and binomial tests use the a priori information. Gonzáles-Debén's π^* and the odds ratio θ (and its logarithm) do not use this information.

(3) To identify such artifacts, one can (a) inspect the Pearson X^2 (or the z) from two-sample CFA without consideration of a priori probabilities, (b) estimate and test θ, or (c) estimate Gonzáles-Debén's π^*. If any or all of the three measures and their significance tests suggest that the configuration under scrutiny does not allow one to discriminate between two groups, the significant X^2 from two-sample CFA with consideration of a priori probabilities is most likely an artifact and this configuration does not constitute a discrimination type. Please note that the reverse can occur too. That is, the tests that do consider the expected cell frequencies can suggest that no discrimination type exists and the tests that do not consider the expected cell frequencies suggest the existence of a discrimination type if a priori probabilities are taken into account. In this case we again go the conservative route and retain the null hypothesis that no discrimination type exists.

Data example 3: CFA of differences in three variables. The following data example is taken from Schneider-Düker (1973; cf. Lienert, 1978; Lautsch & von Weber, 1995). In a study on the covariation of mood and cognitive performance around the menstrual cycle, a sample of 72 female students rated themselves in Mood (M), Concentration (C), and Endurance in cognitive tasks (E). The ratings were performed once before and once after each respondent's most recent period. In the following paragraphs, we analyze the changes in these three variables. When observations are conducted at only two observation points, the a priori probabilities of change patterns are uniformly distributed. In this case, it can be an interesting question whether the observed marginals differ from those expected from a uniform probability distribution, and whether types and antitypes (CFA clusters and anticlusters) result from this deviation. Therefore, we analyze the cross-classification of the difference variables

ΔM, ΔC, and ΔE in two ways. First, we perform a zero order CFA. Results from this analysis will tell us whether deviations from a uniform distribution resulted in types and antitypes. In the second analysis we perform a first order CFA to find out whether these types and antitypes result only from deviation from the uniform distributions or whether they also reflect local variable associations.

Table 70 displays the observed frequencies of the cross-classification of the three first differences variables. Each variable is scaled such that a + indicates an increase in scores and a - indicates a decrease. We used the binomial test and the Bonferroni-adjusted $\alpha^* = 0.00625$. The center of the table panel displays the results from zero order CFA, and the right-hand panel displays the results from first order CFA.

Table 70: Zero order and first order CFAs of Schneider-Düker's mood and cognitive performance data (first differences)

ΔM ΔC ΔE	n_{ijk}	Zero order CFA			First order CFA		
		E_1	p	Type ?	E_2	p	Type ?
+ + +	1	9	.0008	A	2.199	.3503	
+ + -	4	9	.0445		3.079	.3707	
+ - +	3	9	.0158		6.134	.1282	
+ - -	12	9	.1836		8.588	.1452	
- + +	8	9	.4468		5.718	.2110	
- + -	6	9	.1887		9.005	.2983	
- - +	18	9	.0028	T	15.949	.3224	
- - -	20	9	.0004	T	22.319	.3254	

Zero order CFA suggests that one antitype and two types exist. The antitype describes the sole respondent whose mood, concentration, and endurance increased from before to after her period. The first type suggests that more respondents than expected based on the uniform distribution base model experienced decreases in mood and concentration and an increase in endurance. The second type suggests that more respondents than expected experienced decreases in all three variables. The second type also describes the largest group in this study.

As is typical of zero order CFA, this antitype and these two types can indicate that deviations from the uniform distribution occur because of variable main effects and/or because of variable interactions of any order. In the present example, we ask whether first or second order interactions between the difference variables can be the reason for the emergence of the one antitype and the two types. This question can be answered in the affirmative if the antitype and the two types (and perhaps others) still exist if main effects are taken into account. Thus, first order CFA is the method of choice, because it takes the main effects of all variables into account.

The results of first order CFA in the right hand panel of Table 70 suggest that when the main effects of all variables are taken into account, the antitype and the two types disappear. We thus conclude that they emerged only because of main effects of the three difference variables.

A note on the selection of base models for CFA of differences. In Table 70, the first model used to analyze the three differences variables was a zero order base model. This model does not even reproduce the main effects of the differences variables. The selection of this model can be justified because none of the variables was observed under a product-multinomial sampling scheme (see Section 2.3), and none of the variables served as predictor. In each of these cases, the main effects of the variables must be reproduced. However, whenever multiple first differences or multiple second or higher differences are analyzed, all but the last differences can be considered predictors of their immediately following differences. Therefore, their marginal frequencies must be reproduced. As a consequence, the zero order CFA base model can never be used in a CFA of differences, if more than one difference of the same order is considered.

8.2.4 CFA of second differences

CFA of second differences shares methods, virtues, and problems with CFA of first differences (and CFA of higher differences). As was indicated in Section 8.1, the different levels of differences reflect different characteristics of a series of measures. The differences between raw scores, that is, the first differences, describe the up-and-down of a series. This characteristic corresponds to the first derivative of a continuous, differentiable function. Second differences describe how the up-and-downs change. Thus, they are comparable to the second derivative of a continuous, twice-differentiable function, and so forth.

If the first differences are unequal to zero and constant, the series can be described using a linear function, that is, a straight line. If the second differences are unequal to zero and constant, the series can be described using a quadratic function. If the third differences are constant, the series can be describe using a third order polynomial. In general, the differences of a particular order become constant, if a polynomial of the same order allows one to perfectly describe a series.

The signs of first differences have a natural interpretation. Negative signs indicate that the slope is negative, that is, the y-scores become smaller as the x-scores increase. This is well known from regression analysis. The signs of second differences also have a natural interpretation:

(1) Positive signs indicate positive acceleration or, the curve is ∪- or J-shaped.
(2) Negative signs indicate negative acceleration or, the curve is ∩- or inversely J-shaped.

Both (1) and (2) apply accordingly to the signs of higher order differences.

A problem that second and higher differences share in common with first differences is that their a priori probabilities can be calculated, and can be very different from each other. An example of a priori probabilities appeared in Table 66. The probabilities in this table differ by a factor of 7:1.

9. CFA of Level, Variability, and Shape of Series of Observations

CFA of differences, covered in Chapter 8, allows one to analyze the ups-and-downs of a series of measures (first differences), the changes in these ups-and-downs (second differences), the changes in these changes (third differences), and even higher order differences. In addition, several kinds of such changes can be analyzed simultaneously. For example, one can ask whether both the first and the second differences of a series allow one to discriminate between a number of groups of respondents. In other words, the method of differences allows one to analyze various aspects of the shape of a series of measures.

Two important characteristics of series of measures are not considered in the method of differences as it was introduced in Section 8.1. These are the level (magnitude) of the measures and the variability of the measures. The following sections introduce readers to methods for the configural analysis of level and variability of a series of measures. In addition, the joint analysis of various characteristics of series of measures is illustrated.

9.1 CFA of shifts in location

Series of measures can differ in elevation (location), variability, and shape. In this section, we discuss the configural analysis of the location of a series.

As in the analysis of the shape of a series, the location can be analyzed by comparing it to some anchor, and then categorizing the result of this comparison. Three types of anchors have been discussed in the context of CFA. First, *anchors can be absolute scores* such as temperatures, or *population parameters* such as the average intelligence. Second, *anchors can be group-specific* (von Eye & Lienert, 1985) such as the average verbal performance score in a group of middle-aged women. Third, *anchors can be ipsative* (Keuchel & Lienert, 1985), that is, specified relative to an individual's score such as the individual's average or the individual's previous score.

These three types of comparisons can be categorized in many ways. Most popular are two transformations that are analogous to the ones introduced in Section 8.2.2 for the categorization of differences between adjacent scores. Let a be an anchor, for example, the median, the mean, some previous score, or some absolute value. Then, the location of the jth score of individual i on variable X, relative to a, can be categorized to be

$$\delta_{ij} = \begin{cases} + & if \; x_{ij} > a \\ 0 & if \; x_{ij} = a \\ - & if \; x_{ij} < a. \end{cases}$$

If the reliability and the resolution level of some observation is high enough, researchers may wish to use only two categories, or

$$\delta_{ij} = \begin{cases} + & if \; x_{ij} > a \\ - & if \; x_{ij} < a. \end{cases}$$

This method of categorization has been discussed by Bierschenk and Lienert (1977) in the context of clustering profiles and learning curves.

In a fashion analogous to the method of differences, the two transformations proposed for configural analysis of series of measures can also reduce the size of the contingency table. To illustrate, suppose that d variables are examined. The cross-classification of these d variables has $c = \prod_{i=1}^{d} c_i$ cells, where c_i is the number of categories of variable i. After categorization, variable i has $c_i^{\,t}$ categories, and the cross-classification the d transformed variables has $c^{\,t} = \prod_{i=1}^{d} c_i^{\,t}$ cells. If the number $c_i^{\,t}$ is smaller than c_i for at least one i, and none of the variables experiences an increase in the number of categories, then $c^t < c$.

In the following paragraphs, we present two data examples. The first example illustrates the analysis of one variable that was observed four times. The second example illustrates the analysis of two variables, each observed twice. In the context of the second example, we discuss the specification of base models that are specific to the goals of analysis in CFA of shifts in location.

Data example I. The following example uses data published by Krauth and Lienert (1973) on a series of mood scores. A sample of 60 subjects completed seven trials of mathematical calculations. After the first, third, fifth, and seventh trial, each subject rated their mood on 5-point Likert scale with 1 indicating good mood and 5 indicating negative mood. These ratings were transformed such that they depict their position relative to the midpoint of the scale, 3. Thus, rating values 1 and 2 were assigned a +, values of 3 were assigned a 0, and values of 4 and 5 were assigned a -.

Crossed, the transformed ratings from the four trials form a 3 x 3 x 3 x 3 table with 81 cells. The cross-classification of the transformed variables has thus 3^4 cells rather than the 5^4 of the cross-classification of the original variables. This represents a savings of almost 80% in the number of cells. However, for a sample of 60 respondents, this table is still too large. Too many cells would remain empty and the expected cell frequencies would be very small. Therefore, the last two trials were pooled and their average scores were used for the transformation. As a result, we analyze three transformed variables instead of four, and the cross-classification has 27 cells, the result of crossing the three variables M1, M2, and M3. Each of these variables has the categories + for for 'good mood,' 0 for 'average mood,' and - for 'negative mood.' We now analyze this cross-classification using first order CFA. We use Anscombe's z-approximation and the Bonferroni-adjusted $\alpha^* = 0.05/27 = 0.001852$. Table 71 presents the results.

The results in Table 71 suggest that three types and two antitypes exist. Reading from the top to the bottom, the first type is constituted by configuration + + 0. These are the participants who were in good, above-average mood up until the fifth trial. Then, their mood dropped to be no better than scale average. The second type is constituted by configuration + 0 -. A fourth of the sample displayed this pattern. These subjects started out in good mood (first trial). However, over the course of the calculations, their mood dropped first to average (third trial), and then soured to below average (trials five and later). The third type, constituted by configuration

Table 71: First order CFA of changes in mood during mathematical calculations

Cell index	Frequencies		Test statistics		Type/ Antitype?
M1 M2 M3	observed	expected	z	$p(z)$	
+ + +	0	3.05	2.521	.0058	
+ + 0	12	4.35	3.106	.0009	**T**
+ + -	0	5.66	3.497	.0002	**A**
+ 0 +	0	2.03	2.019	.0218	
+ 0 0	1	2.90	1.199	.1152	
+ 0 -	15	3.77	4.487	< α*	**T**
+ - +	0	1.69	1.821	.0343	
+ - 0	0	2.42	2.223	.0131	
+ - -	1	3.14	1.324	.0927	
0 + +	13	2.21	5.155	< α*	**T**
0 + 0	0	3.15	2.567	.0051	
0 + -	0	4.10	2.952	.0016	**A**
0 0 +	0	1.47	1.678	.0466	
0 0 0	0	2.10	2.057	.0198	
0 0 -	1	2.73	1.108	.1340	
0 - +	0	1.23	1.506	.0660	
0 - 0	2	1.75	0.313	.3772	
0 - -	5	2.28	1.674	.0471	
- + +	0	1.05	1.370	.0854	

/ cont.

Cell index	Frequencies		Test statistics		Type/ Antitype?
M1 M2 M3	observed	expected	z	$p(z)$	
- + 0	1	1.50	0.296	.3835	
- + -	1	1.95	0.632	.2639	
- 0 +	0	0.70	1.047	.1476	
- 0 0	1	1.00	0.172	.4318	
- 0 -	0	1.30	1.561	.0593	
- - +	1	0.58	0.726	.2341	
- - 0	3	0.83	2.036	.0209	
- - -	3	1.08	1.682	.0463	

[a] < α* indicates that the tail probability is smaller than can be expressed with four decimal places.

0+ +, shows a somewhat surprising pattern. These participants started out in a rather neutral mood. However, these ratings improved to indicate good mood already at the third trial, and remained in good mood through the rest of the calculations.

The two antitypes indicate which patterns were observed less often than expected based on chance. The first antitype is constituted by configuration + + -. It describes good mood ratings through the fourth trial that are followed by negative mood ratings. This pattern was not observed at all, whereas almost six incidences were expected. Similarly, more than four incidences were expected for pattern 0 + - which describes an improvement in mood from the first to the third trial, followed by a decline to negative mood for the rest of the calculations. No case was observed with this pattern.

Data example II. In the second data example, we use data from the same experiment as in the first (Krauth & Lienert, 1973a). A sample of 60 subjects participated in an experiment in which they had to perform calculations in seven trials. In addition to mood (see Table 70), the flicker fusion threshold (F) and performance in calculations (R) were observed after the first and the seventh trials. Both variables were dichotomized at their median with 1 indicating above and 2 indicating below the median. In the following paragraphs we analyze the 2 x 2 x 2 x 2 cross-classification of the four variables flicker threshold after the first (F1) and the seventh trials (F7), and performance in mathematics after the first (R1) and the seventh trials (R7). We use Pearson's X^2 component test and the Bonferroni-adjusted $\alpha^* = 0.003125$. The first base model is the model of variable independence, that is, global first order CFA. Table 72 displays the results of this analysis.

First order CFA of shifts in location in the variables flicker threshold and calculation performance identified two types and no antitype. The first type is constituted by Configuration 1111. It describes those subjects who have above-average flicker fusion thresholds and perform above average in calculations both in Trials 1 and 7. The second type is constituted by Configuration 2121 and describes those subjects who perform below average in calculations in Trial 1 but above average in Trial 7. The flicker thresholds of these respondents are also below average in trial 1 and above average in trial 2.

As was discussed in the section on the admissibility of CFA base models (Section 2.2), one of the most important criteria for a meaningful base model is that there be only one way to deviate from it. In principle, this is the case for the present base model. The two types emerge because of interactions among the four crossed variables. However, in the present example, the two-way interactions can be grouped in meaningful clusters, and it can be asked whether taking a grouping of the two-way interactions into account helps explain the types. Specifically, there are three groups of two-way interactions. The first involves the interaction between the two measures of the flicker threshold. The second involves the interaction between the two measures of calculation performance, and the third involves the interactions of flicker threshold variables with calculation performance variables.

In longitudinal CFA, researchers often ask whether autocorrelations lead to types and antitypes. Therefore, it may be interesting to take the cross-sectional interactions into account. These are the interactions between R1 and F1 and between R7 and F7. If these interactions are considered in the estimation of the expected cell frequencies, types and antitypes can emerge only from correlations between any variables at Trial 1 and any variables at Trial 7, and from higher order associations.

Table 72: First order CFA of shifts in location in flicker threshold and performance in calculations

Cell index		first order CFA				regional CFA	
R1 R7 F1 F7	n_{ijkl}	E_1	X^2	$p(X^2)$	Type?	E_2	X^2
1111	11	3.72	14.262	.0002	T	9.33	.546
1112	2	5.58	2.294	.1299		2.80	-.478
1121	1	2.00	0.502	.4788		1.67	-.516
1122	0	3.00	3.003	.0831		0.50	-.707
1211	3	4.86	0.713	.3984		1.87	.830
1212	12	7.29	3.038	.0813		14.0	-.535
1221	0	2.62	2.618	.1057		0.33	-.577
1222	4	3.93	0.001	.9706		2.50	.949
2111	1	3.04	1.371	.2417		3.67	-1.393
2112	4	4.56	0.069	.7921		1.10	2.765
2121	7	1.64	17.553	$< \alpha^*$	T	5.33	.722
2122	0	2.46	2.457	.1170		1.60	-1.265
2211	0	3.98	3.978	.0461		0.73	-.856
2212	6	5.97	0.000	.9892		5.50	.213
2221	1	2.14	0.609	.4352		1.07	-.065
2222	8	3.21	7.132	.0076		8.00	.000

[a] $< \alpha^*$ indicates that the tail probability is smaller than can be expressed with four decimal places.

The right hand panel of Table 72 displays the results from the analysis that takes into account the R1 x F1 and the R7 x F7 interactions. The base model for this analysis is

$$\log E = \lambda_0 + \lambda_i^{R1} + \lambda_j^{F1} + \lambda_k^{R7} + \lambda_l^{F7} + \lambda_{ij}^{R1\,F1} + \lambda_{kl}^{R7\,F7}.$$

This equation describes a regional base model in which R1 and F1 form one group and R7 and F7 form another group of variables. Thus, this model can be labeled *longitudinal ISA of shifts in locations*. To identify possible types and antitypes, we use, as for the first order CFA in the middle panel of Table 72, Pearson's X^2 component test and the Bonferroni-adjusted α^* = 0.003125.

The X^2 components in the last column of Table 72 are too small for configurations to constitute types or antitypes. We thus conclude that the two types that emerged from first order CFA resulted because of interactions between flicker threshold and calculation performance at Trial 1 and Trial 7.

9.2 CFA of variability in a series of measures

Thus far, we have considered characteristics of shape and level of series of measures. In many investigations, variability may be of interest in itself or in addition to level and shape. Often, level or shape carry little information about the smoothness of a series. Series identical in level and shape can differ greatly in smoothness, that is, variability. Therefore, Krebs and collaborators (1996) proposed assessing the variability in a series of measures using von Neumann's (1941) variance measure. For a series of T measures x_t, von Neumann's variance V is

$$V = \frac{1}{T - 1} \sum_{t > 1} (x_{t-1} - x_t)^2,$$

for t = 2, ..., T. This measure of variance relates time-adjacent scores to each other. Specifically, this measure is the average of the squared differences between all time-adjacent pairs of measures.

For use in CFA, this measure can be categorized in a fashion parallel to differences and level information. In particular, for a given cut-off a, one can create the transformed variance values where the superscript ^{3} indicates that three categories are created. The cut-off can be chosen to be, for example, a band around the median of V-scores in a sample.

$$V_{ij}^{3} = \begin{cases} + & if\ x_{ij} > a \\ 0 & if\ x_{ij} = a \\ - & if\ x_{ij} < a\ , \end{cases}$$

Alternatively, one can use the transformation

$$V_{ij}^{2} = \begin{cases} + & if\ x_{ij} > a \\ - & if\ x_{ij} < a\ , \end{cases}$$

where the superscript [2] indicates that two categories are created.

Data example. In the following data example, we re-analyze the data published by Krebs et al. (1996). The researchers hypothesized that rats respond to stress initially with a reduction in food intake. After this reduction, they gradually habituate to stress and respond by increasing their food intake again, until their intake is the same as that of rats not exposed to stress. To test this hypothesis, two samples of 18 rats each were studied. The experimental group was exposed to 95 db of white noise, and the control group to 60 db of white noise. Food pellet consumption was observed over 5 days (the raw data are reproduced in Krebs et al., 1996, p. 197).

In the following analyses, we study the two variables S, the number of pellets consumed at Day 1, and V, the volatility of intake over 5 days. Both S and V were dichotomized at their grand medians with + indicating above median and - indicating below median. We cross the experimental condition, G, with S and V, and compare the two experimental groups, e and c, using two-sample CFA. This method allows us to compare the two samples in their volatility and intake at Day 1. We Bonferroni-adjust α to obtain $\alpha^* = 0.0125$, and we use the odds ratio θ to compare the samples (see Section 7.2.2). Table 73 presents results.

Table 73 suggests the existence of two discrimination types. The first type is constituted by Configuration + -. These are the rats that start the 5-day experiment consuming above-average numbers of food pellets, and show below-average volatility in their consumption during the experiment. This pattern was mostly observed in the control group. Configuration - + constitutes the second discrimination type. This configuration describes those rats who consume below average numbers of food pellets at Day 1 and display high volatility in their consumption over the following 5 days. This pattern was observed mostly in the experimental group.

Table 73: Two-sample CFA of food intake of rats under two experimental noise conditions

Cell index SVG[a]	Observed frequencies	θ	$p(\theta)$	π^*	Discrimination Type?
+ + e	3				
		0.470	.3159	0.167	
+ + c	2				
+ - e	2				
		-2.531	.0023	0.409	✔
+ - c	11				
- + e	10				
		1.833	.0102	0.350	✔
- + c	3				
- - e	3				
		0.470	.3159	0.167	
- - c	2				

[a] e indicates the experimental group, and c indicates the control group.

We now ask whether these two configurations constitute a bi-discrimination type (Section 6.2.2.2; Havránek & Lienert, 1986). Table 74 displays the contrasted frequencies.

Inserting in the X^2 equation for biprediction tests given in Section 6.2.2.2 yields

$$X^2 = \frac{[13(13\cdot 2 - 18\cdot 11) - 13(18\cdot 10 - 18\cdot 3)]^2}{13\cdot 13\cdot 18\cdot 18(13 + 13)(18 + 18)/36} = 10.54.$$

For $df = 1$, the tail probability of this X^2 score is $p = 0.0012$. We thus conclude that the two configurations + - and - + allow one to discriminate between the experimental and the control groups.

From these results, we know that the experimental and the control groups differ in the number of food pellets eaten at the first day of the experiment and the volatility of the consumption patterns during the 5 days of the experiment. We now ask whether the two groups also differ in their trends, as was hypothesized. To test trend hypotheses, we expand the cross-

Table 74: Contingency tale for the bi-discrimination types + - and - +

Configuration	Experimental group	Control group	Row totals
+ -	2	11	13
- +	10	3	13
others	6	4	10
Column totals	18	18	36

classification in Table 73 by the signs of the linear trend in food pellet consumption. We thus create a 2 (S; Consumption on first day) x 2 (T; Trend) x 2 (V; Volatility) x 2 (Groups) cross-classification. This table appears in Table 75, along with the results from two-sample CFA. The Bonferroni-adjusted α is $\alpha^* = 0.00625$. We use the standard X^2 test without continuity correction.

Table 75: Two-sample CFA of food intake of rats under two experimental noise conditions; variables included are Consumption on first day, Trend, Volatility, and Experimental group

Cell index STV	Observed frequencies	X^2	$p(X^2)$	π^*	Discrimination Type?
+ + + e	2				
+ + + c	0	2.118	.1456	.50	
+ + - e	2				
+ + - c	1	0.364	.5465	.25	
+ - + e	1				
+ - + c	2	0.364	.5465	.25	

/ cont.

Cell index STV	Observed frequencies	X^2	$p(X^2)$	π*	Discrimination Type?
+ - - e	0				
+ - - c	10	13.846	.0002	.50	✔
- + + e	9				
- + + c	1	8.862	.0029	.44	✔
- + - e	2				
- + - c	1	0.364	.5465	.25	
- - + e	1				
- - + c	2	0.364	.5465	.25	
- - - e	1				
- - - c	1	0.000	1.0000	.00	

The results in Table 75 also suggest two discrimination types. The first is constituted by Configuration + - -. This pattern describes those rats who start consuming above-average amounts of food, reduce their food intake during the experiment, and show below-average volatility in this trend. This pattern was observed predominantly in control group rats. The second discrimination type, constituted by Configuration - + +, describes those rats who start the experiment consuming below average amounts of food, increase their food intake during the experiment, and show above-average volatility in this trend. These are mostly rats from the experimental stress group.

9.3 Considering both level and trend in the analysis of series of measures

Section 9.1 covered methods for the configural analysis of shifts in location. Section 9.2 was mostly concerned with the configural analysis of the variability on a series of measures. Section 9.2 also combined these two approaches in the data example in Table 75. In this example, the amount of food intake was the level information, and trend was conceptualized as linear trend. The present section introduces readers to a different methodology that allows one to simultaneously consider level, linear trend, quadratic trend, and so forth. Specifically, this section presents methods that allow one to first parameterize level and the various trends. After estimating the parameters for the various characteristics of a series of measures, the estimates can be categorized and then subjected to a configural analysis. Two methods are discussed. These are first the estimation of polynomial parameters for equidistant points in time, and second the estimation of polynomial parameters of non-equidistant points in time. We discuss these methods in the context of longitudinal CFA (cf. Krauth, 1973, 1980; Krauth & Lienert, 1975, 1978; Lienert, 1980; von Eye & Hussy, 1980; von Eye & Nesselroade, 1992). However, these methods can also be used to estimate parameters for any other series of measures.

9.3.1 Estimation and CFA of polynomial parameters for equidistant points on X

Polynomials are functions that can be used to describe series of measures. Consider a series of pairs of x_j and y_j scores where variable X is the predictor of variable Y. Using a general regression approach, the scores y_j can be approximated using the polynomial

$$\hat{y} = f(x) = P_I(x) = b_0 x^0 + b_1 x^1 + b_2 x^2 + \dots + b_I x^I = \sum_{i=0}^{I} b_i x^i,$$

where the x are the observation points, the b_i are the regression weights, and the superscripts *i* are exponents. The value *I* is termed the *degree of the polynomial* or the *order of the polynomial*. $P_I(x)$ is thus the polynomial of degree *I*. For instance, the expression

$$\hat{y} = b_0 x^0 + b_1 x^1$$

represents a first order polynomial or a linear regression equation, and the expression

$$\hat{y} = b_0 x^0 + b_1 x^1 + b_2 x^2$$

represents a second degree polynomial, a quadratic regression equation, or a quadratic curve.

In the description of series of measures, polynomials have a number of very desirable characteristics. Most notably, any series of scores can be perfectly described using a polynomial. Here, the expression "perfectly described" means that the polynomial goes exactly through the x-y points. In many instances, to describe a series of t measures perfectly, a polynomial of t - 1st degree is needed. This is the highest degree polynomial that can be constructed for t pairs of scores. In most instances, lower degree polynomials suffice. Differences (see Chapter 8) and polynomials share this characteristic. For reasons of scientific parsimony, researchers typically try to find the lowest degree polynomial that sufficiently describes a series of scores.

Consider the following data example. Researchers attempt to approximate the series of six x-y pairs, {1, 1}, {2, 3}, {3, 5}, {4, 4}, {5, 3}, {6, 2} using polynomials as simple as possible. They begin using a simple regression approach in which they regress Y on X using a straight regression line. The standardized regression coefficient for these data is $b_1 = 0.151$. This value is not significant ($t = 0.306$, $p = 0.775$) and $R^2 = 0.023$. The researchers conclude that this approximation deviates significantly from the observed data. In a next step, the researchers try a quadratic polynomial. For this model the estimated parameters are $b_1 = 4.616$ and $b_2 = -4.561$. Both values are significant and $R^2 = 0.891$. Figure 6 displays both the linear and quadratic and approximation curves.

The approximation polynomials in Figure 6 suggest that the linear regression line fails to describe the data well. The quadratic curve does a much better job. It is still not perfect in the sense that it does not go through the data points exactly. However, researchers will rarely be tempted to improve a solution that is as good as the one presented by the quadratic polynomial in Figure 6. Rarely is the reliability of empirical data points so high that one can justify the selection of a polynomial that goes exactly through each data point.

There exist many methods for estimating the polynomial parameters b_i. Examples of such methods include the least squares method, which is well known from regression analysis, and the method of differences (see Chapter 8). These methods are described in detail elsewhere (Abramowitz & Stegun, 1972; Neter et al., 1996). Computer programs for the least squares method are included in practically all general purpose statistical software packages, for example SYSTAT 10.0.

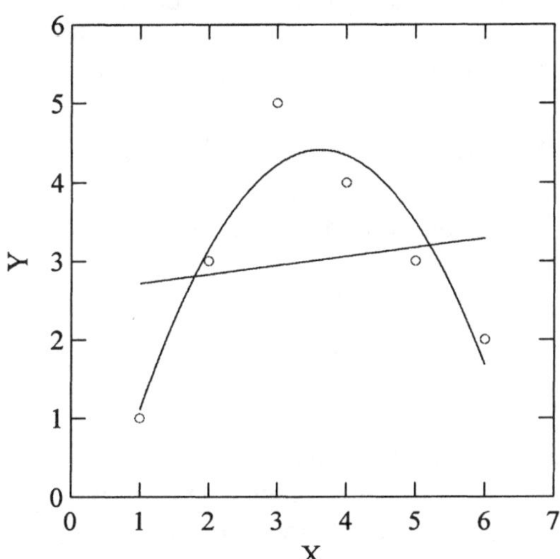

Figure 6: Linear and quadratic approximation polynomials

While very useful, standard polynomials come with a number of characteristics that keep researchers from using them more often. First, the polynomial parameters can depend upon each other. Including higher order parameters can change the estimate of lower order parameters. In the above numerical example, the parameter for the linear slope changed from 0.151 to 4.616 after the quadratic component was introduced. Second, while capable of fitting any series, very high degree polynomials may be needed for certain types of series, for example, cyclical series or series that approximate asymptotes. The first of these problems can be solved by using a special type of polynomial, *orthogonal polynomials*. The second problem is of lesser importance here, in the context of relatively short series of measures. Therefore, we now describe orthogonal polynomials (Section 9.3.1.1) and their use in the context of CFA (Section 9.3.1.2).

9.3.1.1 Orthogonal polynomials

We now approach the same problem as in Section 9.3.1, that is, the problem of approximating a series of *x-y* measures using a polynomial. In contrast to Section 9.3.1, however, we now use a family of polynomials rather than standard polynomials. This family can be described by

$$f(x) = a_0\varphi_0(x) + a_1\varphi_1(x) + a_2\varphi_2(x) + \dots + a_I\varphi_I(x) = \sum_{i=0}^{I} a_i\varphi_i(x),$$

where $\varphi_i(x)$ is an ith degree polynomial and $i = 0, 1, \dots, I$. The family thus described is termed *system of polynomials* $P_I(x)$. It has degree $[P_I(x)] = I$. This system is called orthogonal on the interval $a \le x \le b$, with respect to the weight function $w(x)$, if

$$\int_a^b w(x)P_n(x)P_m(x)\,dx = 0,$$

where $P_n(x)$ and $P_m(x)$ are polynomials of degree n and m, respectively, and $n \neq m$ and $n, m = 0, 1, 2, \dots$ (see Abramowitz & Stegun, 1970). The two polynomials $\varphi_i(x)$ and $\varphi_j(x)$ are orthogonal to each other (with $i \neq j$), if they fulfill the orthogonality condition

$$\sum_{k=1}^{n} \varphi_i(x_k)\,\varphi_j(x_k) = 0,$$

where k indexes points on x and n is the number of points.

As is the case for standard polynomials, there are many ways to estimate the parameters of systems of orthogonal polynomials. In the following paragraphs, we describe estimation for the case in which the distances of the points on x are constant, that is, for *equidistant* points on x.

The calculations begin by setting $\varphi_0(x) = 1$ and $a_0 = \bar{y}$. Then, we replace the equidistant scores on x by orthogonal polynomial coefficients, x'. These coefficients can be found, for instance, in textbooks of analysis of variance (e.g., Kirk, 1995, Table E.10) or in volumes containing statistical tables (e.g., Fisher & Yates, 1948). The polynomial parameters can be estimated using

$$a_j = \frac{\sum_{i=1}^{n} y_i \varphi_j(x_i')}{\sum_{i=1}^{n} \varphi_j^2(x_i')},$$

where j indicates the order of the polynomial (e.g., $j = 1$ indicates that the weight is estimated for the first order polynomial), i indexes the points on x, that is, the polynomial coefficients, and n is the number of points on x. Specifically, one obtains for the first order polynomial

$$a_1 = \frac{\sum_{i=1}^{n} y_i \varphi_1(x_i')}{\sum_{i=1}^{n} \varphi_1^2(x_i')},$$

and so forth.

Once the parameters are determined, the final step of the approximation procedure consists of determining the explicit equation for the polynomial. The parameter estimates a_j are the weights in $\sum_{j=0}^{J-1} a_j \varphi_j(x')$. The polynomials $\varphi_i(x')$ can be determined using the three-term recursion

$$\varphi_{j+1}(x') = (2x' + \alpha_j)\ \varphi_j(x') + \beta_{j-1}\varphi_{j-1}(x'),$$

where j is the degree of the polynomial, and $\varphi_0(x') = 0.5$ and $\varphi_{-1}(x') = 0$. The α_k and the β_{k-1} can be calculated using

$$\alpha_j = \frac{-2\sum_{i=1}^{n} x_i \varphi_j^2(x_i')}{\sum_{i=1}^{n} \varphi_j^2(x_i')}$$

and

$$\beta_{k-1} = \frac{-\sum_{i=1}^{n} \varphi_j^2(x_i')}{\sum_{i=1}^{n} \varphi_{j-1}^2(x_i')}$$

(see Selder, 1973, p. 266).

The *interpretation* of the resulting polynomials is straightforward. The zero degree polynomial is a constant, specified to be the arithmetic mean of the *y*-scores. The first degree polynomial is the linear slope of the y-scores over the range of the corresponding *x*-scores. As will be illustrated below, the parameter of the linear polynomial is identical to the regression parameter, if the *x*-scores are centered. The same analogy applies to the parameter of the zero degree polynomial and all other parameters.

The second order polynomial describes the curvature of the slope. If the parameter for this polynomial, a_2, is negative, the curve describes a "hill." That is, the curve has one peak, and to both sides of this peak, the y-scores decrease. If the parameter a_2 is positive, the curve describes a "valley." That is, the curve has one minimum, and to both sides of this minimum, the *y*-scores increase.

The interpretation of the higher order curves can be performed accordingly. Third degree polynomials have one change in curvature, called the inflexion point. If the parameter of the third degree polynomial, a_3, is negative, the maximum of the curve is located to the right of the one minimum and the *y*-scores to the right of this maximum decrease. If a_3 is positive, the minimum is located to the right of the one maximum and the y-scores to the right of this minimum increase.

Data example. The following data example shows how a series of measures is approximated using orthogonal polynomials. Table 76 displays the *x*-scores and the corresponding *y*-scores, both from a learning experiment (von Eye & Hussy, 1980) and the polynomial coefficients for 1st and 2nd degree polynomials. The experiment examined the retroactive inhibitory effect of the duration of a break before an interfering activity. The x-score indicate the duration of the break. Y-scores indicate the number of items recalled by the participants.

Coefficient a_0 is calculated as $a_0 = \bar{y} = 9.29$. Inserting into the above equations yields $a_1 = 0.15$ and $a_2 = 0.00076$, and the second degree polynomial $\hat{y} = 9.29 + 0.15\varphi_1 + 0.00076\varphi_2$. For estimation,

Table 76: Approximation of a series of five data points by polynomials of 1st and 2nd degree

i	x	$x_{centered}$	y	$\varphi_1(x')$	$\varphi_2(x')$
1	1	-14.2	6.28	-2	2
2	5	-10.2	8.50	-1	-1
3	10	-5.2	8.67	0	-2
4	20	4.8	10.06	1	-1
5	40	24.8	12.94	2	2

extrapolation, and interpolation purposes, the β_j can be estimated and values for y for any x-score between 1 and 40 can be estimated. Inserting the polynomials φ_j yields the interpolation polynomial $\hat{y} = 6.65 + 0.21x - 0.0015x^2$. For instance, the recall rate after a 25 minute break between the end of a learning trial and the interfering activity is $\hat{y} = 6.65 + 0.21 \cdot 25 - 0.0015(25 \cdot 25) = 11.00$.

Figure 7 displays the observed data and the quadratic approximation polynomial. The polynomial in Figure 7 suggests that the second degree polynomial fails to describe the data particularly well. A third degree polynomial may come closer.

We now demonstrate that the polynomial parameters estimated using the methods presented above are identical to regression parameters if the x-scores are (1) equidistant and (2) centered. To do this demonstration, we insert the centered x-scores (Column 3 in Table 76) and the y-scores into a regression program. We obtain the following, slightly edited output from SYSTAT 10.0:

```
Dep Var: RECALL    N: 5    Multiple R: 0.977    Squared multiple R: 0.954
Adjusted squared multiple R: 0.909   Standard error of estimate: 0.739
Effect       Coefficient  Std Error  Std Coeff  Tolerance  t         P
CONSTANT           6.645      0.681      0.000          .     9.755  0.010
TIME               0.217      0.096      1.377      0.061     2.251  0.153
TIME*TIME         -0.002      0.002     -0.419      0.061    -0.684  0.564
```

The coefficients of this analysis are identical to the ones calculated above.

Approximation of learning curve by quadratic polynomial

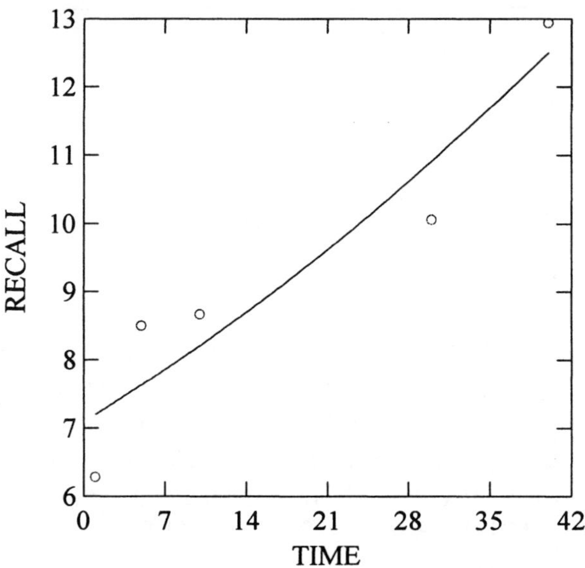

Figure 7: Approximation of y-scores by quadratic polynomial

9.3.1.2 Configural analysis of polynomial coefficients

The polynomial parameters that are estimated using the least squares methods described in Section 9.3.1.1 have several desirable characteristics. For instance, they share the efficiency and minimum variance characteristics of all least squares estimators. Because of these characteristics, the estimates have frequently been used in subsequent analyses in the context of multilevel models (Goldstein, 1987).

However, the estimators described in Section 9.3.1.1 also share some of the less desirable characteristics, such as the susceptibility to outliers. One method that has repeatedly been discussed for dealing with outliers is categorizing data. There can be no doubt that categorization often results in loss of information. However, often categorization also allows researchers to express variable relationships more clearly (for a discussion of these issues see Vargha, Rudas, Delaney, & Maxwell, 1996).

Therefore, we present in this section an application in which dichotomized estimates of polynomial parameters are subjected to configural analysis.

Data example. The following example involves a re-analysis of the data published by von Eye and Nesselroade (1992). A sample of 42 college students was presented a state anxiety questionnaire four times (cf. Nesselroade, Pruchno, & Jacobs, 1986). Each time, all respondents completed both parallel forms, A and B, of the questionnaire. Thus, two time series of four measures each resulted for each student. Von Eye and Nesselroade approximated each of these time series using the second order orthogonal polynomial

$$y = \bar{y} + b_1 t + b_2 t^2,$$

where $\bar{y}$ is the arithmetic mean of the scores in the series, and t counts the equidistant observation points ($t = 1, 2, 3, 4$). Thus, for each student, two sets of parameters ere estimated. For example, one student's parameter estimates were $\bar{y} = 37.8$, $b_1 = 1.0$, and $b_2 = -0.8$ for parallel form A. These estimates suggest that this student's scores increase with time (linear trend, b_1). However, the negative quadratic trend b_2 also suggests that toward the end, this student's scores become lower, outweighing the linear trend in this case.

For the present re-analyses, we only use the mean and the linear trend information (von Eye & Nesselroade, 1992, used the linear and the quadratic trend estimates). The means were dichotomized at their medians, with 1 indicating above the median and 2 indicating below the median. The b_1 parameter estimates were dichotomized at zero, thus separating the positive from the negative slopes. Positive slopes, indicated by 1, indicate increasing state anxiety, negative slopes, indicated by 2, suggest decreasing state anxiety. Thus, zero is a natural cut-off point. From this dichotomization, the four variables A_0, A_1, B_0, and B_1 resulted. A_0 and B_0 were the scales for the means, and A_1 and B_1 the scales for the linear slopes. Crossed, these four dichotomous variables form a 2 x 2 x 2 x 2 contingency table.

We now use first order CFA for the analysis of this cross-classification. We use Anscombe's z-test and adjust α using Bonferroni's method. The resulting threshold is $\alpha^* = 0.003125$. Table 77 displays the results. For the base model we calculated an overall goodness-of-fit Pearson $X^2 = 64.82$. For $df = 11$, this value suggests that the base model must be rejected ($p < 0.05$) and we have reason to expect types and antitypes.

Table 77: First order CFA of means and trends of two state anxiety test parallel forms

Cell index	Frequencies		Test statistics		Type/ antitype?
A_0 A_1 B_0 B_1	observed	expected	z	$p(z)$	
1111	10	3.097	3.223	.0006	T
1112	0	2.106	2.060	.0197	
1121	1	2.558	1.011	.1560	
1122	0	1.740	1.850	.0322	
1211	5	3.749	0.702	.2415	
1212	4	2.549	0.945	.1724	
1221	0	3.097	2.544	.0055	
1222	1	2.106	0.735	.2311	
2111	2	3.097	0.572	.2837	
2112	1	2.106	0.735	.2311	
2121	5	2.558	1.457	.0726	
2122	0	1.740	1.850	.0322	
2211	0	3.749	2.817	.0024	A
2212	1	2.549	1.006	.1572	
2221	2	3.097	0.572	.2837	
2222	10	2.106	4.090	$< \alpha^*$	T

[a] $< \alpha^*$ indicates that the tail probability is smaller than can be expressed with four decimal places.

First order CFA of the responses to the state anxiety questionnaires suggest the existence of two types and one antitype. The first type is constituted by Configuration 1111. It describes those students who score above the median in both parallel test forms, and showed an increasing trend over the four measurement occasions. Only about 3 respondents were expected for this pattern from the base model. However, 10 displayed this pattern. The second type, constituted by Configuration 2222, describes those 10 individuals who scored below the median in both parallel test forms and also exhibited a trend toward lower scores over the four occasions. Only about 2 individuals were expected to show this pattern.

From these two types we can already conclude that these two questionnaire forms are parallel indeed, because more respondents than expected show patterns that suggest comparable responses to the two forms. This interpretation is reinforced by the antitype, which is constituted by Configuration 2211. Almost 4 respondents were expected to show this pattern of results in which the two forms of the questionnaire indicate contradictory patterns of state anxiety. However, not a single respondent produced this pattern. We thus conclude that it is unlikely that the two parallel forms yield conflicting response patterns.

9.3.2 Estimation and CFA of polynomial parameters for non-equidistant points on X

In many instances, observation points are not equidistant. For instance, when a client comes to see the consultant Monday and Wednesday every week, the intervals between observations differ. When students are observed at the beginning and the end of each semester, the intervals differ too. In these and all other cases in which observations are non-equidistant, the algorithms described in Section 9.3.1 cannot be used because they do not take the length of the intervals between observations into account. Therefore, we describe in this section methods for the estimation of polynomial parameters for non-equidistant series of measures. The polynomials are also orthogonal, and the parameters can be interpreted in the same way as for the methods in Section 9.3.1 (Krauth, 1980; Krauth & Lienert, 1975; von Eye & Hussy, 1980).

Consider n non-equidistant scores y_j with j = 1, ..., n. The values x_j correspond to the y_j. The parameters a_i of the orthogonal polynomials can then be estimated as

$$a_i = \frac{\sum_j \varphi_i(x_j)\, y_j}{\sum_j \varphi_i^2(x_j)} ,$$

where i indicates the degree of the polynomial for which the parameter is estimated, and $\varphi(x)$ is the orthogonal polynomial. The polynomials $\varphi_i(x)$ can be calculated using the recursive equation

$$\varphi_{i+1}(x) = (2x + \alpha_i)\varphi_i(x) + \beta_{i-1}\varphi_{i-1}(x) ,$$

where x denotes the values on X, and the starting values for the $\varphi_i(x)$ are set to be $\varphi_{-1}(x) = 0$ and $\varphi_0(x) = 0.5$. The values for α_i and β_{i-1} are

$$\alpha_i = \frac{-2\sum_{j=1}^{n} x_j\ \varphi_i^2(x_j)}{\sum_{j=1}^{n} \varphi_i^2(x_j)}$$

and

$$\beta_{i-1} = \frac{-\sum_{j=1}^{n} \varphi_i^2(x_j)}{\sum_{j=1}^{n} \varphi_{i-1}^2(x_j)} .$$

Data example. In the following example, we re-analyze a data set published by von Eye and Hussy (1980). The authors report results from an experiment on the effects of the length of time between a learning activity and recall of nonsense syllables on recall performance. The length of Time was X = {1, 5, 10, 20, 40} minutes. The averaged recall rates were Y = {6.28, 8.50, 8.67, 10.06, 12.94} syllables. We now approximate this series of measures using a second degree orthogonal polynomial. Inserting into the equation for α_0 yields

$$\alpha_0 = \frac{-2(1\cdot 0.5^2 + 5\cdot 0.5^2 + 10\cdot 0.5^2 + 20\cdot 0.5^2 + 40\cdot 0.5^2)}{5\cdot 0.5^2} = -30.40.$$

To create the coefficients of the first degree orthogonal polynomial, we calculate

$$\varphi_1(x) = (2x + \alpha_0)\ \varphi_0(x) + \beta_{-1}\varphi_{-1}(x).$$

The second summand in this expression equals zero, because $\varphi_{-1}(x) = 0$. Inserting α_0 yields the first order polynomial

$$\varphi_1(x) = (2x - 30.40)\cdot 0.5 = x - 15.20.$$

We now can calculate α_1 and β_0 and obtain $\alpha_1 = \frac{-52109.28}{970.80} = -53.68$

and $\beta_0 = \dfrac{-970.80}{5 \cdot 0.25} = -776.64$. We thus obtain for the coefficients of the second degree polynomial

$$\varphi_2(x) = (2x - 53.68)(x - 15.20) - 776.64 \cdot 0.5$$
$$= 2x^2 - 84.08x + 427.62.$$

The estimates for the polynomial parameters are then

$$a_0 = \frac{0.5(6.28 + 8.50 + 8.67 + 10.06 + 12.94)}{5 \cdot 0.25} = 18.58,$$
$$a_1 = [(1 - 15.20) \cdot 6.28 + (5 - 15.20) \cdot 8.5 + (10 - 15.20) \cdot 8.67$$
$$+ (20 - 15.20) \cdot 10.06 + (40 - 15.20) \cdot 12.94]/970.80 = 0.1527 \text{ , and}$$
$$a_2 = 0.00076 .$$

Based on these calculations, the second order approximation polynomial is

$$\varphi_2(x) = 6.69 + 0.21x - 0.0015x^2.$$

Figure 8 displays the raw data (circles) and the first (straight line) and the second order (curved line) approximation polynomials. Obviously, increasing the degree of the approximation polynomial from 1 to 3 did not result in a major improvement. Indeed, the R^2 of the linear, first degree polynomial is 0.944, and the multiple R^2 of the second degree polynomial is 0.954. This represents an improvement of no more than 1%. Readers are invited to increase the degree of the polynomial to 3 and to evaluate the increase in the portion of variance covered.

Two-sample CFA of orthogonal polynomial parameters. The configural analysis of polynomial parameters based on non-equidistant points on X can be conducted in a fashion parallel to the analysis of polynomial parameters based on equidistant points on X. In each case, the parameters are categorized and then analyzed using some form of CFA. The following data example, taken from von Eye and Lienert (1987; cf. Krauth, 1980a; Zerbe, 1979) combines the analysis of polynomial parameters with two-sample CFA. A sample of 20 obese patients (o) was compared with a sample of 13 controls (c) in eight plasma measurements in inorganic phosphate. The plasma samples were taken immediately after an oral glucose application, that is, after 0 min, and then after 30, 60, 90, 120, 180, 240, and 300 min. Each participant's response curve was approximated by a second degree orthogonal polynomial, estimated based on the non-equidistant points on the time axis. Thus, the eight raw phosphate scores were substituted by the three parameter estimates a_0, a_1, and a_2.

For the following CFA, a_0 was dichotomized at the median, thus creating an above median (+) and a below median (-) group. The parameter estimates a_1 and a_2 were dichotomized at 0, thus discriminating between positive (+) and negative (-) slopes and positively (+) and negatively (-) accelerated curvature, respectively. Crossed with the grouping variable, G, the three dichotomized parameters form a 2 x 2 x 2 x 2 table. We now analyze this cross-classification under the base model of a two-sample CFA. The base model is [a_0', a_1', a_2'][G], where the apostrophes indicate

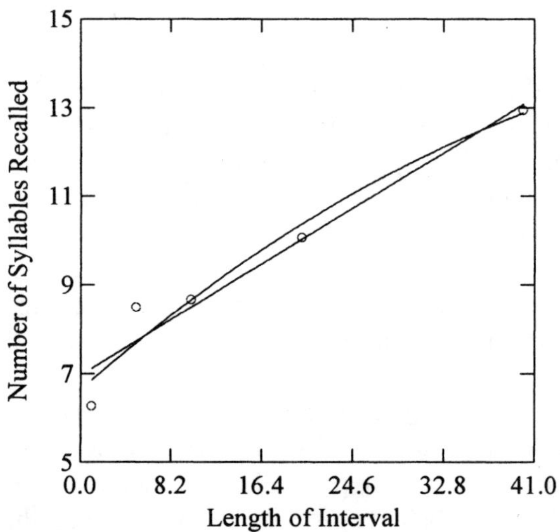

Figure 8: Linear and quadratic approximation polynomials of learning curve

that we use the dichotomized variables. We use the z-approximation of the binomial test and the Bonferroni-adjusted $\alpha^* = 0.00625$. Table 78 summarizes the results.

Table 78: Two-sample CFA of time series of plasma measures

Cell index STVG	Observed frequencies	z	$p(z)$	π^*	Discrimination Type?
+ + + c	1				
+ + + o	3	-.628	.2648	.192	
+ + - c	2				
+ + - o	1	1.014	.1553	.409	
+ - + c	1				
+ - + o	3	-.628	.2648	.192	
+ - - c	0				
+- - o	6	-2.183	.0145	.394	
- + + c	4				
- + + o	0	2.646	.0041	.606	✔
- + - c	4				
- + - o	2	1.511	.0653	.409	
- - + c	1				
- - + o	4	-.964	.1676	.242	
- - - c	0				
- - - o	1	-.819	.2065	.394	

The results in Table 78 suggest that the group of the obese patients and the control group differ only in configuration - + +. This configuration describes the time series of participants who start with below median inorganic phosphate plasma levels. However, these participants display both positive slopes and positively accelerated curvatures. In other words, starting below the median, these respondents augment their inorganic phosphate plasma levels at increasing rates. The group comparison shows that this pattern is observed only in the control group, but not in the group of obese patients.

It is interesting to note, that the complement pattern + - -, which was only observed in the obese patients but not in the controls, did not constitute a discrimination type. This is due to the fact that the frequencies with which this pattern was observed, reflect, in part, the sample size differences. Still, one might suspect that the two complementary patterns - + + and + - - form a biprediction type. Readers are invited to test this hypothesis.

9.4 CFA of series that differ in length; an example of Confirmatory CFA

Series of measures can differ in length for a number of reasons. For example, data can be missing at any position in a series because respondents did not provide answers. In psychotherapy, the number of sessions needed before a patient is considered free of symptoms varies from patient to patient. In problem-solving experiments and in chess, the number of moves needed to solve a problem varies. In this section, we deal with series of measures that (a) have no missing elements, and (b) differ only in length. These typically are series that describe processes that come to a natural end, as in a learning study in which participants reach some criterion after different numbers of learning trials.

Series that differ in length can be approached from various perspectives. For instance, using methods of structural modeling, one can ask whether series differ only in length (Jöreskog & Sörbom, 1993). In the context of CFA, one can ask whether there exist types and antitypes in spite of the differences in length between the series (Lienert & von Eye, 1986).

The characteristics of series that can be considered, are the same as the characteristics of complete series of measures. For instance, one can estimate polynomial parameters up to degree I for series the shortest of which has $I + 1$ scores. Then, the polynomial parameter estimates can be analyzed as shown in Section 9.3, above. These methods will not be repeated in this section. Rather, this section shows, how other characteristics of series of scores than the ones discussed thus far can be analyzed using CFA. All of these characteristics can be used to describe complete and incomplete series. The minimum number of scores in a series is two. Specifically, we consider the three characteristics of series:

(1) *Monotonic trend criterion*. This criterion describes one aspect of the linear trend in the data. Consider a series of length t. This series is said to display a monotonic trend if the inequality $y_{i+1} \geq y_i$ holds for all i = 1, ..., t-1 (weak monotonicity). If this inequality is violated once or more often, that is, if for at least one pair of time-

adjacent measures $y_i > y_{i+1}$, the series is said not to display a monotonic trend.

(2) *Early completion criterion.* When processes are observed until some criterion has been reached, one can expect individuals to differ in the number of steps they need to complete the process. Thus, one can set a threshold and categorize individuals based on this threshold (an example follows below). This criterion can be used even if there is only one response, for example, if a subject solves a problem in the first trial.

(3) *Qualitative criterion.* In addition to the two criteria listed under (1) and (2), one can consider any other criterion. Examples of such criteria include qualitative characteristics of series such as the efficiency of the steps taken to solve a problem, the elegance of a solution, or the availability or use of particular means when solving a problem.

Each of these criteria can be categorized to create variables that enable researchers to employ CFA and to answer questions that can only be answered using CFA.

Data example. The following example, taken from Lienert and von Eye (1986), involves data from a learning experiment. A total of 85 participants (48 males and 37 females) processed a paired-association learning task. Twelve pairs of nouns were presented up to eight times using a memory drum. After each presentation, the stimulus words were presented and subjects were asked to respond with the target word. To complete the task, the respondents had to give eight correct responses. For the following analyses, we create three variables that describe the resulting learning curves which differ in length, and we ask whether female and male participants differ in these variables.

The first variable is the *monotonic trend criterion* (M). A series was assigned a + if the inequality $y_{i+1} \geq y_i$ holds for all $i = 1, ..., 7$. If for at least one pair of time-adjacent measures $y_i > y_{i+1}$, a - was assigned. The second variable is the *early success criterion* (S). A + was assigned if a subject reached the criterion before the eighth trial, and a - was assigned if a subject needed all eight trials. The third variable is the *number-of-errors criterion* (F). The number of wrong associations was counted in addition to the number of hits. A + was assigned if a subject produced more errors than the grand median, and a - was assigned if a subject produced fewer errors. Table 79 displays the (2 x 2 x 2) x 2 cross-classification of M, S, and F, with Gender, G.

Instead of performing a standard two-sample CFA, we now employ a prediction test as presented for biprediction CFA in Section 6.2.2.2. Specifically, we compare females with males in configuration - - - of the three variables M, S, and F. The test is

Table 79: Cross-classification of the monotonic trend (M), early success (S), and number of mistakes (F) in two samples of males and females

Configuration	Comparison groups		Totals
MSF	males	females	
+ + +	12	12	24
+ + -	2	3	5
+ - +	3	2	5
+ - -	6	6	12
- + +	5	6	11
- + -	3	2	5
- - +	2	2	4
- - -	15	4	19
Totals	48	37	85

$$X^2 = \frac{N(ad - bc)^2}{ABCD}.$$

Inserting yields

$$X^2 = \frac{85(15 \cdot 33 - 4 \cdot 33)^2}{19 \cdot 66 \cdot 48 \cdot 37} = 5.029.$$

For $df = 1$, this value has a tail probability of $p = 0.0249$. Thus, we can reject the null hypothesis, according to which configuration - - - does not allow one to discriminate between males and females. Note that α does not need to be adjusted, because we performed only one test.

In contrast to routine *exploratory* CFA, testing only a subset of configurations is part of *confirmatory* or *explanatory* CFA. In the example in Table 79 we only asked whether males and females differ in regard to the pattern non-monotonic slope - no early success - above median number of errors. This hypothesis was largely fueled by an inspection of the frequencies in Table 79. In substantive applications, theory and prior results are needed to justify the selection of configurations for confirmatory analysis. The main advantage of confirmatory CFA is that the number of tests is smaller than in exploratory CFA. The protection of the family-wise or experiment-wise α only needs to take into account this smaller number. Thus, the α* that results in confirmatory CFA can be far less prohibitive than the α* in exploratory CFA. The next section presents additional examples of confirmatory applications of CFA.

9.5 Examining treatment effects using CFA; more confirmatory CFA

This section presents methods for a rather detailed configural examination of treatment effects. These methods are presented for pre-post designs without control group in Section 9.5.1 and with control group in Section 9.5.2.

9.5.1 Treatment effects in pre-post designs (no control group)

In evaluative and experimental research researchers typically pursue specific, a priori formulated hypotheses. Data are examined in regard to these hypotheses. The analyses involve data exploration only in a secondary step, if at all.

In this section, we exemplify application of confirmatory CFA in an evaluation study. Lienert and Straube (1980) treated a sample of 75 acute schizophrenics with neuroleptic drugs for two weeks. Before and after this treatment, the patients were administered the Brief Psychiatric Rating Scale (Overall & Gorham, 1962). Three of the seventeen symptoms captured by this instrument are used for the following analyses: W = emotional withdrawal; T = thought disturbances; and H = hallucinations.

Each of the symptoms was scaled as either present (+) or absent (-). Table 80 displays the data.

Table 80: Evaluation of treatment of schizophrenics with neuroleptic drugs in a pre-post study

Number of symptoms before treatment	Configurations	Number of symptoms after treatment				Totals
		3	2	1	0	
	WTH	+ + +	+ + - + - + - + +	+ - - - + - - - +	- - -	
3	+ + +	1	10	4	0	15
2	+ + - + - + - + +	6	11	17	4	38
1	+ - - - + - - - +	1	4	7	4	16
0	- - -	0	1	2	3	6
	Totals	8	26	30	11	75

We now ask whether the number of patients who display fewer symptoms after the treatment is greater than the number of patients with more symptoms. Table 80 has been arranged such that a count that leads to an answer can easily be performed. Instead of the usual arrangement of configurations in which all permutations are created using a routine scheme in which the last variable is the fastest changing one, the second last variable is the one changing next, and so on, the arrangement in Table 80 groups configurations based on the number of + signs. That is, configurations are grouped based on the number of symptoms displayed by the patient. Looking at the rows, the top configuration includes the patients who suffer from all three symptoms (Row 1). Then come three configurations with two symptoms. These three configurations are considered one category, the category of two symptoms. The following three configurations are also considered one category, the one with one symptom. The last category includes the patients who show none of the three symptoms under study. All this applies accordingly to the columns in Table 80.

The patients who suffer from fewer symptoms after the treatment can be found in the upper right triangle of Table 80, excluding the diagonal. For example, the 10 patients in the second cell in Row 1 are those who suffered from all three symptoms before the treatment and from only two symptoms after the treatment. The first row also indicates that no patient was freed from all three symptoms. The total number of patients freed from one or two symptoms is 10 + 4 + 0 + 17 + 4 + 4 = 39. No patient was freed from all three symptoms.

The patients who suffer from more symptoms after the treatment than before can be found in the lower left triangle of the cross-classification in Table 80, again excluding the diagonal. For example, the table shows that one patient suffered from only one symptom before the treatment but from all three symptoms after the treatment (Row 3, Column 1). The total of patients with an increase in the number of symptoms is 6 + 1 + 4 + 0 + 1 + 2 = 14.

To compare these two frequencies, the one that indicates the number of improved patients and the one that indicates the number of deteriorated patients, we posit as the null hypothesis that there is no difference. That is, discrepancies between these two frequencies are random in nature. There is a number of tests that can be used to test this null hypothesis. Examples include the binomial test given in Section 3.2 and its normal approximations, given in Section 3.3; symmetry tests (see below); and the *diagonal-half sign test.* For the latter, let b denote the number of patients who improved, and w the number of patients who disimproved. Then, the null hypothesis of no difference between b and w can be tested using

$$z = \frac{b - w}{\sqrt{b + w}}.$$

The test statistic is approximately normally distributed. Alternatively, in particular when the samples are small, the binomial test can be used with $p = 0.5$.

To illustrate these two tests we use the data in Table 80. We insert in the z-test formula and obtain

$$z = \frac{39 - 14}{\sqrt{39 + 14}} = 3.434 \ ,$$

and $p = 0.0003$. We thus conclude that the neuroleptic drugs reduce the number of symptoms in schizophrenic inpatients. The same probability results from the normal approximation of the binomial test.

More detailed hypotheses can be tested by focusing on individual symptoms. Two methods of analysis are suggested. First, one can create a pre-intervention x post-intervention cross-tabulation for each symptom and analyze the resulting $I \times I$ table using the Bowker test (1948; cf. von Eye & Spiel, 1996), where I indicates the number of categories, or the McNemar test (1947), when $I = 2$. The test statistic for both tests is

$$X^2 = \sum_i \sum_j \frac{(N_{ij} - N_{ji})^2}{N_{ij} + N_{ji}} \ ,$$

for $i > j$ and $i, j = 1, ..., I$. This test statistic is approximately distributed as χ^2 with $df = \binom{I}{2}$. For $I = 2$, this equation simplifies to

$$X^2 = \frac{(b - w)^2}{b + w} \ ,$$

with $df = 1$ or, with continuity correction,

$$X^2 = \frac{(|b - w| - 0.5)^2}{b + w} \ ,$$

also with $df = 1$, where b and w denote the cell frequencies N_{12} and N_{21}, respectively.

Consider the following example. The cell frequencies for the symptom hallucinations in the neuroleptic drug treatment study are + + = 8, + - = 21, - + = 9, and - - = 32. For these values we calculate

$$X^2 = \frac{(21 - 9)^2}{21 + 9} = 4.80.$$

For $df = 1$, the tail probability of this value is $p = 0.0285$. We thus can reject the null hypothesis that the neuroleptic drug treatment only leads to random changes in hallucinations.

9.5.2 Treatment effects in control group designs

Control groups are often considered an indispensable necessity in research on treatment effects. Control groups allow researchers to distinguish between spontaneous recovery or spontaneous changes on the one hand and treatment effects on the other hand. CFA allows one to compare experimental groups and control groups with two-sample CFA (see Sections 7.1 and 7.2). When there are more than two groups, multi-sample CFA can be employed (see Section 7.3).

In this section, we show how two samples can be compared in regard to the change from one configuration to another. Consider the following scenario. Pattern A is observed before treatment. Pattern B is the desired pattern, and is observed after the treatment. Both observations are made both in the treatment and the control groups. Then, the two groups can be compared in regard to the change from Pattern A to Pattern B based on the 2 x 2 tabulation that is schematized in Table 81.

Table 81: 2 x 2 table for the comparison of two groups in one pattern shift

Patterns	Comparison groups		Totals
	Treatment	Control	
A/B	b	b'	$N_{A/B}$
all others combined	$a + c + d$	$a' + c' + d'$	$n + n' - N_{A/B}$
Totals	n	n'	$n + n'$

The middle columns in Table 81 separate the treatment and the control groups. The frequencies of the treatment group can be considered taken from a 2 x 2 Table of the format given in Table 82. The frequencies of the control group can be considered taken from an analogous 2 x 2 table.

Frequency b in Table 82 is the number of treatment group cases who switched from symptom Pattern A to symptom Pattern B. The remaining three cells contain cases who stayed stable or switched from Pattern B to Pattern A. The cell labels in Table 81 indicate that the same frequencies are used as in Table 82. Thus, cell frequency b in Table 81 is the same as cell frequency b in Table 82. This applies accordingly to the control group, for which a cross-classification parallel to the one in Table 82 can be constructed. The frequencies in Table 81 can be analyzed using the methods described in Sections 7.1 (Table 47) and 7.2.

Table 82: 2 x 2 table of pattern change in treatment group

Patterns Pre-treatment	Post-treatment		Totals
	A	B	
A	a	b	$a + b$
B	c	d	$c + d$
Totals	$a + c$	$b + d$	n

Data example. The number of respondents in Lienert and Straube's (1980) investigation on the effects of neuroleptic drugs who switched from Pattern + + + to Pattern + + - was $b = 9$. The frequency $a + c + d$ is then 66. Now suppose that in a control group of size 54 only 2 patients showed pattern + + +/+ + -. From these frequencies, the cross-classification in Table 83 can be created.

Table 83: Two-sample comparison with respect to change pattern + + +/+ + -

Patterns	Comparison groups		Totals
	Treatment	Control	
+ + +/+ + -	$b = 9$	$b' = 2$	$N_{+++/++-} = 11$
all others combined	$a + c + d =$ 66	$a' + c' + d' =$ 52	$n + n' - N_{+++/++-} =$ 118
Totals	$n = 75$	$n' = 54$	$n + n' = 129$

Using the exact Fisher test described in Section 7.1, we calculate a probability of $p = 0.086$. Using the X^2-test without continuity correction, we calculate $X^2 = 2.77$ and $p = 0.096$ ($df = 1$). The conclusion made in Section 9.5.1, that is, the conclusion that the neuroleptic drugs improve hallucination problems in schizophrenics, must thus be qualified. While there is a significant improvement in units of the number of hallucinations from the first to the second observation, this improvement cannot be considered caused by the drug treatment. The control group patients experience improvements that are not significantly different than those experienced by the patients in the treatment group. This result again illustrates that the use of control groups can prevent researchers from drawing wrong conclusions.

9.6 CFA of patterns of correlation or multivariate distance sequences

Thus far, we have covered CFA of the following characteristics of series of measures:

(1) slope, curvature and higher order characteristics of series in the forms of differences and polynomial parameters;
(2) location/elevation in the form of means of ipsative scores relative to some reference;
(3) variability of series of measures as assessed by von Neumann's variance.

A fourth characteristic of series of measures is their autocorrelation structure. Repeated observations typically are strongly correlated with each other (autocorrelation). It can be of interest to researchers to identify types and antitypes of autocorrelations. Changes in the correlational structure can be as interesting and important as changes in the mean or slope characteristics. A fifth characteristic of series of measures can be captured by multivariate distances. In Section 9.1, we only considered univariate distances in the form of first, second, and higher order differences. Multivariate distances reflect differences between vectors of measures. This section is concerned with CFA of autocorrelations and multivariate distances.

9.6.1 CFA of autocorrelations

Consider the data box (Cattell, 1988) in Figure 9. This box describes the data that are collected from a number of individuals in a number of variables on a number of occasions. The $r_{1.12}$ and $r_{1.23}$ on the right-hand side of the box are correlations. $r_{1.12}$ indicates that, at the first occasion (first subscript), Variables 1 and 2 (last two subscripts) are correlated using all subjects (period in the subscript). $r_{1.23}$ indicates that, at the first occasion (first subscript), Variables 2 and 3 (last two subscripts) are correlated using all subjects (period in the subscript). Using all three occasions, for instance, the correlations $r_{1.12}$, $r_{1.13}$, $r_{2.12}$, $r_{2.13}$, $r_{3.12}$, and $r_{3.13}$ can be estimated.

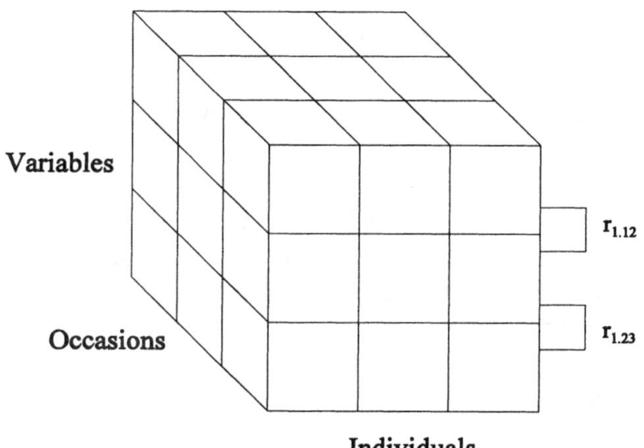

Figure 9: Cattell's data box

In general, six correlation matrices can be created from a data box as the one depicted in Figure 9. Each of these correlation matrices corresponds to one of the six elementary factor analytic techniques described by Cattell (1988). The first correlation matrix is of the individuals x variables type. The factor analytic R technique is used to extract factors of variables from this matrix. The second matrix is of the variables x individuals type, yielding factors of people (Q technique). The third matrix, occasions x variables, uses the P technique to create factors of variables. The fourth matrix, variables x occasions, yields factors of occasions (O technique). The fifth matrix, occasions x individuals, yields factors of people (S technique), and the sixth matrix, individuals x occasions, yields occasions factors (T technique).

Each of these matrices can also be subjected to a CFA. The matrices that contain correlations that vary across occasions are the most interesting ones in the present context of methods of longitudinal CFA. Which of these is selected for a particular analysis is determined by the researchers' research topic. None of the options is a priori superior.

CFA of such a correlation matrix proceeds in the following steps:

(4) Creating of the correlation matrices of interest, e.g., the individuals x variables matrix, separately for each occasion;
(5) Categorizing correlations;
(6) Creating cross-classification of the categorized correlations;
(7) Performing CFA.

It should be mentioned that a very large number of correlation measures has been proposed. Correlations can be calculated between categorical measures, continuous measures, or measures that differ in scale level. Any of these measures can be used for CFA of autocorrelations.

Data example. The following data example, taken from von Eye (1990), illustrates these four steps. A sample of 148 individuals participated in a study on fatigue and mood changes caused by a memory experiment. In the experiment, subjects had to read and recall narratives. Immediately before and after the experiment, subjects were presented with a questionnaire that measured anxiety, arousal, and fatigue. The subjects went through two of these routines, thus filling the questionnaire a total of four times.

In the first step, time-adjacent vectors of scores were correlated separately for each individual. The 3 x 4 matrix of raw scores for each subject was thus transformed into a vector of three correlations. These correlations compare the first with the second, the second with the third, and the third with the fourth responses to the questionnaire.

In the second step, these correlations were categorized. The distribution was bimodal with one mode at around $r = -0.80$ and the other mode at around $r = 0.99$. There were more positive than negative correlations. The median was located at $r = 0.9$. Still, the cutoff was chosen to be at $r = 0.5$. This value identifies the minimum of the frequencies between the two modes. Correlations above the mode were assigned a 1, correlations below the mode were assigned a 2.

In Step 3, the three dichotomized correlations were crossed to form a 2 x 2 x 2 tabulation. This tabulation appears in Table 84, along with the results of CFA. We used Lehmacher's test with Küchenhoff's continuity correction, and Holm's procedure to protect α which led to $\alpha_1^* = 0.00625$.

Table 84: CFA of correlations between four observations of fatigue and mood

Corr.	Frequencies		Test statistics		Holm procedure		
$r_{12}\,r_{23}\,r_{34}$	obs.	exp.	z	p	Rank (p)	critical p	Type ?
111	65	56.41	2.61	.005	1	.006	**T**
112	12	13.16	-0.27	.393	6	.017	
121	31	38.46	-2.34	.010	3	.008	
122	9	8.97	-0.21	.418	7	.025	
211	8	14.95	-2.55	.005	2	.007	**A**
212	3	3.49	0.01	.497	8	.05	
221	16	10.19	2.23	.013	4	.01	
222	4	2.38	0.79	.213	5	.0125	

The results in Table 84 suggest that one type and one antitype exist. The type, constituted by Pattern 111, describes those subjects who have above cutoff correlations throughout. Thus, the strength of the autocorrelation of these subjects' mood and fatigue scores does not seem to be affected by the two experimental trials. The antitype is constituted by Pattern 211. These are subjects who display a low or negative correlation between the mood and fatigue scores observed before and after the first trial. The correlations between the measures after the first and before the second trial are above the cutoff, and so are the correlation between the measures before and after the second trial.

9.6.2 CFA of autodistances

It is well known that distances and correlations are independent of each other. Correlations can be high or low regardless of distance and vice versa. Therefore, researchers often consider both correlations and distances in their analyses rather than only one of the measures. In this section, we call the multivariate distances between time-adjacent observations *autodistances.* This term can be viewed parallel to the term *autocorrelations*.

Many measures of distance have been proposed. The best known is the Euclidean distance

$$s = \sqrt{\sum_i (y_{i+1,j} - y_{i,j})^2},$$

where i indexes the observations and j indexes the variables (or other units of analysis). The Euclidean distance and many other measures of distance can be derived from the Minkowski metric

$$d_r = \left[\sum_j |y_{i+1,j} - y_{i,j}|^r\right]^{1/r}.$$

For instance, setting $r = 2$ yields the Euclidean distance, and $r = 1$ yields the city block metric. (Here, r is a distance parameter, not a correlation.)

It is important to take into account that the Euclidean distance uses raw scores. Thus, if scales are not commensurable (same scale units), there may be a weighting such that the scales with large numbers dominate the distance measurement to the extent that the scale with the smaller numbers become irrelevant. Before using distances, researchers are therefore advised to make sure their scales are commensurable.

CFA of autodistances proceeds in the same four steps as CFA of autocorrelations:

(1) Creating the distance matrices of interest, for example, the individuals x variables matrix, separately for each occasion;
(2) Categorizing distances;
(3) Creating the cross-classification of the categorized distances;
(4) Performing CFA.

Data example. To illustrate that CFA of autocorrelations and CFA of autodistances can yield different patterns of types and antitypes, we use the same data as in Section 9.6.1. The data were collected in a memory experiment in which 148 subjects read and recalled narratives in two trials. Before and after each trial, the subjects provided information on mood and fatigue.

For the following CFA, the distances between the mood and fatigue scores adjacent in time were calculated. The dichotomized variables were scored as a 1 when their raw scores increased and a 2 when their raw scores decreased. The cross-classification of the three dichotomized distances appears in Table 85, along with the results of CFA. To make results comparable with those in Section 9.6.1, we used Lehmacher's test with Küchenhoff's continuity correction and Holm's adjustment of α which led to $\alpha_1^* = 0.00625$.

Table 85: CFA of distances between four observations of fatigue and mood

Distance	Frequencies		Test statistics		Holm procedure		Type ?
$s_{12}\ s_{23}\ s_{34}$	obs.	exp.	z	p	Rank (p)	critical p	
111	17	26.25	-2.674	.0037	4	.01	A
112	18	17.40	0.033	.4867	7	.025	
121	38	24.87	3.905	< α*	2	.007	T
122	12	16.49	-1.357	.0874	5	.013	
211	16	19.46	-0.965	.1673	6	.017	
212	25	12.90	4.228	< α*	1	.006	T
221	18	18.43	0.023	.4908	8	.05	
222	4	12.22	-2.857	.0021	3	.008	A

[a] < α* indicates that the tail probability is smaller than can be expressed with four decimal places.

Table 85 suggests the existence of two types and two antitypes. The first type is constituted by Pattern 121. It describes those subjects whose mood and fatigue scores increased during the trials and decreased between the trials, indicating change toward better mood and less fatigue between the trials and change toward worse mood and more fatigue during the experiment (fatigue had been reverse scored). This pattern not only defined a type, it also was the most frequently observed pattern. The second type, constituted by Pattern 212, describes those subjects whose mood and fatigue scores decreased during the trials and increased between the trials, indicating a change toward better mood and less fatigue during the trials and toward worse mood and more fatigue between the trials.

The first antitype is constituted by Pattern 111. These are the subjects whose mood and fatigue scores increased between each assessment, indicating change toward better mood and less fatigue both during and between the trials. The second antitype, constituted by Pattern 222, describes subjects with just the opposite change pattern as the first antitype. These are the four subjects whose mood and fatigue scores decreased between each assessment, indicating change toward worse mood and more fatigue both during and between the trials. Both of these implausible patterns were observed significantly less often than expected from the base model of variable independence.

9.7 Unidimensional CFA

Thus far, CFA has been presented as a method of analysis of multivariate categorical data. However, there are instances in which univariate arrays are of interest. CFA can also be used to analyze univariate data. Consider the case in which Variable A is subjected to a configural analysis. The number of base models that can be considered in this situation is minimal. The first model that can be implemented is the base model of zero order CFA, $\log E = \lambda_0$. This model allows researchers to answer the question whether and where the observed frequency distribution differs from an expected uniform distribution (see Section 5.1). The next higher model, the main effect model for variable A, is already the saturated model, $\log E = \lambda_0 + \lambda_i^A$. The only additional option involves using a predictor in form of a covariate in the zero order CFA base model (for more detail on covariates see Section 10.5). The base model would then be

$\log E = \lambda_0 + \gamma(x)$, where γ is the parameter for predictor X.

The number of predictors that can be used depends on (a) the number of degrees of freedom needed for the predictor, and (b) the number of available degrees of freedom. For a variable with c categories, the zero order base model leaves c - 1 degrees of freedom. The model is saturated if all of these are used. Therefore, there are c - 2 degrees of freedom that can be invested in predictors.

Data example. The following example uses data published by von Eye, Indurkhya, and Kreppner (2000). The data describe results from a study on family development (Kreppner, 1989; Kreppner, Paulsen, & Schütze, 1982). The variable *Topic of Interaction in a Dyad* was observed in families that just had had a second child. Of the ten categories of this variable, we use here the following four: 0 = interaction partners pursued different topics; 1 = physical closeness; 2 = nursing; and 3 = family/development/education.

We now analyze the frequencies with which these interaction categories were observed in one particular family, named BLEI, at the end of the two-year observation period. We consider two base models. The first is the base model of zero order CFA. The second is the base model with the predictor weights given in the following equation.

$$\log \begin{bmatrix} m_0 \\ m_1 \\ m_2 \\ m_3 \end{bmatrix} = \begin{bmatrix} 1 \\ 1 \\ 1 \\ 1 \end{bmatrix} \lambda_0 + \begin{bmatrix} 1 \\ 5 \\ 6 \\ 7 \end{bmatrix} \gamma .$$

This equation includes two parameters, λ_0 and γ. The former is the well known constant in log-linear models. The second is the parameter for the predictor weights used in this analysis. These weights suggest that the individual categories if the interaction variable are anticipated to occur at rates that differ from a uniform distribution, and as indicated in the weight vector.

It is important to derive these weights from prior results or from theory, because results vary greatly depending on the selection of weights. In the present example, the low weight is justified from the assumption that as the newborn grows older, he/she becomes increasingly able to communicate. Therefore, the number of situations in which the members in a dyad do not really communicate will be smaller than the number of situations in which there is a shared topic of communication. (For more information on the development of these four behavior categories see Table 87 in Section 9.8.) Table 86 displays the results of the standard zero order CFA in its top panel, and the results of zero order CFA enriched with predictive weights in its bottom panel. For both analyses, we used the X^2-component test and the Bonferroni-adjusted α* = 0.0125.

Table 86: Zero order, univariate CFA

	Topic of communication			
	0	1	2	3
	Standard zero order CFA			
observed	71	154	333	403
expected	240.25	240.25	240.25	240.25
X^2	119.23	30.96	35.81	110.25
p	< α*	< α*	< α*	< α*
Type/Antitype ?	**A**	**A**	**T**	**T**
	Zero order CFA with predictive weights			
observed	71	154	333	403
expected	58.95	211.29	290.73	400.03
X^2	2.46	15.53	6.15	0.02
p	.1165	< α*	.0137	.8820
Type/Antitype ?		**A**		

[a] < α* indicates that the tail probability is smaller than can be expressed with four decimal places.

The goodness-of-fit for the standard zero order CFA base model was LR-X^2 = 324.26 (df = 3; p < 0.01), and for the zero order CFA with predictive weights it was $LR\text{-}X^2 = 25.37$ ($df = 2$; $p < 0.01$). Obviously, including the predictor improves model fit dramatically ($\Delta X^2 = 298.89$; $df = 1, p < 0.01$). Still, we can expect types and antitypes to emerge from both approaches. The results in the top panel of Table 86 suggest that each configuration constitutes a type or antitype. The results in the bottom panel suggest that only behavior Category 1, physical closeness, appears significantly less often than expected if the predictive weights are taken into account, and thus constitutes an antitype. The other three behavior categories do not deviate significantly from their expected frequencies any longer.

10.3 Within-individual CFA

In the center of the focus of person-oriented research (Bergman & Magnusson, 1997; see Section 1.2) lies the analysis of the individual. Individuals can be compared or aggregated only if single individuals are validly described. To be able to describe an individual using CFA, a cross-classification (or a count on a single variable; see Section 10.2) must be created. Thus, a repeated observation study must be conducted that yields a number of observations per individual that is large enough for configural analysis.

This section presents a CFA application in which a dyad is considered the unit of analysis (von Eye et al., 2000). Specifically, we ask how a dyad develops over time in one variable. The data used are from Kreppner's family development study again (see Section 10.2), in which families were observed beginning right after the birth of their second child. We study again the variable *Topic of Interaction in a Dyad*. In Kreppner's study, families were observed seven times after the birth of their second child. The first four observations took place during the first year in the life of the second child, and the last three observations in the second year.

For the following CFA, we select family BLEI for the within-individual analyses. The family serves as the individual. We cross the four categories of the interaction variable with the seven observations, thus creating a 7 x 4 contingency table. This table describes, how often each of the four behavior categories was observed on each of the seven occasions. The counts appear in Table 87.

Table 87 shows that the distribution of Topic of Interaction in Dyads in Family BLEI clearly undergoes change over time. Specifically, it seems that the frequencies of Category 0 (no shared topic) decrease rapidly. Only at the sixth observation, this category seems to re-appear. The frequencies of Category 1 (physical closeness) show a slight increase at the beginning, and stay low over the rest of the two-year observation period. A similar pattern can be seen for Category 2 (nursing). In contrast, Category 3 (topics concerning the family, development and education) seems to increase. In the following paragraphs we ask whether these trends manifest in CFA types and antitypes.

We perform a first order CFA using Lehmacher's (1981) test and the Bonferroni-adjusted $\alpha^* = 0.0018$. The log-linear base model is $\log E = \lambda_0 + \lambda_i^{Time} + \lambda_j^{Topic}$. Table 88 displays the results of the analysis.

Table 87: Time x Topic of Interaction in a dyad cross-classification; family BLEI

		Topic of Interaction in Dyads				Total
		0	1	2	3	
Observation Point	1	11	3	50	91	155
	2	12	10	64	75	161
	3	3	4	46	107	160
	4	0	2	49	106	157
	5	0	5	48	101	154
	6	9	4	39	108	160
	7	0	5	24	101	130
	Total	35	33	320	689	1077

The results in Table 88 suggest that there exist four types and two antitypes. The first two types, constituted by Configurations 10 and 11, indicate that lack of shared topic was observed more often than expected from chance in the first half year of observation. Later, this pattern does not occur more often than expected from chance. The third type, constituted by Configuration 22, suggests that nursing also occurs more often than

Table 88: First Order CFA of the cross-classification of Time (T) x Topic of Interaction (I) in dyads in family BLEI

Cells	Frequencies		Test statistics		Type/ Antitype ?
T I	observed	expected	*z*	*p*(*z*)	
10	11	5.037	2.918	.0017	**T**
11	3	4.749	-0.881	.1892	
12	50	46.054	0.749	.0701	
13	91	99.160	-1.475	.2268	
20	12	5.232	3.260	.0006	**T**
21	10	4.933	2.511	.0060	
22	64	47.837	3.021	.0013	**T**
23	75	102.998	-4.982	< α*	**A**
30	3	5.200	-1.062	.1440	
31	4	4.903	-0.448	.3269	
32	46	47.539	-0.288	.3865	
33	107	102.358	0.828	.2038	
40	0	5.102	-2.483	.0065	
41	2	4.811	-1.408	.0796	
42	49	46.648	0.444	.3285	
43	106	100.439	1.000	.1587	
50	0	5.005	-2.456	.0070	
51	5	4.719	0.142	.4435	
52	48	45.757	0.427	.3347	

/ cont.

Table 88, Panel 2/2

Cells	Frequencies		Test statistics		Type/ Antitype ?
T I	observed	expected	z	$p(z)$	
53	101	98.520	0.449	.3266	
60	9	5.200	1.835	.0332	
61	4	4.903	-0.448	.3269	
62	39	47.539	-1.600	.0548	
63	108	102.358	1.006	.1571	
70	0	4.225	-2.227	.0130	
71	5	3.983	0.552	.2906	
72	24	38.626	-2.992	.0014	**A**
73	101	83.166	3.473	.0003	**T**

[a] $< \alpha^*$ indicates that the tail probability is smaller than can be expressed with four decimal places.

expected at the second wave of observations. The fourth type, constituted by Configuration 73, indicates that interaction topics concerning family/development/education appear more often toward the end of the second year of the new child. Before, this topic was less prominent, as is indicated by the antitype 23. The second antitype, 72, suggests that nursing also appears less often than expected based on chance toward the end of the second year. For a comparison of Family BLEI with the entire sample, see von Eye et al. (2000).

Part IV: The CFA Specialty File and Alternative Approaches to CFA

In the following part of this volume on CFA, we present applications that are unique and new in the sense that none of the existing texts on CFA has covered any of them. Most of these applications were developed and proposed between 1995 and 2002. Some of the topics to be covered here have not been published before. Only a few go back to earlier discussions. Each of the applications covered here allows one to answer specific questions, or approaches the goal of identifying types and antitypes from a particular perspective. All together, the topics covered in this part of this book contribute to the development of CFA as a multifaceted, flexible method that allows one to answer virtually all questions of importance in research from the Person Perspective (see Section 1.2). The new facets of CFA will be presented in two groups. The first includes new facets of the approach to CFA described thus far in this book, that is, frequentist CFA. In particular, this group covers the following topics: (1) structural zeros in CFA; (2) the parsimony of CFA base models; (3) CFA of groups of cells; (4) CFA and the exploration of causality; (5) covariates in CFA; (6) CFA for ordinal variables; (7) graphical display of CFA results; and (8) the aggregation of types or antitypes. In addition, there is a section on the use of CFA in tandem with cluster analysis and discriminant analysis. The second group, presented in Chapter 11, includes conceptually new approaches to CFA, both from the frequentist and the Bayesian domains, that is, Bayesian CFA and Victor and Kieser's approach to CFA.

10. More facets of CFA

In this chapter, we present methods that enrich the arsenal of CFA. In addition, we present applications that illustrate the flexibility of CFA.

10.1 CFA of cross-classifications with structural zeros

In many cross-classifications, in particular when there are many cells and the sample is relatively small, there are cells with zero counts. An example of a contingency table with zeros can be found in Table 77, in which Configurations 1112, 1122, 1221, 2122, and 2211 were not observed at all. Each of these configurations had a true probability greater than zero, that is, it could have been observed under different conditions, for example, had the sample been large enough. The zero frequencies in this kind of case are called *sampling zeros*.

However, there are instances in which configurations are theoretically impossible. The zero frequencies in such cells are called *structural zeros* (also called *structural voids*; Wickens, 1989), and tables that contain structural zeros are called *incomplete tables*. Consider the following example. In a study on cancer, one of the stratification variables is the gender of patients. In this study, the zero for the Configuration "female and prostate cancer" is a structural zero. Note that sampling decisions and the attempt to fit a model to part of a table are also reasons to declare cell frequencies structural zeros.

The reason for the distinction between sampling zeros and structural zeros is that sampling zeros are possible counts for events that have greater than zero probabilities. As such, the zeros do make a contribution to the model fitting process and, of importance in CFA, to the process of estimating expected cell frequencies. In contrast, structural zeros do not make a contribution to this process, because the expected frequency for a cell with a structural zero is zero too.

When estimating the expected cell frequencies for incomplete tables, it is important not to assign expected frequencies to cells with structural zeros. In CFA, it is also important not to include configurations with structural zeros in the search for types and antitypes and the process of protecting α.

Quasi-independence log-linear models can be used to estimate frequencies for incomplete tables. These models have the same form as the models used thus far for CFA. However, they contain additional terms that prevent the estimates for structural zero cells from being different than zero. Consider, for example, the case of the $I \times J$ cross-classification of the two variables A and B. The standard log-linear model of independence of A and B is

$$\log E_{ij} = \lambda_0 + \lambda_i^A + \lambda_j^B .$$

Now suppose that this table contains one structural zero. Then, the log-linear model of quasi-independence is

$$\log E_{ij} = \lambda_0 + \lambda_i^A + \lambda_j^B + \delta I ,$$

where δ is a parameter and I is an indicator variable, comparable to a column vector in the design matrix X, that indicates the location of the structural zero. The number of terms for structural zeros can be increased when there is more than one structural zero. Typically, the number of sampling zeros is much larger than the number of structural zeros. Once the expected cell frequencies are estimated using a log-linear quasi-independence base model, CFA proceeds as usual, excluding, however, the configurations with structural zeros.

Data example. The following example involves a re-analysis of data published by M. Riley, Cohn, Toby, and J. Riley (1954; cf. Feger, 1994). The authors discuss a "Dyad Scale of Intimacy" that consists of the three items Person X communicates with Person Y (Item A), Person X actually associates with Person Y (Item B), and Person X wishes to associate with

Person Y (Item C). The items were scaled as 1 if a respondent did not endorse the statement with respect to a particular target person Y, and as 2 if the respondent did endorse the item. A sample of 2673 respondents were administered the scale. Table 89 displays the 2 x 2 x 2 cross-classification of the three items, and the results from two approaches to CFA. The first approach is standard first order CFA, that is, CFA with the base model $\log E_{ijk} = \lambda_0 + \lambda_i^A + \lambda_j^B + \lambda_k^C$. Results from this analysis appear in the top panel of the table.

The second approach considers the structural zero in this table. Riley et al. (1954) had proposed that intimacy takes place only when individuals know each other. Therefore, Pattern 111 does not need to be considered, because it describes relationships of distance. Therefore, Riley et al. did not use this pattern in their analyses and we can declare it a structural zero. Thus, the log-linear CFA base model for the present data is $\log E_{ijk} = \lambda_0 + \lambda_i^A + \lambda_j^B + \lambda_k^C + \delta\, I$. More specifically, the log-linear base model for this CFA is

$$\log \begin{bmatrix} m_{111} \\ m_{112} \\ m_{121} \\ m_{122} \\ m_{211} \\ m_{212} \\ m_{221} \\ m_{222} \end{bmatrix} = \begin{bmatrix} 1 & 1 & 1 & 1 \\ 1 & 1 & 1 & -1 \\ 1 & 1 & -1 & 1 \\ 1 & 1 & -1 & -1 \\ 1 & -1 & 1 & 1 \\ 1 & -1 & 1 & -1 \\ 1 & -1 & -1 & 1 \\ 1 & -1 & -1 & -1 \end{bmatrix} \begin{bmatrix} \lambda_0 \\ \lambda^A \\ \lambda^B \\ \lambda^C \end{bmatrix} + \delta \begin{bmatrix} 1 \\ 0 \\ 0 \\ 0 \\ 0 \\ 0 \\ 0 \\ 0 \end{bmatrix}.$$

The design matrix and the λ parameter vector in this base model are known from standard first order CFA (Section 5.2). New to CFA with structural zeros is the δ parameter and the vector I that indicates which cell contains the structural zero. This vector contains a 1 for the cell with the structural zero, and zeros in all other cells. Each structural zero is specified by a δ parameter and a vector I. Results from CFA that takes into account the structural zero in Cell 111 appears in the bottom part of Table 89. For both analyses, we used the z-test and Bonferroni-adjusted α wich yielded $\alpha^* = 0.00625$.

Table 89: CFA of Riley's data without and with structural zero

Cell index	Frequencies		Test statistics		Type/ Antitype
ABC	observed	expected	*z*	*p*	?
First order CFA without consideration of structural zero in Cell 111					
111	0	134.885	-11.614	< α*	**A**
112	1019	693.962	12.339	< α*	**T**
121	290	163.089	9.938	< α*	**T**
122	522	839.064	-10.946	< α*	**A**
211	93	62.028	3.933	< α*	**T**
212	98	319.124	-12.378	< α*	**A**
221	52	74.998	-2.656	.0040	**A**
222	599	385.850	10.851	< α*	**T**
First order CFA under consideration of structural zero in Cell 111					
111	0	0	-	-	-
112	1019	791.187	8.099	< α*	**T**
121	290	240.363	3.202	< α*	**T**
122	522	799.450	-9.813	< α*	**A**
211	93	96.813	-0.388	.3492	
212	98	322.000	-12.483	< α*	**A**
221	52	97.824	-4.633	< α*	**A**
222	599	325.363	15.170	< α*	**T**

[a] < α* indicates that the tail probability is smaller than can be expressed with four decimal places.

The goodness-of-fit X^2 of the standard CFA model in the top panel of Table 89 is 799.18 ($df = 4$; $p < 0.01$). The goodness-of-fit X^2 of the standard CFA model in the bottom panel of Table 89 is 579.71 ($df = 3$; $p < 0.01$). The model that takes the structural zero into account is thus significantly better than the one that ignores it. Still, in the present example, both X^2 values are large enough for types and antitypes to emerge. Rather than interpreting the types and antitypes in Table 89 in detail, we compare the patterns of types and antitypes in the two panels of the table. First, we find that those configurations that constitute types and antitypes when the structural zero is taken into account, do also constitute types and antitypes when the structural zero is not taken into account. It should be noted that this is not necessarily the case. It is also possible that configurations that constitute types or antitypes in one analysis emerge as inconspicuous under the other. This is the case in the present example. Configuration 211 constitutes a type when the structural zero is not taken into account, but does not differ significantly from its expectancy when the structural zero is part of the model. Note also that the difference between the observed and the expected cell frequency of Configuration 211 is positive in the top part of the table, and negative in the bottom part.

We conclude that

(1) taking into account structural zeros typically brings the estimated expected cell frequencies closer to the observed frequencies than when structural zeros are ignored;

(2) although it reduces the chances of identifying types and antitypes, taking structural zeros into account is strongly recommended;

(3) the pattern of types and antitypes without taking into account structural zeros, does not allow one to predict the pattern of types and antitypes when the structural zeros are taken into account.

10.2 The parsimony of CFA base models

Along with uniqueness of interpretation and consideration of sampling scheme, parsimony is a criterion for the selection of CFA base models (see Section 2.2). Thus far, base models have been selected using the first two criteria, but it has not been discussed whether a particular base model that satisfies these criteria can be made more parsimonious.

The topic of parsimony of CFA base models can be important, particularly when considered in the context of α-protection. The methods used to protect the Type I error from becoming inflated can lead to very conservative statistical decisions. This applies in particular to the most popular Bonferroni procedure. However, if more parsimonious base models can be found, less variability will be covered by the base model, and more variability will be available for the detection of types and antitypes. This portion of variability will be too small to change the fit characteristics of the base model, because the more parsimonious models will be retained only if they are not significantly worse than the less parsimonious models. However, this portion of variability may increase the power available in the search for types and antitypes.

Some CFA base models cannot be made more parsimonious. For example, the zero order CFA base model cannot be reduced, for one obvious reason: It already considers no effects. In addition, when sampling is multivariate product-multinomial (see Sections 2.3.2 and 2.3.3), the marginal frequencies must be reproduced exactly, which typically prevents researchers from finding more parsimonious CFA base models. However, some of the CFA base models are saturated in groups of variables even when sampling is multinomial. These are the candidates for more parsimonious modeling. Examples of such models can be found in Interaction Structure Analysis (ISA; Section 6.1), Prediction CFA (P-CFA; Section 6.2), and k-sample CFA (Chapter 7). Other candidates are second- and higher order global CFA models.

Schuster and von Eye (2000) compared three approaches to two-sample CFA. The first approach was standard two-sample CFA as described in Section 7.1. The second approach involved estimating expected frequencies based on the saturated log-linear model, using maximum likelihood theory. The third approach involved estimating expected cell frequencies using the homogeneous association model for a base model. The results of the comparison of these three approaches suggest that there can be considerable differences in the size of the test statistics. That is, the three approaches differ in the probability of finding types and antitypes.

In this section, we pursue a different route. Rather than comparing standard two-sample CFA with a priori specified models, we look at CFA base models that are saturated in one or more groups of variables, and ask whether there exist base models that are more parsimonious yet not significantly worse than the base model that is partially saturated. The search for more parsimonious models will be exploratory.

The search for a more parsimonious model involves the following three steps:

(1) Identifying that part of the CFA base model that is saturated and can be reduced without violating the constraints imposed by the sampling scheme used for data collection. Variables observed

under a multinomial sampling scheme can typically be subjected to the search for more parsimonious models. Variables observed under univariate product-multinomial sampling schemes can be subjected to this search with the constraint that the univariate marginal frequencies must be reproduced. Variables observed under a bivariate product-multinomial sampling scheme can be subjected to this search with the constraint that their bivariate marginal frequencies must be reproduced. Thus, the two-way interactions of these variables must be part of the model. This applies accordingly to variables observed under more complex product-multinomial sampling schemes.

(2) Collapsing the cross-classification of all variables over those variables that are not involved in the model search. *Collapsing* means removing variables by summing over all of their categories. For example, if a two-sample CFA includes four discriminating variables and one grouping variable, the collapsing is performed by summing over the two categories of the grouping variable. It should be noted that this step does not violate the implications of collapsibility theorems. Specifically, one of these implications is that variables that are independent of all other variables "may be removed by summing over its categories without changing" any parameters (Bishop, Fienberg, & Holland, 1975, p. 47; cf. Clogg, Petkova, & Shihadeh, 1992). In the present context, the damage done by collapsing over the variables not included in the saturated part of the model will be undone by unfolding the table again for the CFA that follows the model search. The collapsing is done solely for the analyses in Step 3, where the variables not included in the saturated part of the model play no role.

(3) Analyzing the model according to three parsimony criteria. First, it must describe the data well so that it can be retained by itself. Second, it must not be significantly worse than the saturated model. Third, it must not violate the constraints imposed by the sampling scheme (see Step 1, above). The search itself can be performed in a number of ways, three of which will be mentioned here. First, relationships among variables that are known from prior research can be made part of the model. All other relationships are not part of the model. In many instances, the base model thus defined already describes the data well. If this model is also substantively meaningful, no additional search is needed. The second method of finding a more parsimonious model involves

using one of the search algorithms available in most general purpose software packages. These algorithms can be viewed parallel to the step-wise search algorithms known from regression analysis. The third method involves estimating all possible models. In the context of CFA, this method is less effortful than it may sound, because the number of variables used in CFA is typically small.

Once a model is identified that meets all the above conditions, one uses the original table for CFA, that is, the un-collapsed cross-classification. Instead of a standard CFA base model, a base model is specified for the estimation of expected cell frequencies that uses the parsimonious model part for those variables that otherwise would be included in the saturated part of the base model.

Data example. The following example presents a re-analysis of data published by Maxwell (1961) and Krauth and Lienert (1973a). In a study on the relationships between the three psychiatric symptoms Depression (D), Feelings of Insecurity (U), and Mood Swings (S) on the one hand and the three psychiatric diagnoses Cyclothymia (C), Anxiety Neuroticism (A), and Neurotic Depression (N). 380 inpatients were diagnosed as either displaying (= 1) or not displaying (= 2) a symptom. Each patient had been diagnosed as falling under C, A, or N. Crossed, these four variables form a 2 x 2 x 2 x 3 contingency table.

We now analyze this table from an Interaction Structure Analysis (ISA; see Section 6.1) perspective. The three symptoms, D, U, and S, form one group of variables, and diagnosis is the sole member of the other group of variables. In standard ISA, the base model would be [D, U, S][G], where G indicates the psychiatric diagnosis. The results from standard ISA appear in Table 90. We used the Pearson X^2 component test and the Bonferroni-adjusted $\alpha^* = 0.00208$. The overall goodness-of-fit $LR\text{-}X^2 = 86.15$ ($df = 14$; $p < 0.01$) is large. Thus, we can expect types and antitypes to emerge.

The results in Table 90 suggest the existence of three types and no antitype. The first type is constituted by Configuration 1112. These are patients who display all three symptoms and had been diagnosed as anxiety neurotics; 19 patients displayed this symptom pattern, but fewer than 9 had been expected.

The second type, constituted by Configuration 1212, describes patients who show only symptoms of depression and mood swings, and had also been diagnosed as anxiety neurotics; 13 patients displayed this pattern,

Table 90: Standard ISA of the Maxwell psychiatry data

Cell index	Frequencies		Test statistics		Type/ Antitype ?
DUSG	observed	expected	X^2	p	
1111	11	12.85	0.267	.6053	
1112	19	8.68	12.254	.0005	**T**
1113	3	11.46	6.248	.0124	
1121	13	10.91	0.402	.5259	
1122	9	7.37	0.361	.5478	
1123	6	9.73	1.428	.2322	
1211	3	6.23	1.676	.1955	
1212	13	4.21	18.348	$< \alpha$*	**T**
1213	0	5.56	5.558	.0184	
1221	4	6.62	1.038	.3084	
1222	12	4.47	12.662	.0004	**T**
1223	1	5.91	4.075	.0435	
2111	30	34.27	0.533	.4654	
2112	14	23.16	3.622	.0570	
2113	44	30.57	5.902	.0151	
2121	38	28.04	3.536	.0600	
2122	11	18.95	3.333	.0679	
2123	23	25.01	0.162	.6877	
2211	18	19.47	0.112	.7384	

/ cont.

Cell index	Frequencies		Test statistics		Type/ Antitype ?
DUSG	observed	expected	X^2	p	
2212	9	13.16	1.314	.2517	
2213	23	17.37	1.826	.1766	
2221	31	29.60	0.066	.7969	
2222	13	20.00	2.450	.1175	
2223	32	26.40	1.188	.2728	

[a] $< \alpha^*$ indicates that the tail probability is smaller than can be expressed with four decimal places.

but only about 4 had been expected. The third type, constituted by Configuration 1222, describes patients who only suffer from symptoms of depression but had also been diagnosed as anxiety depressed; 12 patients suffer only from depression symptoms, but only slightly more than 4 had been expected from the base model. In the following paragraphs, we report the results from the three steps of the search for a more parsimonious base model.

Step 1: Identification of the part of the base model that can be reduced. In the present example, the base model was [D, U, S][G]. This model is saturated in the variables D, U, and S. First, we have to determine whether any of the variables or subgroups of variables have been observed under a product-multinomial sampling scheme. This is not the case for any of the three variables depression, insecurity, and mood swings. Each of these variables is an observed variable. The investigators did not determine the number of observations a priori. In addition, these variables are not used as predictors. Therefore, we can assume multinomial sampling, and there are no constraints on the model simplification process.

Step 2: Collapsing the table over the variables not involved in the model simplification process. The diagnosis variable is not involved in the model simplification process. Therefore, the collapsing reduces the D x U x S x G tabulation with 24 cells to the D x U x S tabulation with 8 cells. This tabulation appears in Table 91.

Table 91: Collapsed D x U x S cross-classification

Configuration	Frequencies	
DUS	observed	expected
111	33	29.9
112	28	24.7
121	16	16.3
122	17	23.0
211	88	91.1
212	72	75.3
221	50	49.7
222	76	70.0

Step 3: Performing the model search. Because we do not entertain any hypotheses about the association structure of the three psychiatric symptoms depression, feelings of insecurity, and mood swings, we employ a model search algorithm. Specifically, we use the model selection option in SPSS10 which can be found under ANALYZE - LOGLINEAR. The program goes through four steps. Table 92 summarizes the results.

Table 92: Results of the model simplification for the variables D, U, and S

Step	Term eliminated	$LR\text{-}X^2$	*df*	*p*
1	[D, U, S]	.631	1	.427
2	[D, S]	.841	2	.657
3	[D, U]	3.200	3	.362

After Step 3, there is no improvement in parsimony. Each of the remaining terms, when eliminated, leads to a significant deterioration of the model. Therefore, the model [U, S][D] is the most parsimonious model. As can be seen from the third column in Table 92, the overall loss is minimal. The $LR\text{-}X^2$ for the final model is 3.200. This value indicates that the simplified model is not significantly worse than the saturated model. Substantively, this result suggests that the three symptoms are largely independent of each other. The only association is between symptoms of depression and feelings of insecurity.

From the perspective of the subsequent CFA, this result indicates that only a small portion of the variability was gained for the search for types and antitypes. This portion is small, and we anticipate therefore, no dramatic changes. It is important to realize that this portion is distributed over the cells in no easily predictable manner. Some CFA tests may come with larger test statistics, others may come with smaller test statistics. Table 93 displays the CFA results. We used, as in Table 90, the X^2 component test and the Bonferroni-adjusted $\alpha^* = 0.00208$.

Table 93: Parsimony ISA of the Maxwell psychiatry data; base model is [US][D][G]

Cell index	Frequencies		Test statistics		Type/ Antitype
DUSG	observed	expected	X^2	p	?
1111	11	11.66	0.037	.8467	
1112	19	7.88	15.692	$< \alpha^*$	**T**
1113	3	10.40	5.265	.0218	
1121	13	9.63	1.778	.2782	
1122	9	6.51	0.952	.3291	
1123	6	8.59	0.781	.3769	
1211	3	6.36	1.775	.1828	

/ cont.

Cell index	Frequencies		Test statistics		Type/ Antitype ?
DUSG	observed	expected	X^2	p	
1212	13	4.30	17.602	< α*	**T**
1213	0	5.67	5.670	.0173	
1221	4	8.96	2.746	.0975	
1222	12	6.05	5.852	.0157	
1223	1	7.99	6.115	.0134	
2111	30	35.47	0.844	.3584	
2112	14	23.97	4.147	.0417	
2113	44	31.63	4.838	.0278	
2121	38	29.31	2.577	.1085	
2122	11	19.81	3.918	.0478	
2123	23	26.14	0.377	.5397	
2211	18	19.35	0.094	.7589	
2212	9	13.07	1.267	.2603	
2213	23	17.26	1.909	.1671	
2221	31	27.26	0.513	.4738	
2222	13	18.42	1.595	.2066	
2223	32	24.31	2.433	.1188	

[a] < α* indicates that the tail probability is smaller than can be expressed with four decimal places.

The results in Table 92 suggest that there are only two types instead of the three in Table 89. Thus, although the power for this second analysis was nominally greater, the number of types turned out to be smaller. The reason for this reduction in the number of types is that while one of the test statistics for the three types increased in magnitude (the one for

Configuration 1112 increased from 12.25 to 15.69), the other two decreased in magnitude, one of them to the extent that it no longer constitutes a type (Configuration 1222).

We therefore conclude that

(1) optimizing base models by making them more parsimonious will always result in a nominal increase in power for the CFA tests;
(2) this power, however, will not always result in an increase in the number of types and antitypes, because it is not predictable that the number of extreme test statistics will increase. Occasionally, the number of types and antitypes can even decrease, as was illustrated in the above example.

In spite of the decrease in the number of types in the example in Tables 90 and 93, it is worth optimizing base models. In many instances, new types and antitypes will emerge, or configurations that were marginal before the optimization will then constitute types or antitypes.

10.3 CFA of groups of cells: Searching for patterns of types and antitypes

Thus far in this book, the focus has been on single cells. The question asked concerned the existence of types and antitypes as defined by a single configuration. However, in many instances it is not the sheer existence of types and antitypes that makes a result meaningful and interesting, but the particular pattern of types or antitypes.

Consider, for example, the analysis of Lienert's LSD data in Table 1. These data resulted from a study on the effects of LSD 50 as measured via the three variables Narrowed Consciousness (C), Thought Disturbance (T), and Affective Disturbance (A; Lienert, 1964). Suppose that, based on prior results and derived from Bonhoeffer's (1917) exogenous response types the pattern of the three mono-symptomatic reactions + - -, - + -, and - - + is expected. Then, rather than testing each configuration individually, it may be interesting to test the pattern as a whole, that is, as a *composite type*.

A test of the existence of composite types or composite antitypes can be described using the *z*-test from Section 3.3.3,

$$z_i = \frac{N_i - Np}{\sqrt{Npq}}$$

where N_i is the observed frequency of Configuration i, Np is the estimated expected cell frequency, and $q = 1 - p$. The sum of the z_i, Σz_i has an expectancy of $E(\Sigma z_i) = 0$ and a variance of t, where t is the number of cells in the summation. We thus can construct a new test statistic for t configurations using the Stouffer z,

$$z = \frac{\sum_{i=1}^{t} z_i}{\sqrt{t}}$$

(von Eye, Lienert, & Wertheimer, 1991; for alternative methods see Darlington & Hayes, 2000; Kristof, 1993). This statistic is approximately normally distributed. When applying this statistic one has to assume that the z_i are independent.

Data example. We now illustrate the test of composite types using Lienert's (1964) LSD data. Table 94 presents these data and the results of standard first order CFA, based on the z-test from Section 3.3.3 and the Bonferroni-adjusted $\alpha^* = 0.00625$.

The results in Table 94 suggest that the z-test does not lead to the detection of individual types or antitypes (for a comparison of results from other CFA tests see Table 13). However, in the present context, we are less interested in types and antitypes of single configurations, but rather in *composite types* or *composite antitypes*. Therefore, we ask whether the three monosymptomatic configurations + - -, - + -, and - - + constitute a composite type. We use the three z-scores from Table 94 and insert in the equation to obtain

$$z = \frac{2.303 + 2.116 + 2.169}{\sqrt{3}} = \frac{6.588}{\sqrt{3}} = 3.804.$$

The tail probability for this z-score is p = 0.00007. This value is smaller than $\alpha = 0.05$, and we retain the hypothesis that the three monosymptomatic reactions to LSD50 constitute a composite type. Readers are invited to test whether the three bisymptomatic reactions + + -, + - +, and - + + form a composite antitype.

Table 94: First order CFA of the three variables Narrowed Consciousness (C), Thought Disturbance (T), and Affective Disturbance (A)

Cell index	Frequencies		Test statistics		Type/ Antitype
CTA	observed	expected	z	p	?
+ + +	20	12.506	2.119	.0170	
+ + -	1	6.848	-2.235	.0127	
+ - +	4	11.402	-2.192	.0142	
+ - -	12	6.244	2.303	.0106	
- + +	3	9.464	-2.101	.0178	
- + -	10	5.182	2.116	.0172	
- - +	15	8.629	2.169	.0150	
- - -	0	4.725	-2.174	.0149	

Two characteristics of this test are of note. First, the significance threshold does not need to be protected if only one test is performed. If several composite types or antitypes are hypothesized, the family-wise α needs to be protected accordingly. Second, for a composite type or antitype to be significant, it is not necessary that each component make a significant contribution, as long as the hypothesis is confirmed overall. Section 10.4.2 will present another example of this second characteristic.

10.4 CFA and the exploration of causality

The investigation of causal processes is typically confirmatory in nature. Researchers adopt a concept of causality, derive causal predictions from this concept, and design experiments and tests of these predictions. Unfortunately, there is no agreement on concepts of causality. Many theorists define causality using Hume's notions of *regularity* and *temporal priority*. The former implies that there exist antecedents that are necessary, sufficient, or both for subsequent events. Temporal priority implies that the antecedent occur prior to the subsequent events. The classical, essentialist perspective of causality proposes that the antecedents be both necessary and sufficient to qualify as causes for subsequent events. Cook and Campbell (1979) cast doubt on the inevitability element involved in this definition and state that this element may be inappropriate for the social sciences. The authors marshal a probabilistic concept of causality that links antecedents and consequences in a probabilistic fashion. In contrast, Sobel (1996) considers probabilistic concepts of causality, in particular Suppes' (1970) theory, not tenable. For a discussion of causality from a philosophical perspective see Stegmüller (1983).

Looking at causality from a data analysis perspective, Bollen (1989) discusses the three criteria *isolation*, *association*, and *direction*. These criteria must be met for a variable or event to qualify as a cause. Of these, *direction* has proven to be the most elusive in the context of statistical analysis (von Eye & Schuster, 1999). Bollen also states that human manipulation, a criterion currently used by many (e.g., Holland, 1986, 1988; Sobel, 1994), is "neither a necessary nor sufficient condition for causality" (1989, p. 41). These are just a few examples that indicate that there is no commonly agreed-upon definition of causality.

Still, causality is a widely discussed and important concept in social science research. Therefore, von Eye and Brandtstädter (1997; cf. von Eye & Brandtstädter, 1998) asked the question whether CFA can be used to *explore* data for loci of possible causal processes or effects. The authors proposed that if causal processes are at work, they must manifest in particular effects in the form of changes in the probability structure. These changes must be specific to the causal processes at work. The authors analyzed the sample cases of the three dependency concepts of the *wedge*, the *fork*, and *reciprocal causation* (see von Eye et al., 1999). The following sections discuss these three concepts in the context of CFA.

10.4.1 Exploring the concept of the *wedge* using CFA

The *wedge* is a concept of dependency that denotes multiple causation. In the least complex case, three events are involved. Consider the three events A, B, and C. The wedge describes the pathways through which C can be reached. In each case, the events A and B are causes, and the event C is the effect. Von Eye and Brandtstädter (1998) distinguish between two forms of the wedge. The *strong wedge* implies that the causes be mutually exclusive and exhaustive. For instance, one can reach the high school diploma via a number of high school tracks. However, each student can complete only one track. Thus, each individual can reach C only either by way of A or by way of B. In contrast, the *weak wedge* does not carry this implication. For example, one can have a headache because of the flu, because one bumped the head, or both. Figure 10 illustrates the dependency concept of the wedge.

CFA is particularly well-suited for the analysis of such concepts as the wedge. The configurations of CFA describe event patterns. Some of these patterns support the notion of a relationship that can be described using the wedge concept. If these patterns form types, the analysis may have detected the location of a causal process. This applies accordingly to antitypes.

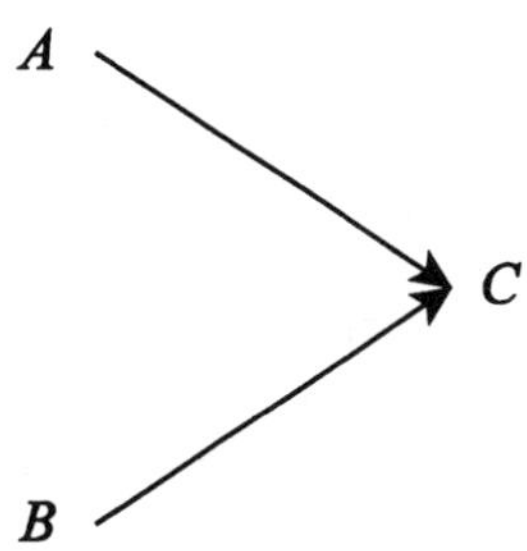

Figure 10: The Wedge

When exploring data that may contain wedge-like relationships with CFA, we use the model of Prediction CFA (P-CFA; see Section 6.2). This model distinguishes between predictors and criteria. In the simplest case, depicted in Figure 10, there will be two predictors and on criterion. Thus, the base model is always

$$\log E = \lambda + \lambda_i^P + \lambda_j^C ,$$

where λ_i^P represents all main effects and interactions among the predictors, and λ_j^C represents all main effects and interactions among the criterion variables. We now ask how CFA can identify wedge-type structures.

The *constituent elements of a wedge* are

(1) two or more antecedent events and
(2) one consequent event.

Note that this and the following considerations can be extended to multiple consequent events. For the sake of simplicity, we stay with one consequent event. The model that we adopt from P-CFA, proposes independence among the antecedent and the consequent events. Thus, types or antitypes indicate predictor-criterion or, in the present case, cause-effect relationships. *A pattern of types or antitypes supports the notion of a wedge-type dependency relationship if one criterion configuration forms types or antitypes with two or more predictor configurations* (see Figure 10). Note that such a relationship involves either only types or only antitypes. The identification of such relationships can be accomplished using the methods of P-CFA and the methods for the identification of composite types or antitypes described in Section 10.3.

Data example: the weak wedge. The following example involves a re-analysis of data presented by Keenan et al. (1996). A sample of N = 213 respondents in a nutrition study answered questions concerning success in their attempts to reduce fat in their diet. The questions concerned support from spouses (F), support from support groups (S), and whether the respondents were able to keep up the changes they had implemented (Q). The answers to the first two questions were scored as no (= 1) versus yes (= 2). The answers to the third question were scored as 1 = sustained change and 2 = changes reversed. In the following analyses, we consider F and S the predictors, and Q the criterion. The P-CFA base model is

$$\log E = \lambda_0 + \lambda_i^F + \lambda_j^S + \lambda_{ij}^{FS} + \lambda_k^Q ,$$

that is, a model that is saturated in the predictors, that considers the main effect of the criterion, and that proposes independence between the predictors and the criterion. Table 95 presents the results of standard P-CFA. We used the *z*-test and the Bonferroni-adjusted $\alpha^* = 0.00625$. We first interpret the results from standard P-CFA and then ask whether the data support the notion of the presence of a wedge-type dependency structure.

The results in Table 94 suggest the existence of two types and two antitypes. Reading from the top to the bottom of the table, the first antitype is constituted by Configuration 111. It indicates that lack of support from spouses and support groups is unlikely to lead to persistent dietary fat reduction. Complementing this result, the first type, constituted by Configuration 112, suggests that total lack of support allows one to predict that dietary changes will be reversed. The second type, constituted by Configuration 221, indicates that support from both spouses and support

Table 95: P-CFA of determinants of success in dietary fat reduction

Cell index	Frequencies		Test statistics		Type/ Antitype ?
FSQ	observed	expected	z	p	
111	7	26.498	-3.788	.0001	**A**
112	61	41.502	3.027	.0012	**T**
121	11	5.455	2.374	.0088	
122	3	8.545	-1.897	.0289	
211	23	31.174	-1.464	.0716	
212	57	48.826	1.170	.1211	
221	42	19.873	4.963	< α*	**T**
222	9	31.127	-3.966	< α*	**A**

[a] < α* indicates that the tail probability is smaller than can be expressed with four decimal places.

groups allows one to predict sustained changes in dietary fat intake. The second antitype complements this type, suggesting that presence of support is unlikely to lead to a reversal of dietary changes (Configuration 222).

While possibly interesting by themselves, these results do not allow us to derive statements concerning patterns that would support the notion of wedge-type structures. We therefore ask whether we can link two predictor levels to one outcome level. For example, we ask whether there are two predictor configurations that allow one to predict sustained dietary change. In the present data, we ask whether the two predictor configurations 12 and 22 both lead to persistent change, that is, criterion category 1. We therefore use the z-scores from Configurations 121 and 221 and insert in the z-formula given in Section 10.3. We obtain

$$z = \frac{2.374 + 4.963}{\sqrt{2}} = 5.188.$$

This score is significant ($p < 0.01$), suggesting that indeed support from support groups alone (Configuration 121) and support from both support group and spouse (Configuration 221) allow one to predict sustained dietary fat reduction[1]. This pattern can be viewed as supporting the notion of a *weak wedge* dependency relation. Figure 11 depicts this result.

We now ask whether it is not only possible to describe the dependency structure of sustained dietary change in terms of the weak wedge, but also a dependency structure of reversal of dietary change. We select Configurations 112 and 212 and calculate

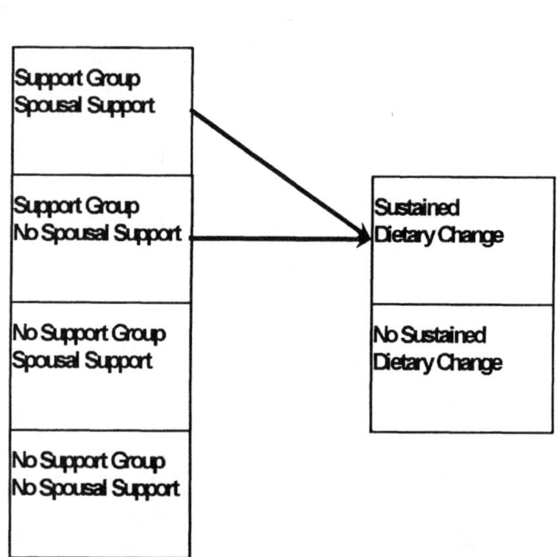

Figure 11: Weak Wedge model of dietary change

$$z = \frac{3.027 + 1.170}{\sqrt{2}} = 2.968.$$

This score is also significant ($p = 0.0015$) and we can conclude that there exists a dependency structure that allows one to describe reversal of dietary change in terms of a weak wedge. Specifically, if an individual has support from neither spouse nor a support group, or if the individual has only spousal support, then a reversal of dietary change can be predicted.

Both results support the notion of a process that can be described in terms of the *weak wedge* concept. All respondents were married and had the opportunity to benefit from support provided by their spouses, by

---

[1]One may ask whether spousal support is necessary at all to achieve persistent dietary change. A logit analysis suggests that spousal support has an effect greater than zero, but that support groups have an effect that is three times as strong (see von Eye & Brandtstädter, 1997).

support groups, or both.

10.4.2 Exploring the concept of the *fork* using CFA

In contrast to the wedge where multiple causes have the same effect, the *fork* is the concept of a process in which *one cause has multiple effects*. Consider the three events A, B, and C. The fork describes the dependency relationship between these three events if, for instance, A causes both B and C. Figure 12 illustrates the fork concept.

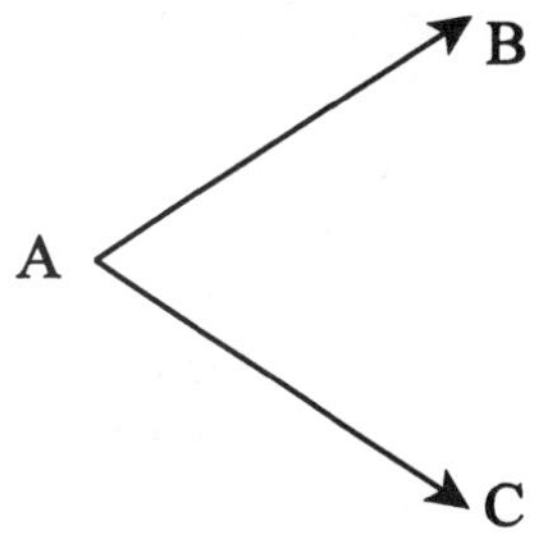

Figure 12: The Fork

As for the wedge, the distinction between a *strong fork* and a *weak fork* can be made. The concept of the strong fork poses the constraint that only one outcome is possible for each individual. For example, there may be enough money to salvage either Social Security or Medicare, but not both. Thus, the strong fork concept may make sense. To illustrate the weak fork concept, consider the flu virus. This virus can cause a running nose, headaches, and weakness, in any combination.

For configural analysis, we again adopt the base model of Prediction CFA (P-CFA), $\log E = \lambda + \lambda_i^P + \lambda_j^C$, where P represents all predictors, and C represents all criterion variables. In the simplest case, illustrated in Figure 12, there are one predictor and two criteria. In this case, this model is saturated in the criteria and takes into account the predictor main effects.

The *constituent elements of a fork* are

(1) two or more consequent events, and
(2) one antecedent event.

Note that this and the following considerations can be extended to multiple antecedent events. For the sake of simplicity, we stay with one antecedent event. The model that we adopt from P-CFA proposes independence among the antecedent and the consequent events. Thus, types or antitypes indicate predictor-criterion or, in the present case, cause-effect relationships. *A pattern of types or antitypes supports the notion of a fork-type dependency relationship if one predictor configuration forms types or antitypes with two or more criterion configurations* (see Figure 12). Note that such a relationship involves either only types or only antitypes. The identification of such relationships can be accomplished using the methods of P-CFA or the methods for the identification of composite types or antitypes described in Section 10.3.

Data example: the strong fork. In the following example, we re-analyze data published by Görtelmeyer (1988). The data were collected in a study on sleep problems in a sample of 273 respondents. The author used first order CFA to define the six types of sleep behavior of respondents who sleep (1) short periods of time early in the morning; (2) symptom-free during 'normal' night hours; (3) symptom-free but wake up too early; (4) short periods early in the morning and show all symptoms of sleep problems; (5) during normal night hours but show all symptoms of sleep problems; and (6) long hours starting early in the evening, but show all symptoms of sleep problems. Of the 273 participants, 107 belonged to one of these types. The remaining 166 did not belong to any type. However, in the following analyses, we treat these 166 individuals as if they belonged to a seventh type.

In the following analyses, we ask whether psychosomatic symptoms allow one to discriminate among the seven sleep behavior categories. Specifically, we cross the seven categories of sleep behavior (S) with psychosomatic symptomatology (P), where 2 indicates above median number of symptoms, and 1 indicates below median number of symptoms. Using P as predictor and S as the criterion, we analyze the resulting cross-classification under the P-CFA base model

$$\log E = \lambda_0 + \lambda_i^P + \lambda_j^S.$$

This model is identical to the base model one would obtain for first order CFA, because there are only two variables. We use Lehmacher's test with Küchenhoff's continuity correction and the Bonferroni-adjusted $\alpha^* = 0.00357$. Table 96 displays the results. We first interpret the results from standard CFA, and then we ask whether the data support the notion of a fork-type relationship present in the data.

The results in Table 96 suggest the existence of four types and four antitypes. Because the criterion variable, P, has only two categories, the types and antitypes can be viewed as forming pairs. The first type,

Table 96: P-CFA of types of sleep behavior as predicted from psychosomatic symptoms

Cell index	Frequencies		Test statistics		Type/ Antitype ?
SP	observed	expected	z	p	
11	19	11.040	3.311	.0005	**T**
12	3	10.960	-3.311	.0005	**A**
21	20	12.044	3.181	.0007	**T**
22	4	11.956	-3.181	.0007	**A**
31	16	9.535	2.832	.0023	**T**
32	3	9.465	-2.832	.0023	**A**
41	5	4.516	-.011	.4956	
42	4	4.484	.011	.4956	
51	4	7.026	-1.383	.0833	
52	10	6.974	1.383	.0833	
61	8	9.535	-.491	.3116	
62	11	9.465	.491	.3116	
71	65	83.304	-4.406	$< \alpha^*$	**A**
72	101	82.696	4.406	$< \alpha^*$	**T**

[a] $< \alpha^*$ indicates that the tail probability is smaller than can be expressed with four decimal places.

constituted by Configuration 11, describes those respondents whose number of psychosomatic symptoms is below average and for which Sleep Pattern 1 can be predicted, that is, sleeping only short periods early in the morning. The corresponding antitype, 12, indicates that this sleep pattern cannot be predicted for respondents with above median numbers of psychosomatic symptoms. The second type-antitype pair, 21 and 22, suggests that for individuals with below median numbers of psychosomatic

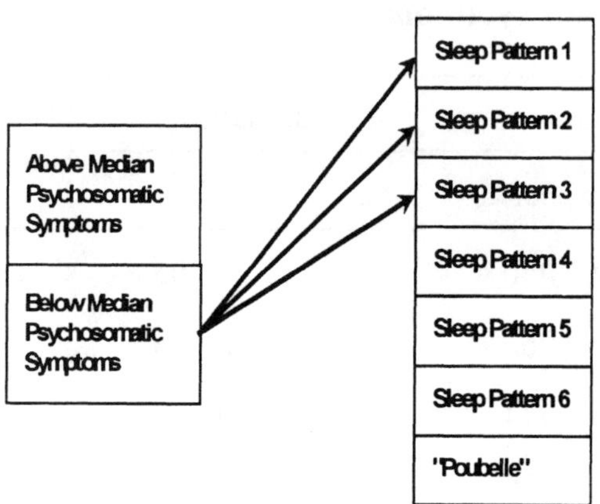

Figure 13: Type-fork structure of sleep patterns

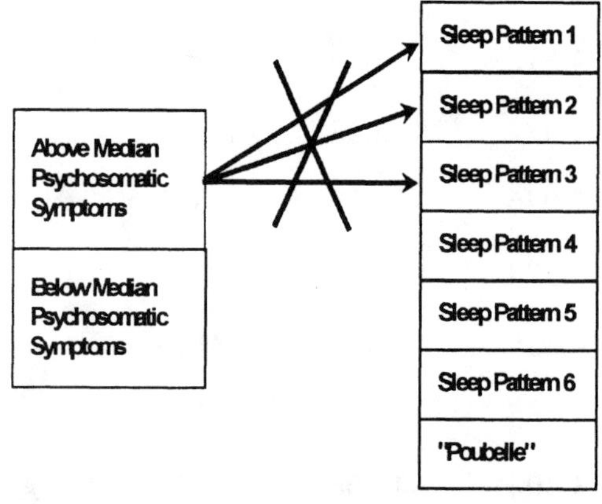

Figure 14: Antitype-fork structure of sleep patterns

symptoms, symptom-free sleeping during regular night hours can be predicted, but not for individuals with above median numbers of psychosomatic symptoms. The third type-antitype pair, 31 and 32, indicates that for individuals with below median numbers of psychosomatic symptoms, symptom-free sleep can be predicted that is shortened by early awakening, but not for individuals with above median numbers of psychosomatic symptoms.

The fourth antitype-type pair was observed for the individuals that do not belong to a particular CFA sleep pattern type. This pair indicates that individuals with below median numbers of psychosomatic symptoms are unlikely to belong to this group (Antitype 71). In contrast, individuals with above median numbers of psychosomatic symptoms are highly likely to belong to this group (Type 72).

We now ask whether these patterns of types and antitypes support an interpretation from the perspective of a fork. Such an interpretation could be justified if one predictor configuration forms types with more than one criterion configuration. This applies accordingly to antitypes.

In Table 95, we find that the first three types form a fork, and the first three antitypes also form a fork. Specifically, the first three types make predictions from predictor Configuration 1. The first three antitypes make predictions from predictor Configuration 2. Therefore, we have a *fork-type* and a *fork-antitype*. Figures 13 and 14 display these two forks. To test the composite fork-type, we use the data from Table 92 and calculate $z = (3.311 + 3.181 + 2.832)/\sqrt{3} = 5.383$. This value is significant ($p < 0.01$) and thus confirms the type pattern. To test the composite fork-antitype, we calculate $z = (-3.311 - 3.181 - 2.832)/\sqrt{3} = -5.383$ and thus again confirm the antitype pattern ($p < 0.01$).

It is important to realize that the antitype-fork structure in Figure 14 is open to more than one interpretation. One interpretation is that the antitype-fork can imply that the antecedent event *fails to produce a consequent event*. A second interpretation is that an antecedent event *prevents a consequent event from materializing*. In exploratory research, it is rarely possible to decide which of these interpretations is supported by the data.

10.4.3 Exploring the concept of *reciprocal causation* using CFA

Reciprocal relations are defined as processes where *two or more agents simultaneously influence each other*. In particular in the social sciences, the concept of reciprocal relations has been a focal point of current theories (e.g., Brandtstädter, 1998; Lerner, 1998; Gottlieb, 1992; Gottlieb, Wahisten, & Lickliter, 1998; von Eye, Lerner, & Lerner, 1998). Consider the following example. Two discussants are engaged in an exchange of arguments. Then, one can assume that each discussant is influenced by the respective other discussant's presence and behavior. Another, more dramatic example of a situation in which two events influence each other reciprocally is an accident in which two cars collide.

The constituent elements of a reciprocal relation are

(1) two or more agents; and
(2) one or more variables that describe both antecedent and consequent events.

Data example. To illustrate the analysis of reciprocal relations with CFA, we use a data set presented by Ohannessian et al. (1994). A sample of 153 young adolescents responded both at the beginning and at the end of their middle school careers to Rohner's (1980) Parental Acceptance and

Rejection Questionnaire (PARQ). For the following analyses with CFA, the sum of the four subscales, warmth and affection, hostility and aggression, indifference and neglect, and undifferentiated rejection, was rank transformed. There are four ranks, with 1 indicating a very poor relationship. Responses are available for both mothers (M) and fathers (F). Table 97 presents the CFA results. The data stem from the first wave of data collection. A first order CFA was performed using the *z*-test and the Bonferroni-adjusted $\alpha^* = 0.003125$. We first discuss the results of standard CFA, and then ask whether the existence of reciprocal relationships can be established.

Standard CFA identifies three types and three antitypes of adolescents' perceptions of their relationships with their mothers and fathers. The types can be interpreted as follows:

- Type 12: adolescents who have very poor relationships with their mothers and poor relationships with their fathers;
- Type 33: adolescents who have good relationships with their mothers and good relationships with their fathers; and
- Type 44: adolescents who have very good relationships with their mothers and very good relationships with their fathers.

The three antitypes can be describes as follows:

- Antitype 14: adolescents who have very poor relationships with their mothers and very good relationships with their fathers; and
- Antitype 32: adolescents who have good relationships with their mothers and poor relationships with their fathers.
- Antitype 42 adolescents who have very good relationships with their mothers and poor relationships with their fathers.

We now ask whether the data in Table 97 support the hypothesis that there exists a reciprocal relationship such that the relationship that an adolescent perceives with his/her mother influences the relationship with his/her father to be perceived as similar, and vice versa. There is support for the notion of a reciprocal relationship if there are types (or antitypes) in those cells that contain cases that confirm the hypotheses of a directed relationship.

Consider, for example, the two variables A and B. A set of hypotheses that involves a reciprocal relationship is

$a_1 \rightarrow b_1$ AND b_2; and

$b_1 \rightarrow a_1$ AND a_2.

Table 97: CFA of adolescents perceptions of their relationships to their mothers and fathers

Cell index	Frequencies		Test statistics		Type/ Antitype ?
MF	observed	expected	z	p	
11	11	6.582	1.722	.0425	
12	40	24.595	3.106	.0009	T
13	2	11.085	-2.729	.0032	
14	0	10.739	-3.277	.0005	A
21	6	5.588	0.174	.4309	
22	27	20.882	1.339	.0903	
23	9	9.412	-0.134	.4466	
24	3	9.118	-2.026	.0214	
31	1	3.850	-1.452	.0732	
32	2	14.386	-3.266	.0005	A
33	16	6.484	3.737	$< \alpha$*	T
34	12	6.281	2.282	.0112	
41	1	2.980	-1.147	.1257	
42	2	11.137	-2.738	.0031	A
43	5	5.020	-0.009	.4965	
44	16	4.863	5.051	$< \alpha$*	T

[a] $< \alpha$* indicates that the tail probability is smaller than can be expressed with four decimal places.

The first of these hypotheses is confirmed by the cases in cells a_1b_1 and a_1b_2. The second of these hypotheses is confirmed by the cases in cells b_1a_1 (= a_1b_1) and b_1a_2. In different words, the cases in cell a_1b_1 confirm both hypotheses, the cases in cell a_1b_2 confirm only the first hypothesis, and the cases in cell b_1a_2 confirm only the second hypothesis. If these three cells form a composite type, there is support for the hypothesis of a reciprocal relationship.

In the present data example, we now test two sets of hypotheses, each of which involves a reciprocal relationship. The first set is

H_1: if the relationship with mother is perceived as very poor (M = 1), the relationship with father is perceived as poor or very poor (F < 3); and
if the relationship with father is perceived as very poor (F = 1), the relationship with mother is perceived as poor or very poor (F < 3).

This set of hypotheses implies that the perception of the relationship with one parent as very poor influences the relationship with the other parent to be perceived as poor or very poor. Cells 11, 12, and 21 contain the cases that confirm this set of hypotheses. Using the *z*-test introduced in Section 10.3 we insert from Table 97 and obtain

$$z = \frac{1.722 + 3.101 + 0.174}{\sqrt{3}} = 2.888 ,$$

a score that is significantly greater than zero ($p < 0.01$). We thus conclude that there is evidence in support of the notion that the perception of the relationship with one parent as very poor influences the relationship with the other parent to be perceived as poor or very poor.

We now ask whether the same is true for positive relationships. We test the second set of hypotheses

H_2: if the relationship with mother is perceived as very good (M = 4), the relationship with father is perceived as good or very good (F > 2); and
if the relationship with father is perceived as very good (F = 4), the relationship with mother is perceived as good or very good (F > 2).

Cells 34, 43, and 44 contain the cases that support this relationship. We insert into the *z*-test formula and obtain

$$z = \frac{2.282 + (-0.009) + 5.051}{\sqrt{3}} = 4.229 ,$$

a score that is also significant ($p < 0.01$). It should be noted that this result is largely carried by the very large test statistic $z = 5.051$ for Cell 44. Neither Cell 34 not Cell 43 make a substantial contribution. However, the concept of a composite type dos not require each individual cell to make a significant contribution for the composite type to be significant (see Section 10.3).

10.5 Covariates in CFA

In the context of analyzing categorical variables, covariates are typically defined as independent variables that affect the joint frequency distribution of the variables that constitute the cross-classification. In the present section, we discuss two approaches to using covariates in configural analysis. (1) In many instances, covariates are categorical variables similar to stratification variables (Graham, 1995). This issue is addressed in Section 10.5.1 (see also Section 6.2.2.1 on Conditional CFA). (2) In other instances, covariates are continuous variables. Researchers also use the cell-specific probabilities of events as covariates or contrast vectors that specify some hypothesis. These issues will be addressed in Section 10.5.2.

The use of covariates typically carries the estimated cell frequencies closer to the observed cell frequencies, because more information is used in the estimation process (von Eye & Niedermeier, 1999). However, as Glück and von Eye (2000) demonstrated, the opposite effect can occur as well. In general, if the correlation between a covariate vector and the residuals of the log-linear base model exists, residuals can be expected to become smaller when covariates are used.

10.5.1 Categorical covariates: stratification variables

Covariates often come in the form of categorical variables. To illustrate, consider the situation in which two raters judge objects using two categories. Aggregating over all objects yields the cross-classification in Table 98.

Now suppose, Table 98 describes the case where two strata were aggregated and the entire sample is analyzed as one group. Using the model proposed by Graham (1995) we now decompose the arrangement in Table 97 to accommodate the two strata (see also Section 6.2.2.1). We obtain Table 99.

Table 98: 2 x 2 Cross-classification of two raters' judgements

		Rater B Rating Categories	
		1	2
Rater A Rating Categories	1	N_{11}	N_{12}
	2	N_{21}	N_{22}

Table 99: 2 x 2 x 2 Cross-Classification of two Raters' Judgements in two Strata

Stratum A		Rater B Rating Categories		Stratum B		Rater B Rating Categories	
		1	2			1	2
Rater A Rating Categories	1	N_{111}	N_{112}	Rater A Rating Categories	1	N_{211}	N_{212}
	2	N_{121}	N_{122}		2	N_{221}	N_{222}

The log-linear base models for the two tables differ in the additional terms that are needed because of the stratification variable. Let the raters be denoted by A and B, and the stratification variable by S. Then, the first order base model for the cross-classification in Table 98 is

$$\log E = \lambda_0 + \lambda_i^A + \lambda_j^B .$$

A more complex base model is not conceivable, because there is only one degree of freedom left in this example. The log-linear base model for the cross-classification in Table 99 is

$$\log E = \lambda_0 + \lambda_i^A + \lambda_j^B + \lambda_k^S .$$

For the cross-classification in Table 99, a number of more complex base models is conceivable. For instance, if the question is whether the stratification variable allows one to predict the raters' judgements, the base model can be

$$\log E = \lambda_0 + \lambda_i^A + \lambda_j^B + \lambda_{ij}^{AB} + \lambda_k^S .$$

As this example illustrates, including a stratification variable and crossing the other variables with the stratification variable not only leads to a higher-dimensional table, but also allows researchers to formulate more elaborate base models and to test a number of different hypotheses. Some of these models, specifically those that allow researchers to distinguish between strata, are more interesting in the context of CFA, other models are more interesting in the context of model fitting (Graham, 1995).

Data example. The following example presents a re-analysis of a cross-classification published by Graham (1995). In the Auckland Heart Study, a community-based case control study of coronary heart disease, members of a random subsample of survivors of myocardial infarction were asked whether their next of kin could also be interviewed about them. Focus of the questions was whether the patients had engaged in any vigorous leisure time activity at least once a week over the last three months. An activity was defined as rigorous if it induced sweating and heavy breathing. Of the variables involved in this part of the study, we look at the following five:

(1) Age of respondent (A): ≤ 55 is labeled as 1; > 55 is labeled as 2;
(2) Next of Kin (N): spouse = 1; non-spouse = 2;
(3) Response of Kin (K): yes = 1; no = 2;
(4) Response of Patient (P): yes = 1; no = 2; and
(5) Gender of Patient (G): male = 1; female = 2.

In the following analyses, we use gender of respondent (= patient) as the covariate. In the first step, we ignore the covariate and perform a routine first order CFA on the 2 x 2 x 2 x 2 cross-classification of A, N, K, and P. Results of this analysis appear in Table 100. We used the *z*-test and the Bonferroni-adjusted $\alpha^* = 0.003125$. The base model for this analysis is

$$\log E = \lambda_0 + \lambda_i^A + \lambda_j^N + \lambda_k^K + \lambda_l^P .$$

Table 100 presents a rich harvest of types and antitypes. We see four types and three antitypes. Reading from the top to the bottom of the table, we find the following types:

1111: younger than 56, spouse reports, both spouse and patient indicate that patient exercises;
1211: younger than 56, non-spouse reports, both non-spouse and patient indicate that patient exercises;
2122: older than 55, spouse reports, both spouse and patient indicate that patient does not exercise;

Table 100 **CFA of the cross-classification of Age (A), Next of Kin (N), Response of Kin (K), and Response of Patient (P; data from the Auckland Heart Study)**

Cell index	Frequencies		Test statistics		Type/ antitype?
ANKP	observed	expected	z	p	
1111	45	7.878	13.225	< α*	T
1112	7	28.047	-3.974	< α*	A
1121	19	34.387	-2.624	.0043	
1122	134	122.416	1.047	.1476	
1211	11	2.505	5.368	< α*	T
1212	2	8.917	-2.316	.0103	
1221	2	10.932	-2.701	.0035	
1222	34	38.918	-0.788	.2152	
2111	12	6.265	2.291	.0110	
2112	1	22.305	-4.511	< α*	A
2121	2	27.347	-4.847	< α*	A
2122	126	97.355	2.903	.0018	T
2211	5	1.992	2.131	.0165	
2212	2	7.091	-1.912	.0279	
2221	4	8.694	-1.592	.0557	
2222	50	30.951	3.424	.0003	T

[a] < α* indicates that the tail probability is smaller than can be expressed with four decimal places.

2222: older than 55, non-spouse reports, both non-spouse and patient indicate that patient does not exercise.

We also find the following antitypes:

1112: younger than 56, spouse reports, spouse indicates that patient exercises, patient indicates that he/she does not exercise;
2112: older than 55, spouse reports, spouse indicates that patient exercises, patient indicates that he/she does not exercise;
2121: older than 55, spouse reports, spouse indicates that patient does not exercise, patient indicates that he/she does exercise.

This type and antitype pattern is quite interesting. All types suggest that it is more likely than expected from the base model that the patient and the next of kin agree in their statements about whether the patient engages in vigorous exercise at least once a week. In contrast, all antitypes suggest that it is less likely than expected from the base model that the patient and the next of kin disagree in their statements about the patient's exercise habits. In addition, there seem to be clear age differences. Readers are invited to test the hypothesis whether all agreement-configurations (these are the configurations in which the last two numbers are the same) form a large composite type, and all disagreement configurations (these are the configurations in which the two last numbers differ) form a large composite antitype.

We now include the gender covariate. We ask, whether this pattern of types and antitypes is the same for males and females. To answer this question, we create the 2 x 2 x 2 x 2 x 2 cross-classification of the variables A, N, K, P, and G and perform a first order CFA. The base model for this analysis is

$$\log E = \lambda_0 + \lambda_i^A + \lambda_j^N + \lambda_k^K + \lambda_l^P + \lambda_m^G .$$

To make results comparable, we used the z-test and the Bonferroni-adjusted $\alpha^* = 0.0015625$. Note, that the comparison of results may suffer from differences in statistical power. The second analysis processes a cross-classification that has twice as many cells as the first analysis, while the sample size is the same. Table 101 displays results.

The results in Table 101 suggest that four types and four antitypes exist. The types are

11111: younger than 56, spouse reports, both spouse and patient indicate that patient exercises, male;
12111: younger than 56, non-spouse reports, both spouse and patient

indicate that patient exercises, male;

21221: older than 55, spouse reports, both spouse and patient indicate that patient does not exercise, male;

22222: older than 55, non-spouse reports, both non-spouse and patient indicate that patient does not exercise, female.

Table 101: CFA of the cross-classification of Age (A), Next of Kin (N), Response of Kin (K), Response of Patient (P), and Gender (G) (data from the Auckland Heart Study)

Cell index	Frequencies		Test statistics		Type/ antitype?
ANKPG	observed	expected	z	p	
11111	38	4.786	15.183	< α*	T
11112	7	3.093	2.222	.0131	
11121	5	17.037	-2.916	.0018	
11122	2	11.010	-2.715	.0033	
11211	15	20.888	-1.288	.0988	
11212	4	13.498	-2.585	.0049	
11221	92	74.363	2.045	.0204	
11222	42	48.045	-0.873	.1913	
12111	8	1.521	5.252	< α*	T
12112	3	0.983	2.034	.0210	
12121	1	5.416	-1.898	.0289	
12122	1	3.500	-1.336	.0907	
12211	1	6.641	-2.189	.0143	
12212	1	4.291	-1.589	.0560	

/ cont.

Cell index	Frequencies		Test statistics		Type/ antitype?
ANKPG	observed	expected	z	p	
12221	15	23.641	-1.777	.0378	
12222	19	15.277	0.952	.1704	
21111	6	3.806	1.125	.1304	
21112	6	2.459	2.258	.0120	
21121	1	13.594	-3.409	.0003	**A**
21122	0	8.756	-2.959	.0015	**A**
21211	1	16.612	-3.830	< α*	**A**
21212	1	10.735	-2.971	.0015	**A**
21221	82	59.139	2.973	.0015	**T**
21222	44	38.216	0.936	.1747	
22111	3	1.210	1.627	.0518	
22112	2	0.782	1.378	.0842	
22121	0	4.308	-2.075	.0190	
22122	2	2.784	-0.470	.3193	
22211	0	5.281	-2.298	.0108	
22212	4	3.413	0.318	.3753	
22221	9	18.801	-2.260	.0119	
22222	41	12.150	8.277	< α*	**T**

[a] < α* indicates that the tail probability is smaller than can be expressed with four decimal places.

The antitypes are:

21121: older than 55, spouse reports, spouse indicates that patient

exercises, patient indicates that he does not exercise, male;
21122: older than 55, spouse reports, spouse indicates that patient exercises, patient indicates that she does not exercise, female;
21211: older than 55, spouse reports, spouse indicates that patient does not exercise, patient indicates that he does exercise, male; and
21212: older than 55, spouse reports, spouse indicates that patient does not exercise, patient indicates that she does exercise, female.

We now ask whether this pattern of types and antitypes suggests gender differences. If the two gender groups do not differ, each type for males, ...1, has a corresponding type for females, ...2. The same would apply for antitypes. The first three types, constitute by Configurations 11111, 12111, and 21221, seem to exist only in the male population. The fourth type, 22222, seems to be a female-only phenomenon. Thus, there is not a single type that appears in both the male and the female populations. In contrast, there are two antitype-pairs. That is, there are two antitypes that appear both in the female and the male populations. These are the antitype pairs 21121 - 21122 and 21211 - 21212. We therefore conclude that the majority of the cells that stand out as types and antitypes suggest gender differences. Taking into account the stratification variable gender thus has led to a far more differentiated picture than aggregating over gender. Readers are invited to perform a two-sample CFA on the data in Table 101, to confirm the gender differences.

10.5.2 Continuous covariates

In many instances, covariates are continuous rather than categorical. Continuous covariates can be treated in two ways. The first involves categorizing the continuous variables and crossing them with the variables that span the tabulation before consideration of the covariates. The analysis proceeds then as described in Section 10.5.1. There are two problems with this procedure. First, categorization can lead to loss of information and has therefore been criticized (for an overview of issues concerning categorization see Vargha et al., 1996). Still, this option is considered viable by many, in particular if there is a natural cut-off point such as, for example, the zero point when regression coefficients are categorized (see the data example in Section 9.3.2).

The second problem concerns statistical power. When categorized (or categorical) covariates are crossed with the variables that span the tabulation before consideration of the covariates, the number of cells in the table increases by a factor of two or more. Therefore, categorizing continuous covariates is an option only if the sample size is very large or if the number of covariates is very small.

The second method of taking covariates into account involves extending the CFA base model by the covariates (Glück & von Eye, 2000; Gutiérrez-Peña & von Eye, in preparation). Consider the standard log-linear CFA base model,

$$\log E = X\lambda ,$$

where X is the design matrix and λ is the parameter vector (see Chapter 2). This model, extended to take into account the covariates, becomes

$$\log E = X\lambda + C\lambda_c ,$$

where C is a matrix that contains the covariates, and λ_c is the parameter vector for the covariates. As this equation indicates, there is one score per covariate for each cell in the table.

The maximum number of covariates that can be included in a model before it becomes saturated, depends on the size of the table and the complexity of the base model. Suppose a table has t cells and the matrix X contains k vectors, the constant vector included. Then, the maximum number of covariates is $t - k - 1$. Consider, for example, a 2 x 3 x 2 cross-classification of the variables A, B, and C which contains $t = 12$ cells. If the base model for a CFA of this table is the first order main effect model of variable independence, the design matrix X has 1 + 1 + 2 + 1 = 5 columns, where the first column represents the constant vector, the second column represents the main effect of variable A, the third and fourth columns represent the main effect of variable B, and the fifth column represents the main effect of variable C. Thus, five degrees of freedom are consumed by the main effect model. A total of 7 degrees of freedom remain available. If all of these are consumed by covariates (one degree of freedom per covariate vector), the model is saturated. Therefore, the maximum number of covariates for this table and the main effect model is six.

If, in contrast, the second order CFA model is the base model, the number of covariates consumed by this base model is (1 + 1 + 2 + 1) + (2 + 1 + 2) = 10, where the degrees of freedom in the second pair of parentheses indicate the degrees of freedom consumed by the A x B, the A x C, and the B x C interactions, in that order. For this model, only one covariate can be used.

The use of continuous covariates involves estimating a score that represents the cases in a cell optimally. Examples of such scores include the mean, the median, the variance, or the maximum score. If this score is a good representative of the cases in a cell, taking into account the covariate can lead to a more informed appraisal of the data than ignoring the covariate. If, however, the covariate poorly represents the cases in a cell, bias can result.

Consideration of a covariate typically carries the expected cell frequencies closer to the observed cell frequencies than possible without the covariate. Therefore, the number of types and antitypes in a CFA with covariates is typically smaller than without covariates. However, it should be noted that a covariate can also have the effect that the type-antitype pattern changes, new types and new antitypes surface, or that the number of types and antitypes even increases. These cases, however, while possible, are rare (see Glück & von Eye, 2000). In the following paragraphs, we give data examples of both.

Data example 1: Covariate makes types and antitypes disappear. The following data example is a re-analysis of data published by Glück and von Eye (2000). A sample of 181 high school students were administered the 24 items of a paper and pencil cube comparison task. After completing each item, the students responded to a questionnaire concerning the perceived difficulty of the item, the strategies they had employed to process the item, and the perceived quality of their strategy (Glück, 1999). The three strategies the respondents used to solve the cube comparison task are mental rotation (R), pattern comparison (P), and change of viewpoint (V). Each strategy was scored as not used = 1 and used = 2.

In the following analyses, we cross the variables R, P, and V and the variable gender (G; 1 = females, 2 = males), because theory and earlier results suggest that performance in spatial task varies with gender. Table 102 displays the results of first order CFA. We used the normal approximation of the binomial test and the Bonferroni-adjusted $\alpha^* = 0.003125$.

The results in Table 102 suggest that a rich pattern of types and antitypes exists, and that the gender groups differ considerably[2].

[2]The results differ from the ones published by Glück and von Eye (2000), because in the earlier analyses, Cells 1111 and 1112 had been blanked out (see Section 10.1).

Table 102: First order CFA of the cross-classification of Rotational Strategy (R), Pattern Comparison Strategy (P), Viewpoint Strategy (V) and Gender (G) without covariate

Cell index	Frequencies		Test statistics		Type/ antitype?
RPVG	observed	expected	z	p	
1111	25	61.295	-4.677	$< \alpha^*$	A
1112	5	103.185	-9.810	$< \alpha^*$	A
1121	17	10.484	2.015	.0219	
1122	42	17.649	5.811	$< \alpha^*$	T
1211	98	88.273	1.048	.1472	
1212	206	148.600	4.811	$< \alpha^*$	T
1221	13	15.098	-0.541	.2942	
1222	64	25.416	7.681	$< \alpha^*$	T
2111	486	398.584	4.651	$< \alpha^*$	T
2112	729	670.919	2.492	.0064	
2121	46	68.167	-2.711	.0034	
2122	95	114.754	-1.875	.0304	
2211	590	573.964	0.732	.2322	
2212	872	966.216	-3.577	.0002	A
2221	39	98.171	-6.057	$< \alpha^*$	A
2222	199	165.251	2.688	.0036	

[a] $< \alpha^*$: tail probability is smaller than can be expressed with four decimal places.

Specifically, there are the following four types and four antitypes.

Types

1122: males that only use the change of viewpoint strategy

1212: males that only use the pattern comparison strategy
1222: males that use both the pattern comparison and the change of viewpoint strategies
2111: females that only use the rotation strategy

Antitypes
1111: females that use no strategy
1112: males that use no strategy
2212: males that use both the rotation and the pattern comparison strategies
2221: females that use all three strategies.

In addition to the four categorical variables used in Table 102, Glück (1999) also asked, whether a number of continuous covariates allows one to predict this pattern of types and antitypes. If this is the case, some or all of these types and antitypes may disappear. Alternatively, the pattern of types and antitypes can change and new types and antitypes may surface. We now present two examples. In the first example (Table 103), we use the covariate handedness, and one type and three antitypes disappear. In the second example (Table 104) we use the covariate item difficulty, and one type disappears and one new type emerges. For both examples, we use the normal approximation of the binomial test and the Bonferroni-adjusted $\alpha^* = 0.003125$. Thus, differences in results cannot be due to differences in the characteristics of the significance test used. Both covariates contribute significantly to the explanation of the frequency distribution in Table 102. The base model for the frequency distribution in Table 102 must be rejected because of the large $LR\text{-}X^2 = 321.68$ ($df = 11$; $p < 0.01$). The base model that takes handedness into account must be rejected too ($LR\text{-}X^2 = 168.14$; $df = 10$; $p < 0.01$). However, it is significantly better than the base model without the covariate ($\Delta LR\text{-}X^2 = 164.21$; $df = 1$; $p < 0.01$). The base model that takes item difficulty into account is also untenable ($X^2 = 296.95$; $df = 10$; $p < 0.01$), but this model too is significantly better than the original base model ($\Delta LR\text{-}X^2 = 22.33$; $df = 1$; $p < 0.01$). Thus, we can expect types and antitypes to emerge for both covariate models.

The results in Table 103 indicate that the three types 1122, 1212, and 2111, and the antitype 1112 still exist. Type 1222 and antitypes 2212 and 2221 have disappeared. We thus can say that taking into account knowledge of handedness makes the configurations that no longer constitute the type and the three antitypes less of a surprise than not taking into account this knowledge. No new type or antitype surfaced. Still, some

Table 103: First order CFA of the cross-classification of Rotational Strategy (R), Pattern Comparison Strategy (P), Viewpoint Strategy (V) and Gender (G) with Handedness (H) as covariate

Cell index	Frequencies		Covariate	Test statistics		Type ?
RPVG	observed	expected	H	*z*	*p*	
1111	25	33.672	.99	-1.502	.0666	
1112	5	87.343	.91	-8.922	< α*	A
1121	17	16.105	.88	0.223	.4116	
1122	42	21.836	.89	4.329	< α*	T
1211	98	106.895	.81	-0.874	.1911	
1212	206	134.852	.83	6.247	< α*	T
1221	13	17.341	.85	-1.045	.1480	
1222	64	51.956	.75	1.683	.0462	
2111	486	418.999	.83	3.487	.0002	T
2112	729	705.236	.81	1.000	.1585	
2121	46	47.402	.92	-0.205	.4187	
2122	95	114.406	.85	-1.844	.0326	
2211	590	646.907	.75	-2.476	.0066	
2212	872	877.095	.76	-0.198	.4213	
2221	39	26.678	.98	2.395	.0083	
2222	199	219.277	.74	-1.414	.0787	

[a] < α* indicates that the tail probability is smaller than can be expressed with four decimal places; covariates keyed in without decimal point.

Table 104: First order CFA of the cross-classification of Rotational Strategy (R), Pattern Comparison Strategy (P), Viewpoint Strategy (V) and Gender (G) with Item Difficulty (D) as covariate

Cell	Frequencies		Covariate	Test statistics		Type?
RPVG	observed	expected	D	z	p	
1111	25	52.327	.64	-3.806	.0001	**A**
1112	5	105.179	.53	-9.917	< α*	**A**
1121	17	9.257	.73	2.540	.0055	
1122	42	15.639	.77	6.681	< α*	**T**
1211	98	80.132	.77	2.019	.0217	
1212	206	162.967	.65	3.452	.0003	**T**
1221	13	15.598	.78	-0.659	.2549	
1222	64	28.885	.74	6.561	< α*	**T**
2111	486	463.014	.40	1.146	.1259	
2112	729	632.333	.62	4.234	< α*	**T**
2121	46	57.752	.79	-1.559	.0595	
2122	95	109.482	.73	-1.406	.0798	
2211	590	535.302	.77	2.567	.0051	
2212	872	979.747	.74	-4.051	< α*	**A**
2221	39	100.601	.81	-6.231	< α*	**A**
2222	199	177.769	.81	1.634	.0511	

[a]< α*: the tail probability is smaller than can be expressed with four decimal places; covariates keyed in without decimal point.

of the changes are dramatic. For example, the expected cell frequency for Configuration 1111 is smaller by almost 50% when the covariate is taken into account, the expected cell frequency for Configuration 1222 more than doubled when the covariate is taken into account (this type disappears), or the expected frequency for configuration 2221 is reduced by over 72% (this antitype disappears).

We now ask whether new types or antitypes can result from taking into account a covariate. We use the covariate item difficulty and perform the same analysis as for Table 103. Results appear in Table 104.

The results in Table 104 are different again. In particular, there is a number of sign changes. Most importantly, the results in Table 104 suggest that the same antitypes exist as in Table 102. However, the type that was constituted by Configuration 2111 is no longer there, and Configuration 2112 now does constitute a type.

We thus conclude

(1) taking into account continuous covariates can be a useful method of explaining types and antitypes;

(2) continuous covariates can play the same role as in experimental research, where they are often used to balance out differences among samples;

(3) continuous covariates can increase the number of types and antitypes, and they can decrease the number of types and antitypes; the effects of a continuous covariate depends on the correlation of the covariate with the residuals of the model without the continuous covariate.

10.6 CFA of ordinal variables

Thus far in this book, and in virtually all applications of CFA, variables were treated as if they were at the nominal level. However, in many instances, variables are at the ordinal or interval levels. The ordinal nature of variables carries information that can be taken into account when estimating expected cell frequencies. Typically, using this information brings the expected and the observed cell frequencies closer together. As a consequence, types and antitypes are less likely to emerge. This has its analogue in log-linear modeling. Without taking the information into account that comes with the ordinal nature of variables, the models that describe the data well can become unnecessarily complex.

In this section, we describe a method that has been developed for log-linear modeling (Fienberg, 1980) and also employed in prediction analysis (von Eye & Brandtstädter, 1988), for use in CFA (von Eye et al.,

2000). The following description follows closely the one given by Fienberg (1980, pp. 62 ff). This description is tailored to two-dimensional tables. Extensions to higher-dimensional tables can be given.

Consider an I x J cross-classifications with cell frequencies N_{ij}. Suppose the categories of the J columns are rank-ordered, and that the ranks, v_j, are known. As was indicated above, if the model of independence of rows and columns does not hold, one can consider (1) the saturated model or (2) a model that takes the scale level of rows and columns into account. The saturated model is of no particular interest in CFA. Therefore, we specify a model that takes the scale level of the columns into account as

$$\log E = \lambda_0 + \lambda_i^{rows} + \lambda_j^{columns} + (v_j - \bar{v})\lambda^{columns\,'},$$

where the λ indicate parameters that need to be estimated and $\lambda^{columns\,'}$ is the set of special parameters that are estimated to reflect the ordinal column characteristics. $\bar{v}$ is the arithmetic mean of the ranks v_j. The expected frequencies for the present case are estimated in an iterative process. The iteration begins by setting the initial expected frequencies $E_{ij}^{(0)} = 1$ for all $i = 1, ..., I$ and $j = 1, ..., J$. Let the iterations be indexed by κ. In the initial iteration step, in which all expected frequencies are set equal to 1, we set $\kappa = 0$. For the subsequent steps κ, one cycles through the three steps

$$E_{ij}^{(3\kappa+1)} = E_{ij}^{(3\kappa)} \left(\frac{N_{i.}}{E_{i.}^{(3\kappa)}} \right),$$

$$E_{ij}^{(3\kappa+2)} = E_{ij}^{(3\kappa+1)} \left(\frac{N_{.j}}{E_{.j}^{(3\kappa+1)}} \right)$$

$$E_{ij}^{(3\kappa+3)} = E_{ij}^{(3\kappa+2)} \left(\frac{\sum_k v_k N_{ik}}{\sum_k v_k E_{ik}^{(3\kappa+2)}} \right)^{v_j} \left(\frac{\sum_k (1 - v_k) N_{ik}}{\sum_k (1 - v_k) E_{ik}^{(3\kappa+2)}} \right)^{1-v_j},$$

for $i = 1, ..., I$, and $j = 1, ..., J$. If these three steps are performed repeatedly, one obtains convergence toward the estimated expected frequencies, E_{ij}. Notice that the exponents for the E_{ij} are just counters. Only the exponents after the parentheses in the third term are real powers. For comments on speed of convergence and alterations to increase this speed see Fienberg (1980, p. 63).

Data example. To illustrate this procedure, we use the data from Section 9.8 again. Table 87 presented the results of a first order CFA of the Time x Topic of Interaction Cross-Classification in dyads in family BLEI. In this earlier analysis, variable Time was treated as nominal level. However, a case can be made that Time is at least ordinal in nature. Therefore, we now re-calculate these results with Time as an ordinal variable. Table 105 presents the observed frequencies for this cross-classification again. Time constitutes the rows. Therefore, we now estimate the expected cell frequencies under the base model

$$\log E = \lambda_0 + \lambda_i^{rows} + \lambda_j^{columns} + (v_i - \bar{v})\lambda^{rows\prime},$$

where $\lambda^{rows\prime}$ is the set of parameters for the ordinal row characteristics. In the present example, we select the v_i to be the natural numbers in ascending order, that is, 1, 2, 3, 4, 5, 6, and 7. The expected cell frequencies from this method appear in Table 105 in *italics*, below the observed frequencies from Table 88.

Obviously, the expected frequencies from the base model that considers the Time variable ordinal are much closer to the expected frequencies in Table 88, where Time was considered a nominal level variable. With only one exception, all types and antitypes disappear. The only remaining type is for Cell 60. The X^2-component for this cell is 19.23 ($df = 1$; $p = 0.00001$; $\alpha^* = 0.0018$), indicating that lack of joint topic in a dyadic interaction surprisingly re-appears in the second year of the second child's life, at the sixth observation.

We conclude that

(1) taking into account the ordinal nature of variables implies using more information when estimating the expected cell frequencies. Therefore, the observed cell frequencies in ordinal models are typically, but not necessarily, closer to the observed frequencies than without consideration of the ordinal nature of variables.

(2) As a consequence, types and antitypes are less likely to surface when the ordinal base models are used than when nominal base models are used. This consequence can be viewed parallel to using covariates (see Section 10.5).

Table 105: Observed and expected (in *italics*) cell frequencies for Time x Topic of Interaction in a Dyad Cross-Classification; family BLEI; Time is ordinal

		Topic of Interaction in Dyads				Total
		0	1	2	3	
Time	1	11 *10.19*	3 *4.94*	50 *56.77*	91 *83.10*	155
	2	12 *7.94*	10 *5.12*	64 *55.31*	75 *92.63*	161
	3	3 *5.87*	4 *5.04*	46 *51.13*	107 *97.96*	160
	4	0 *4.25*	2 *4.86*	49 *46.34*	106 *101.55*	157
	5	0 *3.06*	5 *4.66*	48 *41.71*	101 *104.57*	154
	6	9 *2.32*	4 *4.70*	39 *39.55*	108 *113.43*	160
	7	0 *1.37*	5 *3.69*	24 *29.19*	101 *95.76*	130
	Total	35	33	320	689	1077

10.7 Graphical displays of CFA results

Thus far, we have presented the results of CFA only in the form of tables, in which we labeled types and antitypes. We have used graphical representations to display curves, polynomials, the data box, or models of causality. Indeed, in the literature, there have been only a few attempts to represent CFA results in graphical form. The reason for this sparsity is that it is hard to create easy-to-understand graphical displays for multidimensional cross-classifications. In this section, we discuss three approaches. The first approach focuses on types and antitypes (von Eye & Niedermeier, 1999). The second approach focuses on the observed frequencies (Aksan et al., 1999). The third approach uses Mosaic displays (Mun, von Eye, Fitzgerald, & Zucker, 2001).

10.7.1 Displaying the patterns of types and antitypes based on test statistics or frequencies

Consider the results in Table 102. In a study on the prediction of performance in a spatial task, Glück (1999) crossed the variables rotation strategy (R), pattern comparison strategy (P), viewpoint strategy (V), and Gender, and performed a first order CFA. The pattern of types and antitypes that resulted from this analysis, appears in the last column of Table 102. Figure 15 presents a bar chart of the *z*-scores of this analysis. Positive *z*-scores rise above the zero line, and negative *z*-scores fall below the zero line. Symmetrically to both sides of the zero line, the figure shows two lines that indicate the magnitude of the *z*-value that needs to be surpassed for a configuration to be significant for $\alpha^* = .05/16 = 0.003125$. This *z*-value is 2.7344.

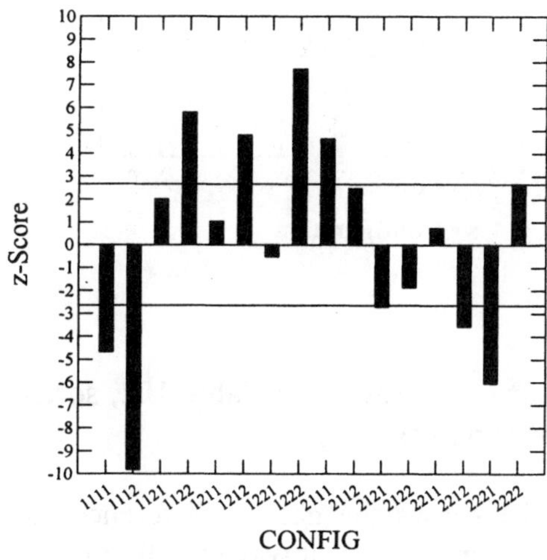

Figure 15: Bar chart of type and antitype pattern for the results in Table 102: order of variables is RPVG.

It is important to note that, because of the nominal level characteristics of the configurations, the order of configurations on the abscissa is arbitrary. Thus, it can be changed to emphasize characteristics of results without changing the validity of the results. For instance, the last digit of the configurations in Figure 15 denotes the respondents' gender. To illustrate the gender differences in a different way than in Figure 15, where the test statistics for the female and male groups are placed next to each other, we now split the bar chart in two halves. The left hand panel of Figure 16 shows the type and antitype pattern for the females, the right hand panel displays the type and antitype pattern for the males.

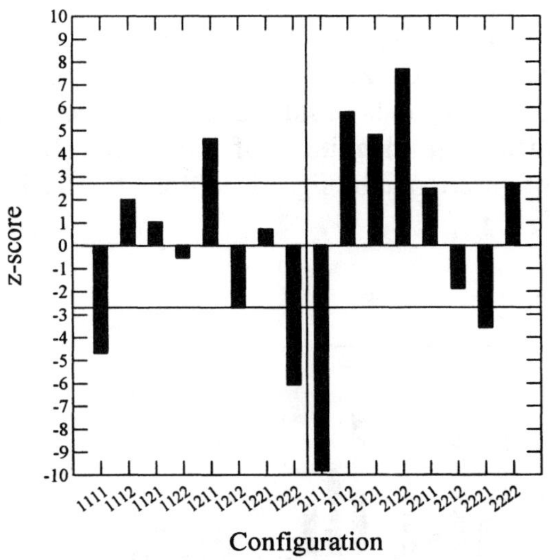

Figure 16: Bar chart of the results in Table 102, sorted by Gender; order of variables is GRPV

The display in Figure 16 shows the gender differences that had been discussed in Section 10.5.2 in a different way than Figure 15. Rather than placing the male and the female test statistics directly next to each other, the statistics are now presented in separate panels.

Both styles, however, while illustrative, lack one important part of the information in a table, the frequencies. From the earlier examples in this book, we know that the magnitude of the z-scores is largely unrelated to the magnitude of the frequencies. Thus, one option is to create a pattern in the bars that varies with frequency. Figure 17 shows the same type/antitype pattern as Figure 16, but the fill pattern is determined by the frequency of a configuration.

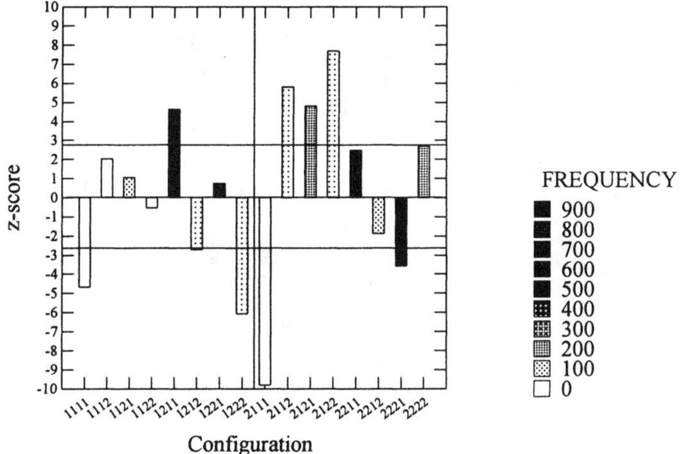

Figure 17: Bar chart of the results in Table 102, sorted by Gender; order of variables is GRPV; fill pattern determined by cell frequency

The content of Figure 17 is interesting because it shows that one of the smallest frequencies comes with the largest test statistics (Configuration 2111; N_{2111} = 5; z_{2111} = -9.81), and the largest frequency is counted for a configuration that constitutes an antitype (Configuration 2221; N_{2221} = 872; z_{2221} = -3.577). However, this display can still be improved. The magnitude of the test statistics is of no concern beyond the type/antitype decision. Researchers do not interpret the magnitude of the test statistics except for using it for decision making. Therefore, a bar chart that shows the cell frequencies and determines the fill pattern depending on the type/antitype decision may be more useful. Figure 18 presents such a bar chart.

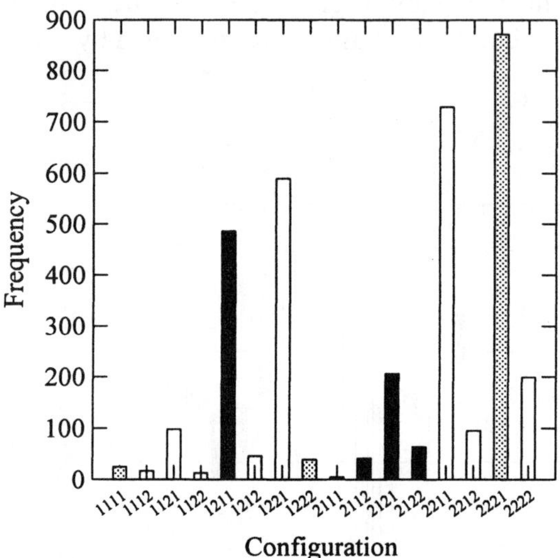

Figure 18: Bar chart of the results in Table 102, fill pattern based on type/antitype decision; order of variables is GRPV (black bars indicate types, gray bars indicate antitypes, white bars are neither types nor antitypes)

The bar chart in Figure 18 displays the observed frequencies from Table 102. It clearly shows the very large differences among the cell frequencies. None of the bar charts in Figures 15 - 17 reflected these differences clearly. In addition, types (black bars) and antitypes (gray bars) are easily located. So, one of the surprising characteristics of this data set, that the largest frequency is found for a configuration that constitutes an antitype, comes out more clearly than in Figure 17. An interesting alternative to the bar chart display in Figure 18, the Mosaic display, is introduced in the next section for use in CFA.

10.7.2 Mosaic displays

Mosaic displays, introduced by Hartigan and Kleiner (1981) and programmed by Wang (1985) and Friendly (1994), represent the cells of a multidimensional contingency table by rectangles. The size of the rectangles is proportional to the cell numbers, typically the cell frequencies. Thus, mosaics are useful for highlighting large differences among cell frequencies.

A mosaic is a pattern of inlaid rectangles that vary in size. There is a correspondence between the rectangles and the cells. This correspondence can be based on the observed frequencies, the expected frequencies, residuals, or whatever cell characteristic is of interest. The order of the variables that span a cross-classification determines the order of the steps in which the rectangles are constructed, and the mosaic pattern.

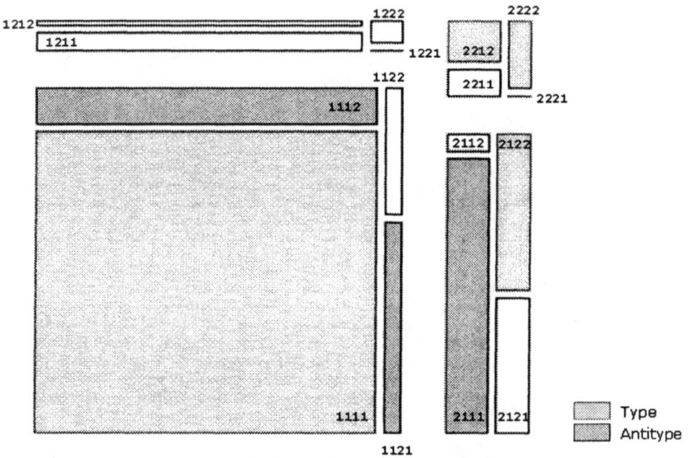

Figure 19: Mosaic display of the CFA of the variables E_1, E_2, I_1, and I_2 (variables explained in text)

Figure 19 displays a sample mosaic. It describes the cross-classification of four variables used in a study by Mun et al. (2001). A sample of 215 boys was rated by their parents when the boys were 3-5 and 6-8 years old. We use the variables Externalizing behavior problems at Wave 1 (E_1); Internalizing behavior problems at Wave 1 (I_1); Externalizing behavior problems at Wave 2 (E_2); and Internalizing behavior problems at Wave 2 (I_2). All four variables had been dichotomized at the clinical cutoffs. A first order CFA of the cross-classification of these four variables showed four types (Configurations 1111, 2122, 2212, and 2222) and three antitypes (Configurations 1112, 1121, and 2111).

The procedure that creates the mosaic displayed in Figure 19, cycles through two iterative steps. Before the first step, a rectangle is created that represents the entire sample. *In the first step, this rectangle is split vertically* in I_1 parts, where I_1 indicates the number of categories of the first variable. The area of these parts is proportional to the marginal frequencies of the first variable. Figure 20 displays this split for the data used for Figure 19. The figure suggests that at the first wave of data collection, when the boys were 3 - 5 years old, many more than half where seen by their parents as not suffering from externalizing behavior problems.

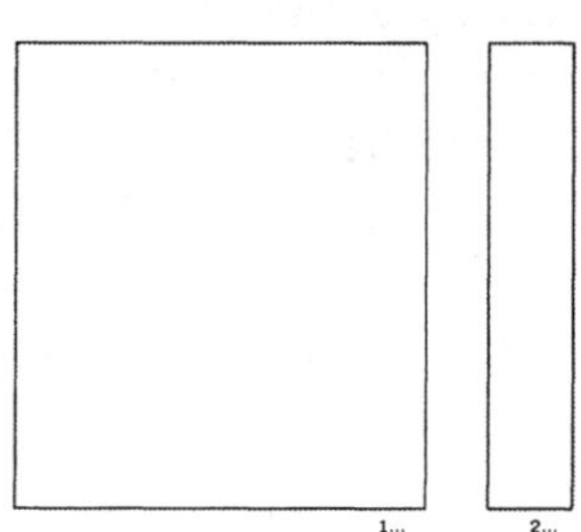

Figure 20: Step 1 of mosaic creation

In Step 2 of the iteration, the I_1 rectangles that resulted from the first step, are split horizontally in I_2 parts, where I_2 indicates the number of categories of the second variable. The area of these parts is proportional to the I_1 x I_2 frequencies of the cross-classification of the first two variables under study. Figure 21 displays the result of this step for the present data example. Externalizing behavior problems at Time 1 is crossed with internalizing behavior

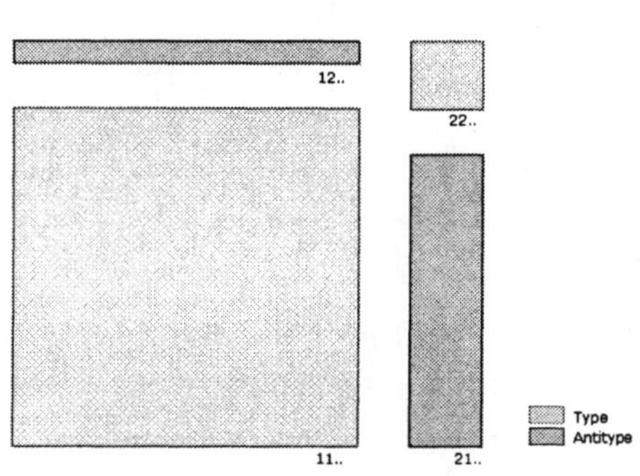

Figure 21: Step 2 of mosaic creation

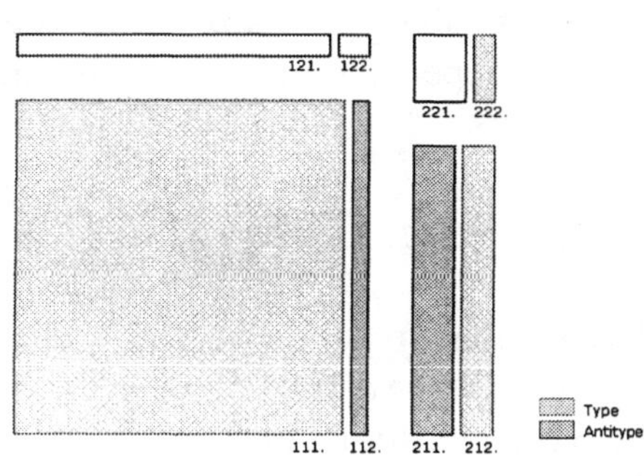

Figure 22: Step 3 of mosaic creation

problems, also at Time 1. The shading of the tiles reflects the results of a CFA of this 2 x 2 table. The results suggest that boys at this young age are perceived by their parents as suffering from either both externalizing and internalizing behavior problems (Type 11) or neither (Type 22). Suffering from only one of these behavior problems is unlikely (Antitypes 12 and 21).

Step 3 of the mosaic building process is an iteration of the first. It involves a vertical split. The resulting I_1 x I_2 x I_3 tiles reflect the cell frequencies of the cross-classification of the first three variables under study. Figure 22 displays the results of this step for the present data example, along with a CFA of this table.

Figure 22 suggests that the CFA of the 2 x 2 x 2 cross-classification of externalizing behavior problems and internalizing behavior problems, both at Wave 1, with externalizing behavior problems at Wave 2 yields three types and two antitypes. The types 111 and 222 suggest again that boys are perceived as consistently showing either all or none of the observed problems. In addition, there seems to be a type of boys with only externalizing behavior problems (Configuration 212). The first antitype (Configuration 112) indicates that it is unlikely that boys develop externalizing behavior problems at age 6-8, if they were not perceived as suffering from these problem at age 3-5. The second antitype suggests that suffering from externalizing behavior problems only at age 3-5 is unlikely too.

The fourth step involves an iteration of the second, involving a horizontal split. Figure 19, above, depicts the results of this step.

In CFA applications, the sequence of steps presented here for illustrative purposes is rarely of interest. Rather, the final step is the main goal of analysis. However, for instance in developmental studies, the changes from one observation point in time to the next may be important. In cross-sectional studies, the order is of lesser interest, in particular when the order of variables is arbitrary. The arrangement of the tiles in a mosaic display depends on the order of variables. The size of the tiles and the shading of types and antitypes is independent of the order of variables.

10.8 Aggregating results from CFA

There are instances in which types or antitypes differ in only one category. Consider, for example, the results from the data example in Section 10.7.2. The analysis resulted in four types, constituted by Configurations 1111, 2122, 2212, and 2222. In this example, the second and the fourth types differ only in the second digit, and the third and the fourth types differ only in the third digit. We now ask whether we can simply these results. We distinguish between two cases: dichotomous variables (Lienert, 1971c), and variables with three or more categories (von Eye & Brandtstädter, 1982).
Dichotomous variables. To simplify results from CFA, we use a theorem from statement calculus that has an analogue on Boolean algebra (Hoernes & Heilweil, 1964). Consider the two statements A and B. The negations of these statements are $\bar{A}$ and $\bar{B}$. In its *disjunctive form*, the theorem states that

$$AB + A\bar{B} = A.$$

In words, the two statements AB and $A\bar{B}$ can be reduced to $A.$ where the period indicates that the aggregation occurred over the second variable. Repeated application of this procedure is called the Quine and McCluskey method (Hoernes & Heilweil,1964). Conditions for proper application of this method are

(1) the elementary statements, that is the statements that involve only one term, e.g., A or B, have only two values (truth values), e.g., A and $\bar{A}$;

(2) the composite statements, that is the statements that involve two or more terms, e.g., AB, can differ in only one term;

(3) the composite statements must share at least one elementary statement.

By way of analogy we now apply the above theorem to the results of CFA. Consider the two types, 11 and 12. These two types share the first digit in common and differ only in the second digit. Application of the above theorem leads to the *reduced type*, 1., where the period indicates that the reduction involved aggregating over the second variable. Reduced types are therefore also called *aggregate types*.

To give another example, consider again the results from Section 10.7.2. There are the four types 1111, 2122, 2212, and 2222. The first of these four types differs from the other three in more than one digit. Therefore, this type cannot be reduced in any form. The second type differs from the fourth in only the second digit. We can therefore create the aggregate type 2122 + 2222 = 2.22. This type suggests that there are more boys than expected based on chance who suffer from externalizing behavior problems at age 3-5 and from both externalizing and internalizing behavior problems at age 6-8, and this regardless of whether these boys suffered from internalizing behavior problems at age 3-5 or not.

Using the same theorem, the third and the fourth types can be reduced to form the aggregate type 2212 + 2222 = 22.2. The period indicates that the aggregation went over the third variable. In a subsequent step, the methods described in Section 10.3 can be used to ensure that the aggregate types or antitypes have a statistical basis.

Variables with three or more categories. Von Eye and Brandtstädter (1982) proposed an extension of the above theorem. The extended version involves the two variables, *A* and *B*. *A* has c_A categories and *B* has c_B categories. Then, the theorem states that

$$A_iB_1 + A_iB_2 + \dots + A_iB_{c_B} = A_{i\cdot}\ ,$$

where all categories of *B* are included. Consider, for example, variable A with the two states A_1 and A_2, and variable B with the three states B_1, B_2 and B_3. Then, the following aggregation is possible: $A_1B_1 + A_1B_2 + A_1B_3 = A_{1.}$.

In general, aggregation of CFA types or antitypes proceeds under the following rules:

(1) Only one variable at a time can be aggregated. However, repeated application of aggregation is possible;

(2) the position of the variable that is aggregated, has no effect; thus, variables can be aggregated regardless of position; however, variables from different positions cannot be aggregated;

(3) aggregation involves either types or antitypes, but not both;

(4) each configuration can be aggregated with any number of other configurations from the same table; Rule 1 always applies;

(5) after aggregation, the methods described in Section 10.3 must be applied to make sure the aggregated type or antitype is still statistically tenable.

Data example. The following example re-analyzes data published by Lienert (1988). A sample of 103 depressed inpatients were assessed in the symptoms anxious (A), tense (T), slowed (S), and suicidal (M). Each symptom was scored as either present (= 1) or absent (= 2). We first analyze these data using a first order CFA. We use the z-test and the Bonferroni-adjusted $\alpha^* = 0.003125$. Table 106 displays the 2 x 2 x 2 x 2 cross-classification of the variables A, T, S, and M, along with CFA results.

Table 106: First order CFA of the variables anxious (A), tense (T), slowed (S), and suicidal (M)

Cell index	Frequencies		Test statistics		Type/ Antitype?
ATSM	observed	expected	*z*	*p*	
1111	5	4.68	0.147	.4417	
1112	9	5.05	1.758	.0394	
1121	4	1.45	2.125	.0168	
1122	4	1.56	1.956	.0253	
1211	2	6.35	-1.726	.0422	
1212	0	6.85	-2.616	.0044	
1221	5	1.96	2.173	.0149	
1222	1	2.11	-0.766	.2220	
2111	6	11.86	-1.702	.0444	
2112	12	12.79	-0.222	.4123	
2121	4	3.66	0.177	.4297	
2122	1	3.95	-1.484	.0689	
2211	22	16.08	1.476	.0699	
2212	25	17.34	1.839	.0329	
2221	3	4.96	-0.881	.1891	
2222	3	5.35	-1.017	.1546	

The results in table 106 suggest no type and no antitype. Still, Lienert discusses the possible existence of the two aggregate types 112. and 221.. To investigate this possibility, we employ the methods for establishing composite types and antitypes presented in Section 10.3. For the possible aggregate Type 112., we calculate

$$z_{112.} = \frac{2.125 + 1.956}{\sqrt{2}} = 2.886 ,$$

a value that suggests that the aggregate type 112. exists (p = 0.002). For the possible aggregate Type 221., we calculate

$$z_{221.} = \frac{1.476 + 1.839}{\sqrt{2}} = 2.344 .$$

The tail probability for this value is $p = 0.009$. Thus, the aggregate type 221. may also exist.

An issue of concern in the present context is the protection of the experiment-wise α. If researchers aggregate types or antitypes, the significance level needs to be protected. Specifically,

(1) if no routine exploratory CFA is performed, the significance level needs to be protected based on the number of aggregate types and aggregate antitypes;

(2) if routine exploratory CFA is performed, the significance level needs to be protected based on the number of tests for types and antitypes plus the number of aggregate types and aggregate antitypes.

Let the number of type/antitype tests be t and the number of aggregate types and antitypes be a. Then, a Bonferroni procedure for the protection of α when testing for aggregate types and aggregate antitypes yields the adjusted significance level

$$\alpha^* = \frac{\alpha}{t + a}.$$

This adjusted level is applied to both the type/antitype tests and the tests in the aggregation step. In the example in Table 106, we performed 16 type/antitype tests and 2 tests for aggregate types. The Bonferroni-adjusted α for the entire analysis is therefore $\alpha^* = 0.05/18 = 0.002778$. Therefore, we can retain the aggregate type 112. but we retain the null hypothesis concerning the aggregate type 221..

10.9 Employing CFA in tandem with other methods of analysis

Rarely, researchers employ just one method of analysis. In most applications, data are looked at from various perspectives and with both exploratory and explanatory goals in mind. For example, when researchers perform regression analysis in an explanatory step, they typically also investigate the correlations among the predictors and the residual distributions. The latter two are exploratory steps of analysis. In a similar fashion, CFA is often employed as an exploratory method before or after other exploratory or explanatory methods. In the following sections, we present two examples. The first example involves using cluster analysis before CFA (Bergman, 2000; Bergman & El-Khouri, 1999). The second example involves using discriminant analysis after CFA (cf. Aksan et al., 1999).

10.9.1 CFA and cluster analysis

Thus far in this book and in virtually all CFA applications, the contingency tables subjected to configural analysis were spanned using categorical variables. However, there are other options. Bergman (2000; cf. Bergman & El-Khouri, 1999) proposed combining cluster analysis and CFA in the following way. First, typical patterns of behavior are created using cluster analysis. The resulting clusters are called *I-states*. They represent an individual's configuration at a particular point in time. Second, methods of CFA are used to analyze the I-states in the space of variables not used in the cluster search, or to analyze the transition patterns when clusters have been formed separately for each point in time. This approach is called *I-States as Objects Analysis* (*ISOA*; Bergman & El-Khouri, 1999).

To describe the role played by CFA in this context, consider a study in which a number of continuous variables are used to form C clusters. Membership in clusters can then be considered a categorical variable with C categories. This variable can be crossed with other categorical variables, and the resulting cross-classification can be subjected to CFA. Suppose, C clusters and d categorical variables are crossed to form a $C \times c_1 \times c_2 \times \ldots c_d$ cross-classification, where c_i is the number of categories of the ith variable, with $i = 1, \ldots, d$. Then, the base model for a first order global CFA of this cross-classification is

$$\log E = \lambda_0 + \lambda^i + \lambda^j + \ldots + \lambda^d + \lambda_k^C,$$

where the superscripts indicate the variables that span the cross-classification and λ_k^C indicates the parameters for the cluster membership variable. Other CFA base models can be specified accordingly. In each of these models, the cluster membership variable is treated in the same way as the other categorical variables.

Data example. The following data example uses the Finkelstein et al. (1994) data on the development of aggressive behavior again (see Sections 3.10.6 and 5.2.2). In this study, the authors administered a questionnaire concerning aggressive behavior in adolescent boys and girls at three points in time. The time intervals were two years each. The questionnaire addressed the four dimensions of aggression Aggressive Impulse, Aggression-Inhibitory Response, Verbal Aggression against Adults, and Physical Aggression against Peers. In addition, the physical pubertal development was assessed using Tanner scores. In the following analyses, we use the data from the first wave of the survey.

In a first step, we clustered the questionnaire data. We used Ward's (1963) method and Euclidean distances. The three-cluster solution was the most interesting. The first cluster contains 52 low-aggression adolescents. They experience infrequent aggressive impulses, low aggression-inhibitory responses, are involved in infrequent physically aggressive acts against peers, and in infrequent verbal aggression. The second cluster contains 36 highly aggressive adolescents. They experience frequent aggressive impulses, average aggression-inhibitory impulses, are involved in very frequent physically aggressive acts against their peers, and in highly frequent verbal aggression. The third cluster contains 24 rather average adolescents. These respondents display a slightly elevated level of aggressive impulses, a very high level of aggression-inhibitory responses, are involved in infrequent physically aggressive acts against their peers, and average levels of verbal aggression.

For the following analyses, we use these clusters as *I-States*, that is, as objects of configural analysis. Specifically, we ask, whether cluster membership can be predicted from the gender of the respondents and their physical pubertal development. The three variables used in this analysis are Cluster Membership (C; categories are 1, 2, and 3), Gender (G; female = 1, male = 2), and Tanner Stage at age 11 (T; pre-pubertal = 1, beginning pubertal = 2, pubertal = 3[3]), and the cross-classification is of size 3 x 2 x 3.

This table is now analyzed using the Prediction-CFA base model

$$\log E = \lambda_0 + \lambda_i^C + \lambda_j^G + \lambda_k^T + \lambda_{jk}^{GT}.$$

This model indicates that Gender and Tanner Stage serve as predictors of Cluster Membership. The analysis used Lehmacher's test and the Bonferroni-adjusted $\alpha^* = 0.0027778$. The observed frequencies and the results of P-CFA appear in Table 107.

The results in Table 107 suggest one prediction antitype and one prediction type. The antitype, constituted by Configuration 212 suggests that female adolescents who display average physical pubertal development are unlikely to belong to the high aggression cluster. The type is constituted by Configuration 222. This type suggests that boys who display average physical development can be predicted to belong to the high aggression cluster.

Discussion. There are two chief reasons why ISOA, that is, *I-states as Objects Analysis* is of importance in the context of CFA. First, as was indicated in the example in Table 107, researchers can use cluster analysis to create typical patterns of behavior that then can be subjected to CFA. When creating these patterns, there is no need to categorize continuous variables which often results in a loss of information. Clusters reflect centers of density in the data space and are comparable to types from zero order CFA. Second, cluster membership for clusters from a series of data waves in repeated observation studies can be crossed and analyzed using CFA (see Bergman, 2000). Resulting types and antitypes indicate most typical and atypical temporal patterns.

---

[3]Note that there was a fourth stage, T = 4. This stage indicates physically mature genital development. This stage was not observed in this sample at age 11.

Table 107: P-CFA of the predictors Gender and Tanner Stage and the criterion Cluster Membership

Cell index	Frequencies		Statistical tests		Type/ Antitype ?
CGT	observed	expected	z_L	p	
111	14	12.07	.862	.1945	
112	17	14.86	.895	.1855	
113	2	3.25	-.974	.1650	
121	13	12.07	.415	.3391	
122	5	9.29	-2.111	.0174	
123	1	0.46	1.074	.1414	
211	4	8.36	-2.079	.0188	
212	4	10.29	-2.803	.0025	**A**
213	2	2.25	-.208	.4176	
221	12	8.36	1.738	.0411	
222	14	6.43	3.982	$< \alpha^*$	**T**
223	0	0.32	-.688	.2456	
311	8	5.57	1.319	.0936	
312	11	6.86	2.102	.0178	
313	3	1.50	1.421	.0777	
321	1	5.57	-2.482	.0065	
322	1	4.29	-1.967	.0246	
323	0	0.21	-.522	.3008	

[a] $< \alpha^*$ indicates that the tail probability is smaller than can be expressed with four decimal places.

10.9.2 CFA and discriminant analysis

In this section, we entertain the question whether results from CFA can be analyzed further in the pursuit of additional questions. For example, one may ask whether the types identified using CFA also differ in the space of variables not used in CFA. If such differences are substantial, the types can be considered *externally valid.*

The parameters inspected in the space of other variables depend on the researchers' interests. For instance, one can examine the covariance structures and ask whether they allow one to distinguish among the types, or the types and the respondents who do not belong to a particular type. Methods of structural equation modeling would be used to answer this question. Stacked or multi-group models are among the suitable approaches (Jöreskog & Sörbom, 1993). In addition or alternatively, one can ask whether the means of members of types differ in other variables. The method to answer this question would be MANOVA. Still a third way of comparing types involves using discriminant analysis. This method allows one to answer the question whether the types are located in different areas of some discriminant space. Finally, one can follow up CFA with other CFA runs. One can ask whether types, antitypes, and non-suspicious configurations can be discriminated in the space of other categorical variables.

In the literature, there exists a number of attempts to follow up CFA with other analyses. For example, Görtelmeyer (1988) created types of sleep disorder using CFA. Then, he asked whether these types can be discriminated in the space of psychosomatic symptoms. To answer this question, Görtelmeyer used CFA again (cf. Table 96, above). Another example is the work of Mahoney (2000). The author used CFA to identify types and antitypes of school-related adjustment behavior in adolescents. In a post hoc step, the author used ANOVA to answer the question whether competence allows one to explain the existence of the outlandish configurations. In the following data examples, we demonstrate the use of discriminant analysis as a follow-up to CFA.

Data example. The following example uses data from a study on successful development among African American and Latino male adolescents (Taylor, Lerner, Villaruel, & von Eye, 2000; cf. Taylor, Lerner, von Eye, Sadowski, Bilalbegovic, & Dowling, 2001). We analyze the data from a sample of 95 male African American adolescents. The respondents were presented with a number of questions in face-to-face interviews. Three of these questions concerned how they settle disputes with their friends (D; 1 = using force, 2 = not using force), whether they go to church (C; 1 = no, 2 = yes), and whether they have sexual relations (S; 1 = yes, 2 = no).

In a first step of analysis, we crossed these three variables and subjected the resulting table to a standard first order CFA. We used Lehmacher's test and the Bonferroni-adjusted $\alpha^* = 0.00625$. The results of this analysis appear in Table 108.

Table 108: First order CFA of the cross-classification of Dispute, Church, and Sex

Cell index	Frequencies		Statistical tests		Type/ Antitype
DCS	observed	expected	z_L	p	?
111	29	17.665	4.318	$< \alpha^*$	**T**
112	9	8.556	0.202	.4200	
121	11	18.041	-2.672	.0038	**A**
122	4	8.738	-2.143	.0161	
211	7	13.998	-2.793	.0026	**A**
212	2	6.780	-2.326	.0100	
221	17	14.296	1.074	.1414	
222	16	6.925	4.388	$< \alpha^*$	**T**

[a] $< \alpha^*$ indicates that the tail probability is smaller than can be expressed with four decimal places.

The results in Table 108 show two types and two antitypes. The first type, constituted by Configuration 111 describes those 29 adolescents who settle disputes with friends using force, do not go to church, and do have sexual relations. The second type, constituted by Configuration 222, describes adolescents with just the opposite profile. These respondents settle disputes with friends peacefully, do go to church, and do not have sexual relationships. Both profiles were observed significantly more often than expected based on chance.

The first antitype is constituted by Configuration 121. These are the respondents who settle disputes with friends using force, but do go to church and do have sexual relations. The second antitype, constituted by Configuration 211, describes those respondents who settle disputes among friends peacefully, do not go to church and do have sexual relations. Both of these profiles were observed significantly less often than expected based on chance.

Although these types and antitypes are interesting in themselves, we now go an additional step and ask whether the two types differ from each other and from all other respondents in the space of additional variables. Specifically, we ask whether the variables Total Assets (defined by the quantity and quality of social support available to an individual), Drug Use, and Safety of Neighborhood allow one to discriminate between the three groups of the non-types (Group 0), Type 111 (Group 1), and Type 222 (Group 2). To answer this question we perform a discriminant analysis. We use the Fisher discriminant criterion which leads to a maximization of the variance between groups and a minimization of the variance within groups. The Wilks Λ for this analysis was 0.2651, indicating that 73.49% of the variance of the criterion, group membership, is explained by the three predictors. This value is significant (df = 3, 2, 92; the F approximation is F = 28.2693; $df1$ = 6, $df2$ = 180; $p < 0.01$). The classification matrix appears in Table 109.

Table 109: Classification matrix for the discrimination among the types and non-types from Table 108

		Respondents grouped into			Original Group Size	percent correct
		Group 0	Group 1	Group 2		
Original group	Group 0	26	21	3	50	52
	Group 1	4	12	0	16	75
	Group 2	4	0	25	29	86
New group size		34	33	28	N = 95	

The classification matrix in Table 109 shows a very interesting pattern. The two types (Groups 1 and 2) are perfectly separated from each other in the space of the variables used for discrimination. None of the re-classified respondents from Type 111 (= Group 1) was re-classified into Type 222 (= Group 2), and vice versa. In addition, the two types were very well reproduced, the % correct scores are 75 for Type 111 and 86 for Type 222. In contrast, the respondents who do not belong to either of these two types, are less well identified. 21 of the originally 50, that is, 42% were grouped into Type 111. That is, in the space of the variables Total Assets, Drug use, and Neighborhood safety these 21 respondents cannot be discriminated from members of Type 111 very well. Three of the originally 50 were grouped into Type 222.

Discussion. Section 10.9 illustrated the application of CFA in the context of other methods of multivariate data analysis. The number of multivariate methods that can be used in tandem with CFA is large. This section discussed cluster analysis and discriminant analysis employed together with CFA. Other examples include the use of CHAID or POIPG in combination with CFA. CHAID, implemented, e.g., in the SPSS package, is the acronym for *Ch*i-squared *A*utomatic *I*nteraction *D*etector. This method analyzes a hierarchy of bivariate cross-classifications in which a dependent variable is related to a hierarchy of independent variables. Lautsch and Ninke (2000) propose using the CHAID method before CFA. Once promising predictor-criterion relationships are established using CHAID, P-CFA can be used to explore these relationships in more detail.

In a similar fashion, Wood (in preparation) proposes using POIPG and CFA in parallel. POIPG is the acronym for "partially oriented inducing path graph." That is, a Bayesian method that allows one to express the dependencies among variables. Wood indicates that representing categorical data by means of directed graphs one the one hand and by CFA on the other may be useful because the two methods allow one to capture different data characteristics. In addition, however, TETRAD can be used to (1) identify subsets of variables implicated in the possible identification of types and antitypes, and (2) check whether types and antitypes sufficiently account for patterns of dependence between several variables.

For a joint application of log-linear modeling and CFA see Netter et al. (2000).

There is a number of benefits to combing methods of analysis. Three benefits are discussed in this section. The first and foremost benefit is that the specific and unique strengths of each method can be exploited. For example, CFA in its present state of development requires that variables be categorical (with the exception of covariates; see Section

10.5.2). Other methods such as cluster analysis, structural equations modeling, regression analysis, MANOVA, or discriminant analysis operate mostly with continuous variables. CFA can produce results similar to some of these methods at the person level. However, it can process continuous variables only if they are categorized which often is paid for by loss of information. Another example is the use of latent variable modeling. Thus far, CFA only processes manifest variables. Therefore, if researchers wish to entertain latent variables hypotheses, structural models are the methodology of choice. In turn, higher order interactions are most easily be dealt with using such methods as log-linear modeling and CFA. Therefore, continuous variables methods are most fruitfully complemented by log-linear modeling or CFA when higher order interactions are of interest. In addition, CFA is the method of choice when analyses are performed at the person level.

The second benefit concerns the size of the cross-classification under study relative to the sample size. CFA is similar to methods of log-linear modeling in common in that it typically uses the cross-classification of all variables. The number of cells in a cross-classification increases with the number of categories in a variable, and increases exponentially with the number of variables. As a consequence, the required number of cases in a sample also needs to increase with the number of variable categories and the number of variables. The limits of *doable* social science research are reached soon when the number of variables to be crossed increases. Therefore, CFA and other methods of multivariate data analysis often complement each other to do justice to the complex multivariate nature of a data set. It should be considered, however, that person level analyses often require the use of CFA. Thus, compromises may be needed.

Third, different methods of data analysis allow one to answer different questions. Therefore, to answer these specific questions, the appropriate methods need to be employed. For example, point estimation as possible in regression analysis, cannot be performed using CFA. Thus, there is no way around regression methods when point estimates are needed. Another example involves the latent variables mentioned above. When testing hypotheses or modeling with latent variables, researchers leave the domain of CFA and resort to using structural equations models (Bartholomew & Knott, 1999).

11. Alternative approaches to CFA

This section presents two alternative approaches to CFA. These two approaches do not just introduce minor cosmetic changes. Rather, they go radically different ways while keeping the main idea of person-level research intact. Both approaches allow researchers to individually test cell frequencies against expected values. The first of the two approaches, proposed by Kieser and Victor (1991, 1999, 2000), uses different methods when estimating expected cell frequencies. Specifically, this approach uses the more general quasi-independence models (cf. Section 10.1, above). The second approach, due to Wood, Sher, and von Eye (1994), and Gutiérrez-Peña and von Eye (2000), views CFA from a Bayesian perspective. This approach allows one to take into consideration prior and subjective information in the search for types and antitypes.

11.1 Kieser and Victor's quasi-independence model of CFA

CFA types and antitypes are statistically defined as local violations of the assumptions specified via the CFA base model. Victor (1989) stressed that the standard CFA base model involves *all* cells of the cross-classification under study. This includes those cells that represent types or antitypes. If types or antitypes exist, deviations from the expected probabilities can result in practically all cells of a cross-classification. This can lead to misinterpretations of the true structure present in a table. Kieser and Victor

(1999) present the following example. In a 3 x 3 table, the cell frequencies are perfectly uniformly distributed, with two exceptions. In Cell 11, there are fewer observations, and in Cell 33, there are more observations than in the rest of the cells. One would, therefore, assume that these two cells are identified as violating the otherwise uniform distribution, and that all other cells are inconspicuous in terms of types and antitypes. However, CFA identifies all cells as types or antitypes, with the exception of Cell 11. Kieser and Victor (1999) present the artificial data example given in Table 110. We use Lehmacher's test and the Bonferroni-adjusted $\alpha^* = 0.005556$.

Table 110: CFA of a table with two violations from independence

Cell index	Frequencies		Tests		Type/ Antitype ?
	Observed	Expected	z_L	p	
11	1	1	0.000	0.5	
12	10	1.429	7.603	$< \alpha^*$	**T**
13	10	18.571	-5.986	$< \alpha^*$	**A**
21	10	1.429	7.603	$< \alpha^*$	**T**
22	10	2.041	5.971	$< \alpha^*$	**T**
23	10	26.531	-9.765	$< \alpha^*$	**A**
31	10	18.571	-5.986	$< \alpha^*$	**A**
32	10	26.531	-9.765	$< \alpha^*$	**A**
33	370	344.898	11.675	$< \alpha^*$	**T**

[a] $< \alpha^*$ indicates that the tail probability is smaller than can be expressed with four decimal places.

Kieser and Victor (1999, p. 969) conclude from this example that CFA "does not appropriately describe deviations from the general population rule." The reasons for this lack of appropriate description lies in the following two characteristics of standard CFA:

1. The hypotheses tested in CFA are logically dependent (see Section 3.10). Because of this dependence, CFA can yield only an a priori determined number of types and antitypes. The authors note three examples[1]. (a) In tables with two or more dimensions, there exists either no local violation at all or at least four. (b) In two-dimensional tables of the minimum size 3 x 3, there cannot be exactly five violations of the base model (Perli, 1984). (c) Sole violations of the CFA base model cannot be modeled at all.
2. When fitting the CFA base model, possible type configurations or antitype configurations are not taken into account. Thus, the base model implies the assumption that types or antitypes do not exist.

For these reasons, Victor (1989) and Kieser and Victor (1999, 2000; cf. Lienert, 1989) proposed an alternative way of estimating expected cell frequencies in *confirmatory CFA*. Specifically, the authors proposed estimating expected cell frequencies using the more general log-linear models of quasi-independence. These are models that allow one to blank out specified cells and to fit the CFA base model to the rest of the table. The blanked out cells are those for which types and antitypes were suspected. In more technical terms, the original CFA log-linear base model was

$$\log E = X\lambda,$$

where X is the design matrix and λ is the parameter vector. The log-linear model of quasi-independence is

$$\log E = X_b\lambda + X_t\tau,$$

where X_b is the design matrix for the original base model, and X_t is the design matrix in which researchers specify the cells for which they expect types and antitypes. τ is a parameter vector analogous to λ. This model is equivalent to the model with structural zeros introduced in Section 10.1, above. In fact, the model proposed by Kieser and Victor treats cells for which types and antitypes are expected as structural zeros. In other words, Kieser and Victor's model blanks out cells that are type or antitype candidates and asks whether the base model fits for the remainder of the table.

---

[1]It should be noted that the reference base model for all three results is the log-linear main effect model.

Using the log-linear model of quasi-independence, Kieser and Victor (1999) propose a new approach to confirmatory CFA according to which a set of configurations, *T*, is declared to constitute types or antitypes if the following two assumptions hold:

(i) the CFA base model holds for all cells that do not belong to *T*, and
(ii) a superimposed models holds for the cells in *T*.

To examine these two assumptions, two hypotheses must be tested:

(1) H_1: the CFA base model reproduces the frequencies of those cells adequately that do not belong to *T*. This hypothesis can be tested using a quasi-independence log-linear model that blanks out those cells for which types or antitypes are anticipated.

(2) H_2: the cell probabilities for the cells that do belong to T deviate from the association structure defined by the CFA base model. This can be shown by testing the parameters that come with each of the vectors for τ. (Note that this procedure is slightly different than the procedure described by Kieser and Victor, 1999, who proposed using conditional likelihood test statistics. For alternative tests see Lienert, Dunkl, & von Eye, 1990.)

Data example. To illustrate Kieser and Victor's approach to confirmatory CFA, we re-analyze the same data as Kieser and Victor (1999). These are Lienert's (1964) LSD data (see also Tables 1, 13, and 94). In a sample of 65 students, the effects of LSD 50 were measured in the three variables Narrowed Consciousness (C), Thought Disturbance (T), and Affective Disturbance (A). Each of these variables was scaled as 1 = present or 2 = absent. To compare results, we use both the classical CFA base model and Victor and Kieser's quasi-independence model. The log-linear base model of quasi-independence is

$$\log E = \begin{bmatrix} 1 & 1 & 1 & 1 \\ 1 & 1 & 1 & -1 \\ 1 & 1 & -1 & 1 \\ 1 & 1 & -1 & -1 \\ 1 & -1 & 1 & 1 \\ 1 & -1 & 1 & -1 \\ 1 & -1 & -1 & 1 \\ 1 & -1 & -1 & -1 \end{bmatrix} \begin{bmatrix} \lambda_0 \\ \lambda_C \\ \lambda_T \\ \lambda_A \end{bmatrix} + \begin{bmatrix} 1 & 0 \\ 0 & 0 \\ 0 & 0 \\ 0 & 0 \\ 0 & 0 \\ 0 & 0 \\ 0 & 0 \\ 0 & 1 \end{bmatrix} \begin{bmatrix} \tau_{111} \\ \tau_{222} \end{bmatrix}.$$

The first design matrix on the right hand side of this equation represents the standard first order CFA base model of variable independence. The second design matrix indicates the cells that are suspected to represent the type (Cell 111) and the antitype (Cell 222). The corresponding parameters are τ_{111} and τ_{222}. In standard CFA, the log-linear parameters are of lesser interest. Here, the τ-estimates are used to determine the status of configurations as types or antitypes.

Table 111 presents results from both standard CFA and Victor and Kieser's confirmatory CFA. Specifically, the table reproduces the expected frequencies that were estimated for both approaches. Note that for reasons of estimability and to replicate the results in Kieser and Victor (1999), the observed zero in Cell 222 was replaced by 0.125 (cf. Agresti & Yang, 1987). Thus, the results in Table 111 and the results in Tables 13 and 94 are not strictly comparable (differences in results are minimal, however).

Table 111: Results from standard CFA and Victor and Kieser's CFA for Lienert's LSD data (Configurations 111 and 222 are expected to constitute a type and an antitype, respectively)

Configuration	Frequencies		
CTA	observed	expected for standard CFA	expected for Victor and Kieser's CFA
111	20	12.506	20
112	1	6.848	1.947
121	4	11.402	3.328
122	12	6.244	11.662
211	3	9.464	2.662
212	10	5.182	9.328
221	15	8.629	15.947
222	0.125	4.725	0.125

To establish the type for Configuration 111 and the antitype for Configuration 222, we first compare the goodness of fit indices of the two base models. We obtain for the standard CFA base model the $LR\text{-}X^2 = 43.916$ ($df = 4$; $p < 0.01$) and for the quasi-independence model the $LR\text{-}X^2 = 0.843$ ($df = 2$, $p = 0.6561$). The difference between these two models is significant ($\Delta X^2 = 43.073$; $\Delta df = 2$; $p < 0.01$). We now ask whether indeed the blanking out of the two designated cells makes a significant contribution to the explanation of the structure in the 2 x 2 x 2 cross-classification. The parameters are estimated to be $\tau_{111} = 3.569$ ($se = 0.760$; $z_{111} = 4.695$; $p < 0.01$), and $\tau_{222} = -6.098$ ($se = 2.870$; $z = -2.125$; $p = 0.0168$). Both values are significant. We thus conclude that the hypotheses that Configuration 111 constitutes a type and Configuration 222 constitutes an antitype can be retained.

Standard exploratory CFA of these data can yield the results presented in Table 13, above. That is, one can interpret all configurations as types or antitypes (based on Lehmacher's tests) or none (based on e.g., X^2-tests), or a selection of configurations. Standard confirmatory CFA would yield results identical to the ones created using Kieser and Victor's CFA. More specifically, when only two cells are tested, the Bonferroni-adjusted α is $\alpha^* = 0.025$. The probability for the z-score for Configuration 111 in Table 13 was $p = 0.017$, and the probability for the z-score for Configuration 222 in Table 13 was $p = 0.015$. Thus, the CFA null hypothesis can be rejected for both configurations.

However, there are two major differences between Kieser and Victor's confirmatory CFA and standard confirmatory CFA. First, it is not always the case that the expected cell frequencies with and without cells blanked out are as similar as in the present example. Therefore, it cannot be expected that the type/antitype decisions from the two CFA approaches are always the same. Second, only based on Kieser and Victor's confirmatory CFA one can conclude that the base model fits in those cells that do not constitute types or antitypes.

Kieser and Victor's approach to CFA can be applied in an exploratory context too. The authors (1999) proposed two stepwise search procedures. The first involves *forward inclusion* which is followed by *all-models-fit*. This procedure requires that the researcher determines the number of configurations that are anticipated to constitute types or antitypes before analysis. The forward inclusion method then identifies the subset T of cells that possibly contain types or antitypes. The subsequent all-models-fit procedure then tries to minimize the number of type/antitype cells while maximizing the number of cells for which the base model fits.

Alternatively, the authors propose a two-stage forward inclusion procedure. Here too, the researchers need to determine the number of configurations that are candidates for type or antitypes before the search starts. The first analytic step is then the same as for the first procedure. It identifies the subset of type/antitype cells. The cells in T are then ranked based on the degree to which they deviate from the base model. In the second step of the search procedure, these cells are included again in the set of non-suspicious cells, beginning with the one that deviates the least from the base model. This second step is repeated until the base model does not fit any longer. A SAS/IML program can be requested from Dr. Kieser[2].

Discussion. The approach to CFA proposed by Victor and Kieser is most useful when researchers assume that the presence of types and antitypes has a masking effect. This effect can manifest in the identification of types and antitypes that, in the population, do not exist, or in the description of configurations as conforming to the base model that, in the population, are types or antitypes. Both kinds of misidentifications can occur in the same cross-classification as was illustrated in Table 110. If, however, researchers believe that no such masking occurs, standard CFA can be the method of choice. Indeed, a comparison of the standard CFA model and Kieser and Victor's model, both given at the beginning of this section, shows that the standard model is a special case of Kieser and Victor's model. The standard model and Kieser and Victor's model are identical if no cells are blanked out.

11.2 Bayesian CFA

Application of Bayesian inference is based on Bayes' Theorem and involves the following four principal steps (Everitt, 1998; Gelman et al., 1995):

(1) Calculate the likelihood, $f(x|\theta)$, that describes the data X in terms of the unknown parameters θ;

(2) Calculate the prior distribution, $f(\theta)$ which reflects the knowledge about θ that existed prior to the collection of data;

(3) Employ the Bayes theorem to calculate the *posterior distribution* $f(\theta|x)$ which reflects the knowledge about θ after having observed

---

[2]Dr. Kieser's e-mail address is Meinhard.Kieser@Schwabe.de. A similar procedure is implemented in SYSTAT (see Section 11.1.2, below).

the data; this step implies that the distribution $f(\theta|x)$ can be updated each time new data come in; and

(4) Derive inference statements and make statistical decisions based on the posterior distribution.

In the following sections, we first review two of the key concepts of Bayesian statistics, the prior and the posterior distributions. These concepts are then applied in the context of CFA[3].

11.2.1 The prior and posterior distributions

Consider for the following review the cross-classification of $d \geq 2$ categorical variables. Let i index all cells, and let π_i be the population probability for Cell i. The vector of the probabilities π_i is π. Assume furthermore that sampling is multinomial (for product-multinomial sampling see von Eye, Schuster, & Gutiérrez-Peña, 2000), in which case F, the vector of observed frequencies can be considered an observation from a $(k - 1)$-dimensional multinomial distribution with $N = \sum_i f_i$, where f_i is the observed frequency for Cell i and k is the total number of cells in the cross-classification, and with unknown parameter vector π.

In Bayesian statistics, all prior beliefs about the values in the vector π are described in the form of a *prior distribution*. The usual conjugate prior is the Dirichlet distribution. Wood et al. (1994) discussed the Dirichlet distribution in the context of CFA. This distribution is described by a parameter vector $\beta = (\beta_1, ..., \beta_k)$ such that

$$E(\pi_i) = \frac{\beta_i}{\sum_j \beta_j},$$

where $i, j = 1, ..., k$ (cf. Gelman et al., 1995).

In many applications, researchers do not possess or do not wish to make prior beliefs part of their statistical analyses. Therefore, in the absence of prior information, an *ignorance prior*, also called *noninformative prior* can be used. One example of such a prior is Dirichlet-distributed with parameter $\beta = (0.5, ..., 0.5)$. This prior has the characteristic of being *conjugate*, that is, closed under sampling. This

---

[3]The following sections borrow heavily from Gutiérrez-Peña and von Eye (2000).

means that the posterior distribution (see below) is of the same family at each stage of sampling. Specifically, the posterior distribution of π is also Dirichlet, with parameter $\beta = (f_1 + 0.5, ..., f_k + 0.5)$. This distribution contains the entire available information about the population proportions π_i, conditional on the observed frequencies.

The base model used in CFA to specify assumptions concerning variable interrelations imposes constraints on the range of possible values of π. In other words, if a base model is implemented, the population probability of Cell i is $\pi_i^* = f_i(\pi)$ for some functions $f(i)$. Consider, for example, a 2 x 2 cross-classification and the base model of variable independence. Then one obtains for the π_i^*

$$
\begin{aligned}
f_1(\pi) &= (\pi_1 + \pi_2) \cdot (\pi_1 + \pi_3) \\
f_2(\pi) &= (\pi_1 + \pi_2) \cdot (\pi_2 + \pi_4) \\
f_3(\pi) &= (\pi_3 + \pi_4) \cdot (\pi_1 + \pi_3) \\
f_4(\pi) &= (\pi_3 + \pi_4) \cdot (\pi_2 + \pi_4) \, .
\end{aligned}
$$

The base model can be tested as a whole. One uses the posterior distribution of the statistic

The resulting quantity δ can be viewed as deviance. It is always zero or greater. It is zero only if the base model is true. In this case, the X^2 values used in frequentist CFA are zero, too. Unfortunately, the posterior distribution of δ is not readily available. Therefore, one resorts to Monte Carlo techniques to estimate the degree to which calculated values of δ are

$$\delta = \sum_i \log\left(\frac{\pi_i}{\pi_i^*}\right) \pi_i \, .$$

extreme. In general, distributions of δ with a mean near zero are interpreted as in support of the base model. Posterior distributions remote from zero allow one to reject the base model. In more technical terms, one uses the following decision rule: if under the H_0 that $\delta = 0$ the value $\delta = 0$ is not contained in the α% most extreme posterior density region, reject the null hypothesis. This test plays a role parallel to the null hypothesis test in conventional, frequentist CFA.

11.2.2 Types and antitypes in Bayesian CFA

Consider Configuration *i*. If $\pi_i > \pi_i^*$, Configuration *i* constitutes a *Bayesian CFA type*. If $\pi_i < \pi_i^*$, Configuration *i* constitutes a *Bayesian CFA antitype*. Using the posterior distribution of π, we can, in principle, calculate the posterior probability of any event that involves the population proportions, π. If the probability $\Pr(\pi_i > \pi_i^*)$ is close to 1, one can state that Configuration *i* constitutes a type. If $\Pr(\pi_i > \pi_i^*)$ is close to zero, one can state that Configuration *i* constitutes an antitype. In practice, one would classify configurations as types and antitypes only if $\pi_i - \pi_i^*$ is significantly different than zero. Therefore, Gutiérrez-Peña and von Eye (2000) proposed the rule that Configuration *i* be classified as constituting a type only if $\pi_i > \pi_i^* + \varepsilon_i$, and that Configuration *i* be classified as constituting an antitype only if $\pi_i < \pi_i^* - \varepsilon_i$. The value ε_i is a suitably chosen threshold, for example, two times the posterior standard deviation of $\pi_i - \pi_i^*$.

11.2.3 Patterns of types and antitypes and protecting α

An interesting possibility is that Bayesian CFA allows one to calculate the posterior probability of *any* specific pattern of types and antitypes in a cross-classification. For a particular CFA base model, the posterior distribution of π implies a probability distribution on the set of all possible patterns. Consider, for example, a 2 x 2 cross-classification. There are $3^4 = 81$ possible patterns of the outcomes T = type, A = antitype, and N = neither type nor antitype. Examples of such patterns include

$$\begin{bmatrix} T & T \\ T & T \end{bmatrix} \cdots \begin{bmatrix} N & A \\ T & N \end{bmatrix} \cdots \begin{bmatrix} T & A \\ A & T \end{bmatrix} \cdots \begin{bmatrix} A & T \\ T & A \end{bmatrix} \cdots \begin{bmatrix} A & A \\ A & A \end{bmatrix}.$$

As was discussed in the context of Kieser and Victor's (2000) approach to CFA in Section 11.1, some of these patterns have probability zero, that is, they are impossible. Examples of such patterns include the first and the last of the above. Other patterns will have very low probabilities. Bayesian CFA reports the *most probable pattern*. This pattern can be identified using a number of strategies. One strategy that guarantees that this pattern will be identified involves calculating the probabilities for all possible patterns. This option, however, can require enormous computational resources. The number of patterns for a cross-classification with t cells is 3^t. So, if a table has, for example, 8 cells, the total number of patterns is already $3^8 = 6561$. This can easily be handled by standard PCs. If, however, a table has 4 x 2 x 4 = 32 cells, the number of patterns is $1.853 \cdot 10^{15}$ patterns. This number implies patience on the researcher's side. Therefore, Gutiérrez-Peña and von Eye (2000) proposed looking only at patterns 'in the neighborhood' of a particular pattern suggested either by exploratory analysis which looks at each cell individually or by theory which makes the analysis focus on a selection of cells only.

There are two consequences from this characteristic of Bayesian analysis. The first characteristic is that *hypotheses about patterns of types and antitypes can be tested* in a way not possible in frequentist CFA. In addition, because there is only one test, the family-wise or global α does not need to be protected or adjusted. The factual significance threshold α will always be equal to the nominal threshold.

11.2.4 Data examples

Data example I: Görtelmeyer's sleep data, exploratory CFA. In this section, we first present a data example from Gutiérrez-Peña and von Eye (2000). The data are the same as in Table 95. In Görtelmeyer's (1988) study on sleep problems, data were collected in a sample of 273 respondents. Using first order CFA, Görtelmeyer defined the six types of sleep behavior of respondents who sleep (1) short periods of time early in the morning; (2) symptom-free during 'normal' night hours; (3) symptom-free but wake up too early; (4) short periods early in the morning and show all symptoms of sleep problems; (5) during normal night hours but show all symptoms of sleep problems; and (6) long hours starting early in the evening, but show all symptoms of sleep problems. Of the 273 participants, 107 belonged to one of these types. The remaining 166 did not belong to any type. However, as in the analyses for Table 96, we again treat these 166 individuals as if they belonged to a seventh type. Table 112 displays the results of Bayesian CFA (for a comparison with the results from frequentist CFA see Table 96). The base model was that of first order CFA. The significance threshold was set to two times the posterior standard deviation of $\pi_i - \pi_i^*$. The prior used for this analysis was noninformative.

Table 112: First order Bayesian CFA of Görtelmeyer's sleep behavior data

Cell index SP	Observed Frequencies	Probabilities Type	Neither	Antitype	Type/ Antitype ?
11	19	.9588	.0442	.0000	**T**
12	3	.0000	.0442	.9558	**A**
21	20	.9321	.0679	.0000	**T**
22	4	.0000	.0679	.9321	**A**
31	16	.8618	.1382	.0000	**T**
32	3	.0000	.1382	.8618	**A**
41	5	.0466	.9415	.0119	
42	4	.0199	.9415	.0466	
51	4	.0000	.6496	.3504	
52	10	.3504	.6496	.0000	
61	8	.0038	.8948	.1014	
62	11	.1014	.8948	.0038	
71	65	.0000	.0046	.9954	**A**
72	101	.9954	.0046	.0000	**T**

The Bayesian CFA of Görtelmeyer's sleep behavior data, summarized in Table 112, yields three interesting results. First, the harvest of types and antitypes is the same as in the frequentist analysis in Table 96. This does not come as a big surprise considering that we used noninformative priors. However, Gutiérrez-Peña and von Eye's (2000) results suggest that Bayesian CFA may have more power than standard frequentist CFA. As soon as informative priors are used, results cannot be expected to be the same any more.

Second, Table 112 shows interesting information not provided by standard, frequentist CFA. Specifically, the table shows the probability with which each configuration constitutes a type, an antitype, or neither. Please note that the probabilities are supposed to sum to 1.0 in each row (differences from 1.0 in Table 112 are due to rounding).

Third, Bayesian CFA allows one to evaluate the solution as a whole. We inspect the posterior distribution of the deviance equivalent δ for the sleep behavior data. This distribution appears in Figure 23.

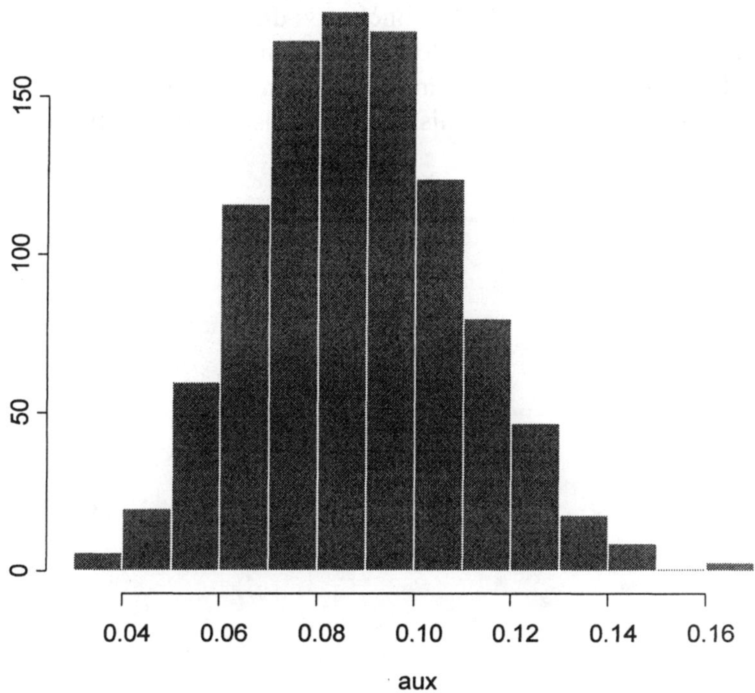

Figure 23: Posterior distribution of deviance equivalent δ

The figure shows that the density mass of the solution and its 'neighbors' is remote from zero. It appears that zero is not part of the distribution. We thus can safely retain the exploratory solution.

Data example II: The causal fork in Görtelmeyer's sleep behavior data. In this section, we employ Bayesian CFA in a confirmatory context. Specifically, we test the hypothesis of a strong fork entertained in Section 10.6.2. This hypothesis implies two hypotheses:

1. The first three types in Table 112 form a fork (see Figure 12 in Section 10.6.2); and
2. The first three antitypes in Table 112 form a fork.

The posterior probabilities of these two hypotheses suggest that
Pr(Configurations 11, 21, and 31 are all types) = 0.762, and
Pr(Configurations 11, 22, and 32 are all antitypes) = 0.762.

Thus, Bayesian CFA supports the conclusion drawn based on frequentist CFA in the existence of the type-fork and the antitype fork that had been suggested in Section 10.6.2. Sample program code for the Bayesian analyses appears in Part IV of this book (see also Gutiérrez-Peña & von Eye, 2000).

Part V: Computational Issues

12. Using General Purpose Software to Perform CFA

In the following sections we discuss computational issues of CFA. Specifically, we show how a CFA can be run on a PC. There are many programs available for CFA. Examples include the following:

- Hammond's CFA program (can be downloaded free from the web): http://www.liv.ac.uk/~pbarrett/programs.htm#CFA
- Funke's program, which is a module of the R package (can be downloaded free and used within the R-package which can also be downloaded free): http://www.stat.ufl.edu/system/man/R/library/cfa/html/cfa.html
- Lautsch and von Weber's program: This program comes on a diskette with Lautsch and von Weber's book on CFA (1995; in German).
- Krauth and Hebben's program: This program comes on a diskette with Krauth's book on CFA (1993; in German).
- Bergman and El-Khouri's program (1998): This CFA program is part of the software package SLEIPNER for pattern-oriented analyses. It can be requested from bei@psychology.su.se
- Dunkl's program (2000): This program comes in the form of a SAS

module. It can be requested from dunkl@imerem.de
- von Eye's program CFA 2002 (2001): This FORTRAN program comes in the form of an executable file. It is gratis and can be requested from voneye@msu.edu

In addition, parts of a CFA can be performed using the log-linear and cross-tab modules in most general purpose software packages. We illustrate the use of two general purpose statistical software packages, SYSYAT (Wilkinson, 1999) and S-plus (see Venables & Ripley, 1994), and CFA 2002, a stand-alone program for CFA (von Eye, 2001). We use SYSTAT and the stand-alone program for frequentist CFA and S-plus for Bayesian CFA.

The following sections present sample code and walk the reader through the steps necessary to perform CFA. We begin with SYSTAT, continue with S-plus, and conclude with CFA 2000. All of the examples have been executed in a Windows 2000 system using an IBM PC. In other Windows or MacIntosh systems, the required commands are analogous (CFA 2002 exists only for Windows systems).

12.1 Using SYSTAT to perform CFA

SYSTAT is a general purpose statistical software package. It can be used to perform a large number of descriptive, graphical, and inferential routines, it has modules for modeling and estimation, and it can be used for simulations. Although there is no particular module for CFA, its two-way cross-tabulation and its log-linear modeling modules can be used to perform some of the steps of CFA. We first present an example using the two-way cross-tabulation module, and then an example using the log-linear module.

There are several ways to input data in SYSTAT. For the sake of simplicity, we assume for the following examples that a cross-classification already exists. If this is not the case, SYSTAT can produce it using the raw data. For the various options, the reader may consult the program manual, in particular the volume DATA.

12.1.1 SYSTAT's two-way cross-tabulation module

SYSTAT contains a module that allows one to count configurations of categorical variables. This module includes three parts. The first part produces univariate counts. The second part produces bivariate counts, that is, two-way cross-classifications. The third part produces three- or higher-way cross-classifications. In the following example, we use the two-way submodule.

For the following illustration, we use the two-way cross-classification from Table 17. This cross-tabulation results from crossing the two variables T1 and T2. These variables describe the Tanner stages in a sample of 83 adolescents, observed in 1983 and 1985. Tanner stages indicate the progress an adolescent has made in his/her physical pubertal development. In 1983, only stages 1 through 3 were observed, and in 1985, only stages 2 through 4 were observed. Thus, the cross-classification has 3 x 3 cells. Using the two-way module, one can perform some of the steps of a global, first order CFA.

Before we employ this module, we need to input the data. As was indicated above, we assume that the cross-tabulation and the counts for the individual cells already exist. Therefore, we only key in (or read from a file) the cell indices and the cell frequencies. To perform this step, we proceed as indicated in the following table. We assume that SYSTAT is running and the command window is open. The following commands are issued.

Command/Input	Effect
click VIEW and DATA	a spreadsheet-type display opens that allows one to input data directly in a rectangular format
click the cell VAR00001	highlights the column for Variable 00001
click DATA and VARIABLE PROPERTIES	opens the boxes that allow one to specify variable name and properties
type T1 in the Variable Name box; hit the ENTER key	labels the first variable T1, the first Tanner observation; carries back to command window
click the cell VAR00002	highlights the column for Variable 00002

click DATA and VARIABLE PROPERTIES	opens the boxes that allow one to specify variable name and properties
type T2 in the Variable Name box; hit the ENTER key	labels the second variable, T2, the second Tanner observation; carries back to command window
click the cell VAR00003	highlights the column for Variable 00003
click DATA and VARIABLE PROPERTIES	opens the boxes that allow one to specify variable name and properties
type FREQ in the Variable Name box; hit ENTER; then place cursor in the first cell of the first column	labels the third variable, FREQ, the observed cell frequencies; carries back to command window; we now have defined the row-variable (T1), the column variable (T2), and the cell frequencies (FREQ); the spread sheet is now ready for data input; first, we key in the row indicators
type 1 and then ↓, 1↓, 1↓, 2↓, ..., 3 and ENTER; then place cursor in the first cell of the second column	inserts the row-indices in the first column
type 2↓, 3↓, 4↓, ..., 4↓ and ENTER; then place cursor in the first cell of the third column	inserts the column indices in the second column

type 31↓, 7↓, ..., 6 and ENTER	inserts the observed cell frequencies in the third column; we now have the 3 x 3 cross-classification input and ready for analysis; we recommend saving the data at this stage. Before analyzing this table, we need to let the program know which of these three variables contains the cell frequencies.
click DATA and FREQUENCY and highlight FREQ; click ADD and OK	indicates that FREQ is the frequency variable. The following steps invoke the two-way submodule.
Click STATISTICS, CROSSTABS, and TWO-WAY	opens the dialog box for the two-way program;
highlight T2 and ADD it to the ROW VARIABLE window	specifies that T1 is the row variable of the two-way table
highlight T1 and ADD it to the COLUMN VARIABLE	specifies that T2 is the column variable of the two-way table; in the list under the TABLES list, we ...
... check Expected and Standardized Deviates	specifies that we wish to see the expected frequencies and the standardized deviates in the output. The latter are defined as $z = \frac{f - e}{\sqrt{e}}$. The box by Frequencies is checked already. That is, the observed frequencies are presented by default.
click OK	and the program performs the calculations.

The following, slightly edited output results from these commands:

```
  Case frequencies determined by value of variable FREQ.

Frequencies
 T1 (rows) by T2 (columns)
```

	2	3	4	Total
1	31	7	0	38
2	14	15	10	39
3	0	0	6	6
Total	45	22	16	83

```
Expected values
T1 (rows) by T2 (columns)
```

	2	3	4
1	20.602	10.072	7.325
2	21.145	10.337	7.518
3	3.253	1.590	1.157

```
Standardized deviates: (Observed-Expected)/SQR(Expected)
T1 (rows) by T2 (columns)
```

	2	3	4
1	2.291	-0.968	-2.707
2	-1.554	1.450	0.905
3	-1.804	-1.261	4.504

```
WARNING: More than one-fifth of fitted cells are sparse (frequency
< 5).
        Significance tests computed on this table are suspect.
```

Test statistic	Value	df	Prob
Pearson Chi-square	43.971	4.000	0.000

Reading from the top to the bottom of the output, we first see the confirmation that SYSTAT interprets variable FREQ as carrying the frequency information. The cross-classifications with the observed and the expected cell frequencies follow. Please notice that the expected frequencies are the same as the ones in Table 17. This confirms that we selected the same base models; in this present example, this is the main effect model of variable independence. The third table shown in this output contains the *standardized deviates*, that is, the square root of the X^2 components.

The tables are followed by the warning that more than 20% of the expected cell frequencies are smaller than $e_{ij} = 5$. Based on the conclusion from the discussion in Section 3.7.1, we ignore this warning because none of the expected cell frequencies is smaller than $e_{ij} = 1$. The final block of information presents the contingency X^2-test which indicates that the base model must be rejected. This result leads one to expect types and antitypes.

Obviously, SYSTAT's two-way module provides information that is necessary for CFA. Two important parts of the information needed for CFA are missing, however. First, there is no protection of the family-wise α. We thus have to hand-calculate some adjusted α or use some tabulated adjusted significance threshold. For the present example, we can use the adjusted scores from Table 17. The second missing element is the tail probability for each of the standardized deviates. Again, we may have to calculate these using some other means. One problem with the presented results is the number of decimal places. This number is certainly big enough for the expected cell frequencies. However, for the standardized deviates, one would wish for more decimal places. In cases that have deviates close to the critical value, decisions may be hard to make based on only three decimal places. Still, SYSTAT's two-way module is a convenient first step toward a CFA of a two-way table, in particular when the data are already available in a SYSTAT system file.

In addition to providing only parts of the information needed for a complete CFA, the two-way module has three characteristics that limit its usefulness as a CFA program. First, the program can handle only two variables. In most instances, more than two variables are analyzed simultaneously. Second, other base models than the first order CFA base model are often of interest. The two-way module only uses the main effect model. Third, covariates or special effects cannot be considered. With only a few exceptions, SYSTAT's log-linear model module allows one to do all this. Therefore, we illustrate in the next section how the same data can be analyzed using the log-linear model module.

12.1.2 SYSTAT's log-linear modeling module

For the description of SYSTAT's log-linear modeling module, we assume again that the cross-classification is already given. In addition, we assume that the cell indices and the cell frequencies have already been keyed in, and that the frequency variable has been specified. The log-linear module itself provides a number of options that are of use in CFA. Four of these options are particularly useful. First, the program allows one to specify CFA base models. This specification is done in the form of a hierarchical log-linear model that one asks the program to fit to the data. Naturally, more than one base model can be fit. Second, the program provides a number of deviance scores. Some of these are of interest in CFA. Third, the program allows one to take into account one covariate. Fourth, the program allows one to take into account structural zeros.

In the following example, we re-analyze the data from Table 17 and from Section 11.1.1 using the log-linear module. The following table contains the commands that need to be issued.

Command/Input	Effect
click STATISTICS, LOGLINEAR MODEL, and ESTIMATE MODEL	invokes the log-linear program; now we first define the cross-classification to analyze
highlight T1 and click CROSS by the DEFINE TABLE box	specifies T1 as the row variable
highlight T2 and click CROSS by the DEFINE TABLE box	specifies T2 as the column variable
highlight T1 and ADD it to the MODEL TERMS box	specifies that the main effect for T1 is part of the base model
highlight T2 and ADD it to the MODEL TERMS box	specifies that the main effect for T2 is part of the base model
we now click the STATISTICS button to tailor the modeling to the needs of CFA	
in the STATISTICS dialogue box we check (or leave checked) the following options: in the TEST STATISTICS box: Chi-square	yields goodness-of-fit X^2 for the base model
in the CELL CONTENTS box we check observed frequencies, expected frequencies, standardized deviates, and Pearson	produces the tables with the observed and the expected cell frequencies, and the table with the standardized deviates, $z = \frac{f - e}{\sqrt{e}}$

all other boxes that may be checked are un-checked, because we don't need this information for CFA	one interesting option is the specification of the number of *outlandish cells*. These cells are defined in the same way as types and antitypes in Kieser and Victor's (1999) CFA. These cells indicate significant deviations of observed from expected cell frequencies. In the present example, we request two outlandish cells and type
2	and make sure the option is checked
we click CONTINUE and, in the next dialogue box, OK	the program responds with the output of the modeling results.

The following, slightly edited output results from these commands:

```
  Case frequencies determined by value of variable FREQ.

Observed Frequencies
====================
T1       |              T2
         | 2             3             4
---------+-----------------------------------
1        |       31.000        7.000        0.000
2        |       14.000       15.000       10.000
3        |        0.000        0.000        6.000
---------+-----------------------------------
Pearson ChiSquare    43.9714  df     4  Probability  0.00000
     LR ChiSquare    45.3216  df     4  Probability  0.00000

Expected Values
===============
T1       |              T2
         | 2             3             4
---------+-----------------------------------
1        |       20.602       10.072        7.325
2        |       21.145       10.337        7.518
3        |        3.253        1.590        1.157
---------+-----------------------------------

Standardized Deviates = (Obs-Exp)/sqrt(Exp)
===========================================
T1       |              T2
         | 2             3             4
---------+-----------------------------------
1        |        2.291       -0.968       -2.707
2        |       -1.554        1.450        0.905
3        |       -1.804       -1.261        4.504
```

```
Pearson Chi-square = (Obs-Exp)^2/Exp
====================================
T1          |              T2
            | 2              3              4
------------+-----------------------------------
1           |           5.247          0.937          7.325
2           |           2.414          2.103          0.819
3           |           3.253          1.590         20.282
------------+-----------------------------------

The  2 most outlandish cells (based on FTD, stepwise):
======================================================

                                                  T1
  ln(MLE) LR_ChiSq  p-value Frequency  |  T2
  --------- -------- -------- ---------  -  -
  -24.232   22.802    0.000        0     1  3
  -20.606    7.253    0.007        0     3  1
```

Reading from the top, this output can be interpreted as follows. After the confirmation that variable FREQ carries the frequency information, the output presents the observed frequency table. This table is followed by the Pearson Chi-square and the likelihood ratio Chi-square goodness-of-fit test information. Both tests suggest that the base model of independence of the two Tanner score assessments must be rejected. The tables with the expected cell frequencies and the standardized deviates are presented next.

Most interesting is the final block of information which contains the two most outlandish cells. The procedure that identifies these cells is a relative of the exploratory procedure proposed by Kieser and Victor (1999). Based on the Freeman-Tukey deviates, these are normally distributed deviates when the data are from a Poisson distribution, the configuration with the largest deviate under the base model is declared a structural zero. The Freeman-Tukey deviate for Cell i is defined as $\sqrt{N_i} + \sqrt{N_i + 1} - \sqrt{4e_i + 1}$. The base model is then fit to the remaining cells of the table. The first $LR\text{-}X^2$ reported in this block of the output is for the model with Cell 1 3 blanked out.

For the model with one cell blanked out, the Freeman-Tukey deviates are determined and the procedure iterates through another cycle. In each iteration step, an additional cell is declared a structural zero, and the model is refitted to the remainder of the cells. In the present example, the second cell blanked out is labeled Cell 3 1. Please note that the program mislabels cells in this part of the output. What is labeled Cell 3 1 is, using the labels in the other parts of the output, Cell 3 2.

Clearly, the log-linear modeling module provides many more options of interest to CFA than the two-way module. However, except for the Kieser-Victor-type procedure that led to the detection of two antitypes, the program performs no complete CFA. Thus, the user is left again with having to hand-calculate the probabilities of individual deviates. In addition, the program cannot be used to estimate a zero order base model. Therefore, more specialized CFA programs are often used. Before illustrating one such program, we briefly show how S-plus can be used to perform Bayesian CFA.

12.2 Using S-plus to perform Bayesian CFA

S-plus (cf. Venables & Ripley, 1994) is better described as a system that provides an environment that allows users to (a) apply existing statistical tools and (b) implement new statistical ideas. Thus, S-plus is as much a statistical software package as it is a language that can be used to formulate new routines and procedures. Most of the S-plus environment is open to the extent that users are provided with the possibility to change design characteristics implemented in the package as it can be purchased. A module for standard, frequentist CFA that works both under S-plus and its relative R was provided by Funke (see the beginning of Chapter 11).

In the present context, we illustrate how S-plus can be used to perform Bayesian CFA (see Section 11.2). Specifically, we show how to perform the example in Table 112 (Section 11.2.4). There is no Bayesian CFA routine available in S-plus. Therefore, code had to be written. In the following paragraphs, we present this code (from Gutiérrez-Peña & von Eye, 2000), along with some comments and a selection of the resulting output information. The following program was developed and tested in S-plus 4.5 under Windows 95/98, Windows 2000, and Unix. In addition, the program was tested in S-plus 3.2. Thus, it should be functional in a wide range of environments.

We begin with the program code. This code can be cut and pasted in the command box in S-plus under Windows 2000. In Windows 95/98, it can be imported using the *source* command:

```
> source("PATH/file name")
```

The command file contains the following lines, where text after a pound mark (#) is commentary and explanation[1]:

```
# External function

rdirich <- function(n, alpha, k)
{
        if(length(alpha) != k + 1)
        stop("alpha vector is the wrong length")
        km1 <- k + 1
        M <- matrix(0, n, km1)
        ve <- vector("numeric", n)
        for(i in 1:km1) {
                M[, i] <- rgamma(n, alpha[i])
                ve <- ve + M[, i] }
        M <- M/ve
        M
}


m_c(19, 3, 20, 4, 16, 3, 5, 4, 4, 10, 8, 11, 65, 101)
total_sum(m)
tabla_matrix(m,7,2,byrow=T)

p.i_1:7
p.j_1:2
for(i in 1:7){ p.i[i]_sum(tabla[i,])/total }
for(j in 1:2){ p.j[j]_sum(tabla[,j])/total }

m_m+0.5

N_1000
sample_rdirich(N,m,13)

p.i.aux_p.i
p.j.aux_p.j
sam.hat_sample


for(n in 1:N)
{
  tabla.aux_matrix(sample[n,],7,2,byrow=T)
  for(i in 1:7){ p.i.aux[i]_sum(tabla.aux[i,]) }
  for(j in 1:2){ p.j.aux[j]_sum(tabla.aux[,j]) }

  sam.hat[n,]_as.vector(t(p.i.aux%o%p.j.aux))
print(n)
}


aux_0*(1:N)
for(k in 1:14)             # Computes the Deviance Equivalent #
{
aux_log(sample[,k]/sam.hat[,k])*sample[,k] + aux
}

win.graph()                # Draw  histogram #
```

---

[1]Thanks go to Eduardo Gutiérrez-Peña (IIMAS, Autonomous University of Mexico City) for making this program available.

```
hist(aux)

P_1:14
for(k in 1:14)
{
P[k]_sum(ifelse(sample[,k] > sam.hat[,k],1,0))/N
}


## Uses the modified definition of  Types  and  Antitypes ##

e_1:14
P.mat_matrix(0,14,3)
patt.bay_1:14
patt_matrix(0,N,14)
for(k in 1:14)
{
e[k]_2*sqrt(var(sample[,k]-sam.hat[,k]))

patt[,k]_ifelse(sample[,k] > (sam.hat[,k] + e[k]),1,0)
patt[,k]_ifelse(sample[,k] < (sam.hat[,k] - e[k]),-1,patt[,k])
patt[,k]_ifelse(abs(sample[,k] - sam.hat[,k]) <= e[k],0,patt[,k])

P.mat[k,1]_sum(ifelse(patt[,k]==1,1,0))/N # Posterior probability that
                                          # the cell is a Type
P.mat[k,3]_sum(ifelse(patt[,k]==-1,1,0))/N # Posterior prob. that
                                            # the cell is an Antitype
P.mat[k,2]_1 - P.mat[k,1] - P.mat[k,3]   # Posterior probability that
                                           # the cell is neither #
if(P.mat[k,1]>max(P.mat[k,2],P.mat[k,3])){patt.bay[k]_1}
if(P.mat[k,3]>max(P.mat[k,1],P.mat[k,2])){patt.bay[k]_-1}
if(P.mat[k,2]>max(P.mat[k,1],P.mat[k,3])){patt.bay[k]_0}
}

patt.other_c(1,-1,1,-1,1,-1,0,0,0,0,0,0,0,0)


P.other_0
P.bay_0
for(n in 1:N)
{
if(sum(ifelse(patt[n,]==patt.other,1,0))==14){P.other_P.other+1}

if(sum(ifelse(patt[n,]==patt.bay,1,0))==14){P.bay_P.bay+1}
print(n)
}

P.bay_P.bay/N       ## Posterior probability of the pattern from the
                    ## Bayesian analysis
P.other_P.other/N   ## Posterior probability of the other pattern ##

## Test the hypotheses of a fork of Types and of a fork of Antitypes
##

fork.type_(sample[,1] > (sam.hat[,1] + e[1]) ) &
          (sample[,3] > (sam.hat[,3] + e[3]) ) &
          (sample[,5] > (sam.hat[,5] + e[5]) )

fork.antitype_(sample[,2] < (sam.hat[,2] - e[2]) ) &
              (sample[,4] < (sam.hat[,4] - e[4]) ) &
              (sample[,6] < (sam.hat[,6] - e[6]) )
```

```
P.fork.type_sum(ifelse(fork.type,1,0))/N
P.fork.antitype_sum(ifelse(fork.antitype,1,0))/N
```

After pasting this program into the command box in S-plus, it runs automatically. After the program has started, the screen displays

```
[1] 1
[1] 2
[1] 3
[1] 4
.
.
.
[1] 1000,
```

as the program goes through the first round of 1000 iterations (see Line 23 of the program). Then, the program draws in a separate window the histogram shown in Figure 23 (Section 11.2.4). The program terminates after completion of the last 1000 iterations. The results are stored in the files P.mat, P.bay et cetera. The content of these files can be sent to the screen by typing the file name, for example,

```
> P.mat
```

Here is the slightly edited content of file P.mat:

```
> P.mat
        [,1]  [,2]  [,3]
 [1,] 0.959 0.041 0.000
 [2,] 0.000 0.041 0.959
 [3,] 0.922 0.078 0.000
 [4,] 0.000 0.078 0.922
 [5,] 0.873 0.127 0.000
 [6,] 0.000 0.127 0.873
 [7,] 0.057 0.934 0.009
 [8,] 0.009 0.934 0.057
 [9,] 0.000 0.666 0.334
[10,] 0.334 0.666 0.000
[11,] 0.004 0.907 0.089
[12,] 0.089 0.907 0.004
[13,] 0.000 0.002 0.998
[14,] 0.998 0.002 0.000
```

This file contains the information reproduced in columns 3, 4, and 5 of Table 112.

12.3 Using CFA 2002 to perform frequentist CFA

CFA 2002 (von Eye, 2001) is an interactive program that largely performs Configural Frequency analyses. It is, in principle, capable of estimating log-linear models. However, other programs such as the ones included in SAS, SPSS, SYSTAT, or S-plus, may be more convenient for log-linear modeling. The following sections first give a description of program characteristics and options, and then some sample applications.

12.3.1 Program description

The current version of CFA 2002 has the following specifications:

1. *Size* of executable program file: 381 KB
2. *Operating Systems*: Windows NT 4.0, Windows 98, Windows 2000 *Number of subroutines*: 34
3. *Input options*: data input either via keyboard or via file; via file, frequency tables and raw data can be read; the maximum number of raw data is 50; variables can be re-ordered and categorized
4. *Output*: written in a file; partial results appear on screen
5. *Number of variables that can be simultaneously analyzed*: 10 (or limited by computer memory)
6. *Number of categories per variable*: up to 9 (or limited by computer memory)
7. *Memory allocation*: dynamic; limits depend on user's PC
8. *Variants of CFA*

- *zero order CFA* (Lienert & von Eye, 1989; see Section 5.1): CFA 2002 compares observed with expected cell frequencies based on the assumption of a uniform frequency distribution. The log-linear base model for zero order CFA is log $F = \mathbf{1}\lambda + e$, where $\mathbf{1}$ is a constant vector. The resulting types and antitypes reflect agglomerations and sparser populated sectors in the data space.
- *first order CFA* (Lienert, 1969; see Section 5.2): considers all main effects when estimating the expected cell frequencies. Many consider this the *classical* method of CFA, even the *only* method of CFA (Krauth, 1993). To illustrate, consider a cross-classification spanned by the three variables, A, B, and C. For this table, the model used for estimation of expected cell frequencies is $\log E = \lambda_0 + \lambda_i^A + \lambda_j^B + \lambda_k^C$, where the λ are the parameters for the main effects.
- *second and higher order CFA* (von Eye & Lienert, 1984; see Sections 5.3 and 5.4): in general, kth order CFA considers all associations up to *k*-1st order when estimating expected cell

frequencies. For instance, consider the three variables A, B, and C. Then, the second order CFA base model is $\log E = \lambda_0 + \lambda_i^A + \lambda_j^B + \lambda_k^C + \lambda_{ij}^{AB} + \lambda_{ik}^{AC} + \lambda_{jk}^{BC}$, where the double subscripted and double superscripted elements indicate pair-wise interactions. Second order CFA types and antitypes can result only if second and/or higher order interactions
exist. This applies accordingly to higher order CFA.

- *two-sample CFA* (Lienert, 1971; see Sections 7.1 and 7.2): k-sample CFA allows one to compare k groups of respondents with each other. To illustrate the case of two-sample CFA, consider a cross-classification that is spanned by the three variables A, B, and C. Suppose that variable C indicates the groups. The log-linear base model for this approach is $\log E = \lambda_0 + \lambda_i^A + \lambda_j^B + \lambda_k^C + \lambda_{ij}^{AB}$. This model is saturated in the variables used to discriminate between the groups. Therefore, types and antitypes can emerge only if there is an interaction between the grouping variable, C, and predictor A, predictor B, and/or both A and B. Sampling is product-multinomial in the variables used for discrimination (von Eye, Schuster, & Gutiérrez-Peña, 2000).
- *prediction CFA* (P-CFA; Lienert & Krauth, 1973; see Section 6.2): In prediction CFA, variables are classified in the two groups of predictors and criteria. The base model is saturated in both the predictors and the criteria, but proposes independence of predictors from criteria. Consider the two predictors A, B, and the two criteria C and D, the base model for a prediction CFA of these four variables is $\log F = \lambda_0 + \lambda_i^A + \lambda_j^B + \lambda_k^C + \lambda_l^D + \lambda_{ij}^{AB} + \lambda_{kl}^{CD}$. Types and antitypes can emerge only if there is an interaction between predictors and criteria. Sampling is product-multinomial in the predictors (von Eye & Schuster, 1998).
- *interaction structure CFA* (limited to two groups of variables; Lienert & Krauth, 1973; see Section 6.1): in a fashion similar to k-sample CFA and prediction CFA, Interaction Structure CFA (ISA) classifies variables in two groups. However, in contrast to both k-sample CFA and P-CFA, ISA does not assign to either group the status of predictors or criteria. Thus, ISA is a method for analyzing the relationships among two groups of variables that do not differ in their status. The log-linear base model for ISA is the same as that for prediction CFA in many situations. For example, for the

four variables A, B, C, and D, the model is the same as given above for P-CFA. However, because the status of the variable groups is the same, sampling is typically multinomial for both groups. Thus, selection of base models is less constrained than the selection of base models for P-CFA (for details see von Eye & Schuster, 1998; von Eye, Schuster, & Gutiérrez-Peña, 2000).

- *longitudinal CFA* (see Part III of this volume): CFA offers a large number of options to analyze longitudinal data. CFA 2002 allows one to create results for almost all of these options. There is currently only one exception. CFA of first, second, and higher differences can be run under a base model that is not log-linear because the cell probabilities are known a priori. This base model cannot be realized in the current program version. Therefore, the estimation module used in program CFA 2002 cannot be used. There is currently no program option to key in expected cell frequencies or cell probabilities (future versions will make these options available). Therefore, CFA of differences can currently not be performed using program CFA 2002. All other options can be used, although some of them may require that the data be processed before feeding them into CFA. Examples of problems that require data processing before CFA include the analysis of categorized polynomial coefficients that have been estimated separately for each individual (von Eye & Nesselroade, 1992).
- *CFA with covariates* (Glück & von Eye, 2000; see Section 10.7): Covariates are typically defined as independent variables that allow one to predict the dependent variables and are not controlled by the experimenter. In CFA, covariates are defined as variables that are not under control of the researchers and may allow one to predict the observed frequency distribution. To accommodate possible covariates, Glück and von Eye (2000) proposed an extension of the CFA base model. The extended model has the form $\log E = X_b\lambda_b + X_c\lambda_c$, where subscript b indicates the usual base model and its parameters, and subscript c indicates the covariates and their parameters. The program CFA 2002 keeps track of the number of covariates and prevents researchers from using more covariates than possible based on the remaining number of degrees of freedom. Covariates come either in form of categorical variables that describe everybody in a particular cell, or in form of measures of central tendency that characterize

everybody in a cell.

- *symmetry CFA* (design matrix needs to be keyed in; see Lienert & Netter, 1986a; von Eye et al., 1996): axial symmetry proposes that the distribution in pairs of cells mirrored about the main diagonal of a square matrix is uniform. Von Eye and Spiel (1996) have illustrated that axial symmetry can be cast in terms of a nonstandard log-linear model. Applied to CFA one can ask which cells deviate significantly from axial symmetry and thus form types and antitypes of asymmetry. This can be of interest in drug control studies or in studies of change.

6. *Statistical tests for global CFA* (see Chapter 3)
    - exact binomial test
    - binomial test using Stirling's approximation of factorials
    - Pearson X^2-component test
    - normal approximation of the binomial test
    - *z*-test
    - Lehmacher's asymptotic hypergeometric test (Lehmacher, 1981)
    - Lehmacher's test with Küchenhoff's continuity correction (Küchenhoff, 1986)
    - Anscombe's z-approximation (described in Upton, 1978).
7. *Statistical tests for 2-sample CFA* (see Sections 7.1 and 7.2, above)
    - X^2-test for 2 x 2 tables
    - X^2-test with continuity correction
    - normal approximation of the binomial test
    - z-test
    - λ, the log-linear interaction plus significance test based on jackknife procedures (von Eye, Spiel, & Rovine, 1995)
    - $\tilde{\lambda}$, the marginal-dependent variant of λ (Goodman, 1991) plus significance test based on jackknife procedures (von Eye et al., 1995)
    - ρ, the correlation in 2 x 2 tables (Goodman, 1991) plus significance test based on jackknife procedures (von Eye et al., 1995)
    - δ, the absolute value of λ, both marginal independent (Goodman, 1991; von Eye et al., 1995) plus significance test based on jackknife procedures
    - θ, the log-odds ratio plus asymptotic significance test
8. *Descriptive measures for global CFA* (see Chapter 4)

- relative risk ratio, *RR* (see DuMouchel, 1999)
- log*P*, that is, the Poisson probability of the observed frequency when compared to the expected frequency, estimated under some chance model (DuMouchel, 1999)
- rank of *RR*
- rank of log*P*

13. *Descriptive measures for two-sample CFA* (see Section 7.2): coefficient π which describes the goodness-of-fit in cross-classifications (here: 2 x 2 tables; Rudas, Clogg, & Lindsay, 1994; Gonzáles Debén & Méndez Ramírez, 1999)
14. *Availability*: The executable program file and a manual can be obtained gratis from the author at VONEYE@MSU.EDU.

12.3.2 Sample applications

The following sections present sample applications of the program CFA 2002.

12.3.2.1 First order CFA; keyboard input of frequency table

This section presents a sample run of first order CFA. We use the same data as in Sections 11.1.1 and 11.1.2, that is, Lienert's (1964) LSD data. The data describe a sample of 65 students who were administered LSD 50. The observed variables are C = narrowed consciousness, T = thought disturbance, and A = affective disturbance. Each symptom was rated as either 1 = present or 2 = absent. The cross-classification of the three symptoms has 2 x 2 x 2 = 8 cells. In the present sample run we enter the frequency table using the keyboard.

For users to replicate this sample run we assume that they have the executable file, **CFA**, of the program on their computer, that the computer runs under Windows 95 or higher, under Windows NT 4.0 or higher, or under Windows 2000, and that there is a program shortcut to the executable file on the screen. If there is no shortcut, the program can be started by double-clicking the program file name within Windows' Explorer. The following steps must be performed to analyze the LSD data with First Order CFA:

Command	Effect
double click shortcut to CFA program	starts CFA program; program window appears on screen; the program responds with a header telling the user that the CFA program was started and asks whether data will be input via file (=1) or interactively, via the keyboard (=2). We select interactive data input and type
2 (Enter)	The program responds by asking for the number of variables. We type
3 (Enter)	The programs asks for number of categories of the first variable. We type
2 (Enter)	This is repeated until the number of categories for each variable is given. The program follows up by prompting the cell frequencies. We respond to the prompt for the first cell, that is, Cell 111, by typing
20 (Enter)	To the prompt for the second cell frequency we respond by typing
1 (Enter)	For the following cells we type
4, 12, 3, 10, 15, and 0	each number followed by Enter (no commas). When all cell frequencies are keyed in, the program responds by presenting the sample size - in the present example N = 65 -, and by asking whether the user wishes to save the data (yes = 1; no = 2). We select to save the data and type
1 (Enter)	The program then asks for the name of the data file. Up to 80 spaces are read. The name must be given in DOS style, that is, including the path. If no path is given, the file will be saved in the currently active directory. In the present example, this is the directory that contains the CFA program file. We type

leuner.dat (Enter)	The program responds by presenting the current program options concerning CFA models. The current version allows one to perform any of the global CFA models, that is, any of the models where the status of all variables is the same. In addition, the program can perform a two-sample CFA. Later, in Section 12.3.2.2, it is shown how P-CFA and ISA can be performed. One indicates the CFA model by typing the order number of the CFA model. For example, for zero order CFA one types 0, for first order CFA one types 1, and so forth. For the present example we select first order CFA and type
1 (Enter)	The program responds by presenting the unidimensional marginal frequencies on the screen and by asking whether the user wishes to include a covariate (yes = 1; no = 2). In the present example we opt not to include covariates and type
2 (Enter)	The program then presents the eight statistical tests currently included in the program. We select the Lehmacher test with continuity correction and indicate our choice by typing
7 (Enter)	The program then requests input of a significance level. We go with the standard $\alpha = 0.05$ and type
.05 (Enter)	The program now requests the name of the output file. We type
leuner.out (Enter)	A total of 80 spaces can be used for the file name. The program responds by performing calculations and writing results to the file leuner.out. Finally, the program asks whether the uses wishes that the design matrix, *X*, be printed. In this example we would like to see the design matrix and type
1 (Enter)	After concluding the analysis the program window disappears.

The above sample run resulted in the following, slightly edited, output file, leuner.out:

```
                         Configural Frequency Analysis
                         ---------- --------- --------
          author of program: Alexander von Eye, 2002


       Marginal Frequencies
       --------------------
       Variable Frequencies
       -------- -----------
           1        37.    28.

           2        34.    31.

           3        42.    23.

     sample size N =       65

    Lehmachers test with continuity correction was used
    Bonferroni-adjusted alpha =  .0062500
    a CFA of order   1  was performed


                                              Table of results
                                              ----- -- -------
    Configuration     fo        fe     statistic        p
    -------------    ----  -------- ---------       -------
         111          20.    12.506      3.183     .00072795    Type
         112           1.     6.848     -2.800     .00255136    Antitype
         121           4.    11.402     -3.198     .00069093    Antitype
         122          12.     6.244      2.819     .00240908    Type
         211           3.     9.464     -2.874     .00202764    Antitype
         212          10.     5.182      2.442     .00730272
         221          15.     8.629      2.887     .00194350    Type
         222           0.     4.725     -2.458     .00698741


                    chi2 for CFA model =    37.9198
                    df =      4       p =  .00000012

                LR-chi2 for CFA model =    45.0749
                    df =      4       p =  .00000000


   Descriptive indicators of types and antitypes
   ---------------------------------------------
        cell  Rel. Risk  Rank        logP    Rank
    --------  ---------  ----        ----    ----
         111     1.599     4        1.549     2
         112      .146     7         .730     7
         121      .351     5         .938     5
         122     1.922     2        1.595     1
         211      .317     6         .787     6
         212     1.930     1        1.421     4
         221     1.738     3        1.536     3
         222      .000     8         .515     8


  Design Matrix
```

```
  1.0   1.0   1.0
  1.0   1.0  -1.0
  1.0  -1.0   1.0
  1.0  -1.0  -1.0
 -1.0   1.0   1.0
 -1.0   1.0  -1.0
 -1.0  -1.0   1.0
 -1.0  -1.0  -1.0
                                    CARPE DIEM
```

Read from the top to the bottom this print out can be interpreted as follows. After the program title and a authorship statement the program presents the marginal frequencies. The order of the variables is the same as the order in which the variables were input. The sample size is given next. In the following line, the program confirms the choice of significance test. In the present example, this was Lehmacher's test with Küchenhoff's continuity correction. Bonferroni adjustment of the test-wise α resulted in the adjusted $\alpha^* = 0.05/8 = 0.00625$. A statement confirming that a first order CFA is performed is followed by the table of results. The columns of this table contain (a) the indices of the cells of the cross-tabulation; (b) the observed cell frequencies, labeled fo; (c) the estimated expected cell frequencies, labeled fe; (d) the values of the selected test statistic[2]; (e) the one-sided tail probabilities of the tests statistic; and (f) if applicable, the designation of a configuration as constituting a type or an antitype.

The present analysis suggests that there exist three types and three antitypes. For purposes of illustration we interpret the first type and the first antitype. The first type has cell-index pattern 111. It suggests that LSD 50 causes more participants than expected from chance to experience all three symptoms, that is, narrowed consciousness, thought disturbances, and affective disturbances. The chance model had been specified in the CFA base model in which we had postulated that the three symptoms are not associated (i.e., the log-linear main effect model or model of variable independence). The first antitype has cell-index pattern 112. It suggests that presence of the first two symptoms and absence of the third symptom co-occur less often than expected from the chance model. (For a substantive interpretation of the complete results see Lienert, 1964.)

Under the frequency table there is information on the goodness-of-fit of the CFA base model. This is given in units of the Pearson X^2 and the likelihood ratio X^2, both followed by the model degrees of freedom and their tail probabilities.

---

[2]When one of the binomial tests is selected, this column is omitted.

The table below the significance test results displays the relative risk ratio, *RR* (see Section 4.1), and Log*P* (see Section 4.2), which are also used for data mining in large sparse contingency tables and in Bayesian analysis of cross-classifications (DuMouchel, 1999; von Eye & Gutiérrez-Peña, in preparation).

It is important to note that the status of *RR* and Log*P* in this context is that of descriptive measures rather than significance tests. Therefore, rather than printing a probability for the *RR* score, the scores are ranked and so are the Log*P*, and the ranks are printed. Thus, two goals can be accomplished. First, when the sample size-to-table size ratio is large enough, the usual CFA significance tests can be employed and interpreted. Second, when the table is sparse and the tests can not be taken seriously any longer, the descriptive measures can be used as indicators of the degree to which the discrepancy between n and e is extreme. Type and antitype decisions can then be based on selecting the α% most extreme discrepancies.

In the above example, the configurations identified as types and antitypes are among the most extreme ones in the rank order of Log*P* values. Note, however, that the most extreme *RR* (Configuration 212) constitutes neither a type nor an antitype. (For details how these measures relate to each other see Section 4.3, DuMouchel, 1999; or von Eye and Gutiérrez-Peña, in preparation.)

The last part of the printout is optional. It presents the design matrix that was used to estimate the expected cell frequencies. The design matrix contains all vectors needed for the main effects and interactions in the model. The effects are expressed in terms of effect coding. The constant vector is implied. Covariates are part of this protocol if they are part of the CFA base model.

CARPE DIEM means SEIZE THE DAY.

12.3.2.2 Two-Sample CFA with Two Predictors; Keyboard Input

Two-sample CFA allows researchers to compare two independent groups of individuals. This variant of CFA can only find *discrimination types* (no discrimination antitypes). The reason is that if there are more cases than expected from the base model in one group, there must be fewer cases than expected in the other group. This is by necessity because CFA typically estimates cell frequencies such that the marginal frequencies are reproduced. The two exceptions to this strategy are von Eye's (1985) CFA of directed relationships and CFA of differences (see Section 8.2). The log-linear base model for two-sample CFA is [P][G], where P comprises all variables used to discriminate between the two groups, and G is the grouping variable.

The following sample run re-analyzes Lienert's suicide data (see Tables 39a and 39b; Krauth & Lienert, 1973a). The data describe suicide patterns in pre- and post-WWII Germany for males (=1) and females (=2). In the years 1952 (=1) and 1944 (=2), the numbers of incidences were counted in which suicide was committed by gassing (=1), hanging (=2), use of drug overdose (=3), drowning (=4), cutting veins (=5), shooting (=6), and jumping (=7). The base model for the following gender comparison is [Year, Means of Suicide][Gender]. This model is saturated in the predictors, that is, it takes into account the main effects for Year and Means of Suicide and the interaction between Year and Means of Suicide. In addition, the model assumes independence between the two predictors and Gender. Therefore, if an interaction exists between the two predictors and Gender, there must be a difference between the gender groups and Means of Suicide, for a given year.

The following example first illustrates how the CFA program can be used to perform two-sample CFA. Second, the example shows how one can perform CFA for regional models, that is for models where variables differ in status. In the present example, there are two predictors and one grouping variable. The interaction between the two predictors is part of the two-sample CFA base model, but the interactions between the two predictors and the criterion are not part of the base model. The following paragraphs illustrate how to estimate expected cell frequencies for a model with two interacting predictors and one independent criterion.

Consider the above model, [P][G]. The cross-tabulation of the two predictors is $P_1 \times P_2$. This cross-tabulation has $I \times J$ cells. It contains all the information available on the interaction between P_1 and P_2. The saturated model also exhausts all available information. In the following analyses, we declare the cells of this table the categories of a *composite predictor*. Suppose, for example, that $I = J = 2$. Then, the indices of the cells of the cross-tabulation of these two variables are 11, 12, 21, and 22. Now, we declare these four cells to be the four categories of a composite predictor and obtain for the indices 1 = 11, 2 = 12, 3 = 21, and 4 = 22. This applies accordingly for three or more variable categories, three or more predictor variables, and two or more criterion variables in Prediction CFA.

By using the CFA program for two-sample CFA, we indicate to the program that we have a dichotomous variable for the grouping. This must be the last in the list of variables, that is, the fastest changing variable. The first variable is either a composite predictor that results from crossing predictors, or a series of one or more predictor variables. Results do not depend on the definition of the predictors. The following table summarizes data and command input via keyboard.

Command	Effect
double click shortcut to CFA program	starts CFA program; program window appears on screen. The program asks whether data will be entered via file (=1) or keyboard (=2). We enter data via keyboard and type
2	The program now asks for the number of variables. We type
2 (Enter)	The program prompts the number of categories for the predictor. We have a *composite predictor* that results from crossing a 2-category with a 7-category variable. Thus, we have a 14-category composite predictor and type
14 (Enter)	For the gender variable we type
2 (Enter)	The program then prompts the cell frequencies. We type
52, 47, 31, 14, 44, 97, 20, 10, 22, 5, 3, 0, 2, 2, 16, 61, 76, 35, 7, 9, 19, 54, 15, 4, 35, 11, 9, 2	Each of these numbers is followed by Enter (commas must not be entered). The first number in this pattern is the frequency with which males committed suicide by a given means and in a given year. The second number is the frequency for this pattern for females. After completion of data input the program asks whether the user wishes to save the data. We type
1 (Enter)	to indicate that yes. After the prompt we give

suicide.dat (Enter)	for the data file name. Up to 80 spaces can be used for the file name. The program then asks what model the user wishes to run. We type
20 (Enter)	to indicate that we want a two-sample CFA. The program then presents the marginal frequencies and requests the significance level. We type
.05 (Enter)	The program then prompts the name for the output file. We type
suicide.out (Enter)	The program responds by presenting the current options for significance tests. Our samples are relatively large. Therefore we can select one of the *z*-tests. We select the *z*-approximation of the binomial test and type
3 (Enter)	The program then asks whether the user wishes to perform a first order CFA using the same data. We indicate no by typing
2 (Enter)	The program closes and the program window disappears.

The following protocol contains the slightly edited result file, suicide.out.

```
                     Configural Frequency Analysis
                     ---------- --------- --------
          author of program: Alexander von Eye, 2002

        Marginal Frequencies
        --------------------
        Variable Frequencies
        -------- -----------
             1       99.   45.  141.   30.   27.    3.    4.   77.  111.
   16.   73.   19.   46.   11.

             2      351.  351.

       sample size N =  702.
             The z-approximation of the binomial test will be performed

     Bonferroni-adjusted alpha =  .0035714

                                       Table of results
                                       ----------------
Configuration    f      statistic        p       pi*   Type?
```

```
      11        52.
      12        47.       .542        .293839     .048
 -----------------------------------------------------------------------
      21        31.
      22        14.      2.620        .004402     .274
 -----------------------------------------------------------------------
      31        44.
      32        97.     -4.993        .000000     .273  Discrimination Type
 -----------------------------------------------------------------------
      41        20.
      42        10.      1.866        .031017     .250
 -----------------------------------------------------------------------
      51        22.
      52         5.      3.336        .000424     .386  Discrimination Type
 -----------------------------------------------------------------------
      61         3.
      62         0.      1.736        .041303     .500
 -----------------------------------------------------------------------
      71         2.
      72         2.       .000        .500000     .000
 -----------------------------------------------------------------------
      81        16.
      82        61.     -5.435        .000000     .369  Discrimination Type
 -----------------------------------------------------------------------
      91        76.
      92        35.      4.241        .000011     .270  Discrimination Type
 -----------------------------------------------------------------------
     101         7.
     102         9.      -.506        .306499     .111
 -----------------------------------------------------------------------
     111        19.
     112        54.     -4.328        .000008     .324  Discrimination Type
 -----------------------------------------------------------------------
     121        15.
     122         4.      2.558        .005257     .367
 -----------------------------------------------------------------------
     131        35.
     132        11.      3.661        .000126     .343  Discrimination Type
 -----------------------------------------------------------------------
     141         9.
     142         2.      2.127        .016697     .389
 -----------------------------------------------------------------------


                 Alternative Measures of Deviation from Independence
                 ----------- -------- -- --------- ---- ------------

 f1  f2  lambda lambdat  rho  delta  theta
 --  --  ------ -------  ---  -----  -----
 (1st line: measures, 2nd line: standard errors, 3rd line: z, 4th
line: p(z))

 52. 47.   .029    .020   .020   .020   .118
           .055    .038   .035   .035   .217
           .537    .539   .578   .578   .542
         .29559 .29506 .28155 .28155 .29391

 31. 14.   .212    .104   .099   .099   .847
           .085    .042   .055   .055   .331
          2.483   2.474  1.782  1.782  2.555
         .00652 .00668 .03736 .03736 .00530

 44. 97.   .245    .196  -.188   .188  -.980
           .051    .041   .030   .030   .201
          4.846   4.811 -6.312  6.312 -4.886
         .00000 .00000 .00000 .00000 .00000
```

```
20. 10.    .181    .073    .070    .070    .723
           .103    .041    .069    .069    .395
          1.755   1.764   1.024   1.024   1.831
         .03959  .03883  .15301  .15301  .03358

22.  5.    .383    .147    .126    .126   1.532
           .138    .053    .075    .075    .501
          2.786   2.764   1.678   1.678   3.056
         .00267  .00285  .04666  .04666  .00112

 3.  0.    .489    .073    .057    .057   1.954
           .000    .000    .000    .000    .000
           ---     ---     ---     ---     ---
           ---     ---     ---     ---     ---
 2.  2.    .000    .000    .000    .000    .000
           .344    .045    .217    .214   1.003
           .005    .005    .000    .001    .000
         .49803  .49790  .50000  .49957  .50000

16. 61.    .371    .232   -.205    .205  -1.483
           .075    .048    .044    .044    .292
          4.948   4.849  -4.705   4.705  -5.075
         .00000  .00000  .00000  .00000  .00000

76. 35.    .229    .167    .160    .160    .914
           .056    .041    .034    .034    .220
          4.107   4.076   4.759   4.759   4.151
         .00002  .00002  .00000  .00000  .00002

 7.  9.    .064    .019   -.019    .019   -.257
           .136    .040    .097    .097    .510
           .472    .486   -.198    .198   -.504
         .31851  .31359  .42160  .42160  .30696

19. 54.    .289    .176   -.163    .163  -1.156
           .071    .044    .044    .044    .278
          4.066   4.012  -3.705   3.705  -4.152
         .00002  .00003  .00011  .00011  .00002

15.  4.    .338    .110    .097    .097   1.354
           .160    .052    .090    .090    .568
          2.119   2.132   1.072   1.072   2.384
         .01704  .01650  .14179  .14179  .00856

35. 11.    .308    .152    .138    .138   1.231
           .092    .046    .056    .056    .354
          3.344   3.301   2.484   2.484   3.473
         .00041  .00048  .00649  .00649  .00026

 9.  2.    .381    .095    .080    .080   1.524
           .261    .062    .122    .122    .785
          1.460   1.520    .656    .656   1.941
         .07221  .06419  .25591  .25591  .02614


                         CARPE DIEM
```

Read from the top, this printout can be interpreted in a fashion parallel to the printout in Section 2.1. The table of results, however, is arranged differently. More specifically, the table of results presents the frequency for a predictor pattern for the two groups always between a pair of lines. Here, the frequencies for the males appear first and the frequencies for the females appear second. The information whether a discrimination type was found always appears in the second line. For example, consider the third pair of lines, that is, the lines with indices 31 and 32. This is the predictor pattern *Suicide by drug overdose in 1952*. The discrimination type suggests that this pattern is observed more often in females ($m_{32} = 97$) than in males ($m_{31} = 44$). In contrast, the next discrimination type suggests that, in 1952, males committed suicide by cutting veins more often than females ($m_{51} =$ 22 versus $m_{52} = 5$). The remaining three discrimination types can be interpreted accordingly.

The column between the tail probabilities and the designation of a pair of cells as discrimination type displays the coefficient π^* (see Section 7.2).

The block of significance test results is followed by a block of the five other measures of deviation from independence, λ, $\tilde{\lambda}$, ρ, Δ, and θ, and their significance tests. For each of these measures a standard error, a z-score, and a one-sided tail probability is printed. For λ, $\tilde{\lambda}$, ρ, and Δ these values are estimated using the jack-knifing procedure described in the appendix of von Eye et al. (1995). The standard error of the log odds ratio is estimated as described, for example, by Christensen (1997, p. 30). If one of the comparison frequencies is zero, the *z*-score and the tail probabilities are not estimated, and the standard errors are printed as zero. The present example suggests that these five measures can lead to quite discrepant appraisals of the two samples. The only exception includes ρ and Δ which differ only in sign (if the correlation is negative).

There is no design matrix included in the protocol of two-sample CFA. However, the design matrix used by the program is created using the same method as the design matrices for the main effect models.

12.3.2.3 Second Order CFA; frequency table input via file

This section illustrates the use of *second order CFA* and data input via file. In contrast to the base model of first order CFA, which does not consider any variable interactions, the base model of *second order CFA* considers all pair-wise interactions. For example, consider the three variables, A, B, and C. In bracket notation, the base model for first order CFA of these variables is [A][B][C]. The base model for second order CFA of these three variables is [AB][AC][BC]. This is a hierarchical log-linear model that implies the lower order terms, that is, in the present example, the main effects of all variables.

To illustrate second order CFA we use Lienert's (1964) LSD data again. We now assume that these data are available in the file named leuner.dat. The following print out displays the contents of this file:

```
  3  2  2  2
    20.
     1.
     4.
    12.
     3.
    10.
    15.
     0.
```

This file shows how data files must be structured to be readable for the CFA program. In a first string, the CFA program expects information about the size of the cross-tabulation to be analyzed. Specifically, the program expects to read the number of variables and then, for each variable, the number of categories. For both the number of variables and the number of categories for each variable, three places are used. In the present example, the first line of the data file indicates that we have three variables with two categories each. Next, the program expects to read the observed cell frequencies. It is important to note that the cells must be in the proper order, with the fastest changing variable being the last in the array. Please notice the periods after the frequencies. The format in which the frequencies are read is (x, f6.0), where the x indicates a blank at the beginning of the row. If the frequencies are presented with the period, they can appear anywhere within the six columns. If the period is omitted, the last digit must be placed in the sixth column of the format, that is, the seventh column of the line.

The following commands must be issued to perform Second Order CFA with the LSD data.

Command	Effect
double click shortcut to CFA program	starts program, opens program window on screen. The program then asks whether the data will be entered via file or keyboard (interactively). We select input via file and type

1 (Enter)	The program then asks whether the file is a raw data file (= 1) or a frequency table (= 2). We have a frequency table and type
2 (Enter)	The program then prompts the name for the input file. We type
leuner.dat (Enter)	The program confirms on the screen that this file is now open and presents the CFA model options. To calculate a second order CFA we type
2 (Enter)	The program confirms the selection, presents the marginal frequencies, and asks whether a covariate will be included. We have no covariate and type
2 (Enter)	The program then presents the available significance tests. Because the sample size is relatively small we can select the binomial test. Thus, we type
1 (Enter)	The program then prompts the significance level. We type
.05 (Enter)	The program then asks for the name of the output file. We type
leuner2.out (Enter)	The program writes the results to the file leuner2.out. We opt to include the design matrix in the output and key
1 (Enter)	This command concludes the run and closes the program window.

The following slightly edited protocol presents the results of second order CFA of the LSD data:

```
                    Configural Frequency Analysis
                    ---------- --------- --------
        author of program: Alexander von Eye, 2002

    Marginal Frequencies
```

```
Variable Frequencies
-------- -----------
    1       37.    28.

    2       34.    31.

    3       42.    23.

 sample size N =        65
Bonferroni-adjusted alpha =  .0062500
a CFA of order   2  was performed
significance testing used  binomial test

                                    Table of results
                                    ----- -- -------
 Configuration     fo       fe            p
 -------------    ----  --------      ---------
      111          20.   14.200      .05983935
      112           1.    6.800      .00652806
      121           4.    9.800      .02423844
      122          12.    6.200      .01898418
      211           3.    8.800      .01798567
      212          10.    4.200      .00860371
      221          15.    9.200      .03572556
      222           0.    5.800      .00229865    Antitype

               chi2 for CFA model =     37.4653
                df =      1       p =  .00000000

            LR-chi2 for CFA model =     44.1061
                df =      1       p =  .00000000


Descriptive indicators of types and antitypes
---------------------------------------------
     cell  Rel. Risk  Rank       logP   Rank
 --------  ---------  ----       ----   ----
      111     1.408     4       1.130     4
      112      .147     7        .719     6
      121      .408     5        .644     8
      122     1.936     2       1.615     2
      211      .341     6        .661     7
      212     2.381     1       1.959     1
      221     1.630     3       1.344     3
      222      .000     8        .770     5


Design Matrix
------ ------

  1.0  1.0  1.0  1.0  1.0  1.0
  1.0  1.0 -1.0  1.0 -1.0 -1.0
  1.0 -1.0  1.0 -1.0 -1.0  1.0
  1.0 -1.0 -1.0 -1.0  1.0 -1.0
 -1.0  1.0  1.0 -1.0  1.0 -1.0
 -1.0  1.0 -1.0 -1.0 -1.0  1.0
 -1.0 -1.0  1.0  1.0 -1.0 -1.0
 -1.0 -1.0 -1.0  1.0  1.0  1.0
                                    CARPE DIEM
```

As is obvious from the model specification, considering all pair-wise interactions carries the model closer to the saturated model than the base model of first order CFA. Thus, the second order CFA estimated expected cell frequencies will typically and on average be closer to the observed cell frequencies than the first order CFA estimated expected cell frequencies. A comparison of the results from the protocol in Section 12.3.2.1 with the present protocol confirms this. The discrepancies between the observed and the expected cell frequencies are smaller, and there is only one antitype left[3]. This is the antitype of those individuals that did not experience any of the LSD effects. Under the second order CFA base model 5.8 cases were expected to show no effects. However, none were observed.

12.3.2.4 CFA with covariates; input via file (frequencies) and keyboard (covariate)

This section illustrates the use of covariates in CFA (see Section 10.7). Consider, for example, the log-linear CFA base model given above. The inclusion of covariates leads to the model,

$$\log E = X_b\lambda_b + X_c\lambda_c ,$$

where X_c is a matrix with the covariates in its columns, and λ_c is the parameter for the covariate. Subscript b refers to the base mode. Using a covariate implies that more information than in standard base models is used when estimating expected cell frequencies. As a result, the expected cell frequencies typically (but not always) are closer to the observed cell frequencies, and it is less likely that types and antitypes will emerge. Covariates in CFA are particularly useful when there is information that may systematically vary over the cells of a cross-classification.

The following example illustrates the use of covariates by re-analyzing the data presented by Khamis (1996; cf. von Eye, Spiel, & Rovine, in press). The data describe the use of Cigarettes (C), Alcohol (A), and Marijuana (M) in a sample of 2,276 high school students. Each drug was scored as either used (= 1) or not used (= 2). These data can be

---

[3]Notice that the binomial test is also less powerful than Lehmacher's test. Thus, differences in power can also contribute to this difference in results. The Lehmacher test is not applicable in Second Order CFA. Therefore, a direct comparison between results from the two base models is not possible when Lehmacher's test is used.

analyzed using, for instance, log-linear modeling (Khamis, 1996) or CFA. Now suppose that, after a first analysis it becomes known in an imaginary re-analysis that all of those students that use both marijuana and alcohol also have police records for traffic violations (V = 1), and none of the others are known for traffic violations (V = 2). One may now ask whether knowledge of this covariate changes CFA results. The following equation gives the CFA base model with covariate for the present example. The base model is a log-linear main effects model, that is, a model that includes all main effects but no interaction.

$$\log \begin{bmatrix} \hat{e}_{111} \\ \hat{e}_{112} \\ \hat{e}_{121} \\ \hat{e}_{122} \\ \hat{e}_{211} \\ \hat{e}_{212} \\ \hat{e}_{221} \\ \hat{e}_{222} \end{bmatrix} = \begin{bmatrix} 1 & 1 & 1 & 1 \\ 1 & 1 & 1 & -1 \\ 1 & 1 & -1 & 1 \\ 1 & 1 & -1 & -1 \\ 1 & -1 & 1 & 1 \\ 1 & -1 & 1 & -1 \\ 1 & -1 & -1 & 1 \\ 1 & -1 & -1 & -1 \end{bmatrix} \begin{bmatrix} \lambda_0 \\ \lambda_C \\ \lambda_A \\ \lambda_M \end{bmatrix} + \begin{bmatrix} 1 \\ 2 \\ 2 \\ 2 \\ 1 \\ 2 \\ 2 \\ 2 \end{bmatrix} [\lambda_V] \; .$$

The vector on the left-hand side of the equation represents the expected cell frequencies, $\hat{m}_{ijk}$. The matrix right after the equal sign is the indicator matrix. The first column in this matrix, a column of constants, is needed for estimation of the 'grand mean parameter,' λ_0. The following three columns contain the indicator variables for the main effects of variables C, A, and M. The second summand in this equation contains the vector for the covariate, multiplied by the one-element vector for the covariate parameter.

Table 113 summarizes the results of standard, first order CFA of these data without the covariate. The results with covariate appear in the following output protocol. CFA was performed using the normal approximation of the binomial test with Bonferroni adjustment of the testwise α. The adjusted α^* was 0.00625.

The application of first order CFA with no covariate suggests that more high school students than expected from the assumption of variable independence use all three drugs, Marijuana, Alcohol, and Cigarettes (Type 111); fewer students than expected use only Cigarettes and Alcohol (Antitype 112); more students than expected use only Marijuana and Cigarettes (Type 121); fewer students than expected use only Cigarettes

Table 113: CFA of Khamis' drug use data

Cell index	Frequencies		Significance tests	
CAM	observed	expected	z	$p(z)$
111	279	64.88	26.97	< α*; T
112	2	47.33	-6.66	< α*; A
121	456	386.70	3.87	< α*; T
122	44	282.09	-15.15	< α*; A
211	43	124.19	-7.49	< α*; A
212	3	90.60	-9.39	< α*; A
221	538	740.23	-9.05	< α*; A
222	911	539.98	18.28	< α*; T

[a] < α* indicates that the tail probability is smaller than can be expressed with four decimal places.

(Antitype 122), only Alcohol and Marijuana (Antitype 211), only Alcohol (Antitype 212), or only Marijuana (Antitype 221); and more students than expected do not use any of the three drugs (Type 222).

Also considering the (hypothetical) citation record creates a different picture (the complete output follows below; cf. Mellenbergh, 1996). The discrepancies between the observed and the expected cell frequencies are, on average, smaller and the overall X^2 is smaller by almost one half (824.16 from 1411.39). In spite of the large sample size, the resulting pattern of types and antitypes is no longer the same. Configuration 112 no longer constitutes an antitype and neither does Configuration 212.

The following table and output illustrate the use of the CFA program for first order CFA with a covariate. We assume that the data are stored in a file named "Khamis2.dat." This file only contains the frequencies of the cross-tabulation. The covariate will be entered via the keyboard. The following output displays the data file:

```
 3  2  2  2
279.
2.
456.
44.
43.
3.
538.
911.
```

The following commands are needed to perform first order CFA of Khamis' drug data with a covariate:

Command	Effect
double click shortcut to CFA program	starts program. The program asks how the data will be entered. We type
1 (Enter)	The program then asks whether the data file contains raw data (= 1) or a frequency table (=2). We have a frequency table and type
2 (Enter)	thus indicating that the frequency table will be read from a file. The program then prompts the name of the data file. We type
khamis2.dat (Enter)	The program responds by confirming that this file has been opened and presents the CFA model options. We type
1 (Enter)	to indicate that we wish to calculate a first order CFA. Next the program asks whether we would like to include a covariate. We type
1 (Enter)	to indicate that we wish to use a covariate. The program requests the values of the covariate for each cell. We type
1, 2, 2, 2, 1, 2, 2, 2	each value followed by Enter (commas must not be entered). The program then asks whether another covariate will be entered. We type

2 (Enter)	thus indicating that we have only one covariate. We then type
4 (Enter)	to indicate our selection of the *z*-test, we type
.05 (Enter)	to indicate the significance level, and we type
khamis2.out (Enter)	to name the output file. Entering
1 (Enter)	includes the design matrix in the output.

The following, slightly edited protocol displays the contents of the output file khamis2.out:

```
                         Configural Frequency Analysis
                         ---------- --------- --------
            author of program: Alexander von Eye, 2002

         Marginal Frequencies
         --------------------
         Variable Frequencies
         -------- -----------
             1       781.  1495.

             2       327.  1949.

             3      1316.   960.

       sample size N =      2276

      the normal z-test was used
      Bonferroni-adjusted alpha =  .0062500
      a CFA of order   1  was performed

                                         Table of results
                                         ----- -- -------
      Configuration    fo        fe    statistic        p
      -------------   ----  --------- ---------      -------
           111        279.   110.493     16.031     .00000000     Type
           112          2.     1.716       .217     .41409490
           121        456.   341.087      6.222     .00000000     Type
           122         44.   327.704    -15.672     .00000000     Antitype
           211         43.   211.507    -11.587     .00000000     Antitype
           212          3.     3.284      -.157     .43767758
           221        538.   652.913     -4.497     .00000345     Antitype
           222        911.   627.296     11.327     .00000000     Type

                      chi2 for CFA model =   824.1630
                      df =      3       p =  .00000000

                  LR-chi2 for CFA model =    939.5626
                      df =      3       p =  .00000000
```

```
 Descriptive indicators of types and antitypes
 ---------------------------------------------
      cell  Rel. Risk  Rank        logP   Rank
  --------  ---------  ----        ----   ----
       111      2.525     1      40.467      3
       112      1.166     4        .298      7
       121      1.337     3       8.890      5
       122       .134     8      83.438      1
       211       .203     7      42.527      2
       212       .913     5        .231      8
       221       .824     6       5.703      6
       222      1.452     2      25.896      4


Design Matrix
------ ------

  1.0   1.0   1.0   1.0
  1.0   1.0  -1.0   2.0
  1.0  -1.0   1.0   2.0
  1.0  -1.0  -1.0   2.0
 -1.0   1.0   1.0   1.0
 -1.0   1.0  -1.0   2.0
 -1.0  -1.0   1.0   2.0
 -1.0  -1.0  -1.0   2.0
                                   CARPE DIEM
```

This protocol can be interpreted as the other protocols, above. Note that Lehmacher's tests are not applicable when covariates are used.

References

Abramowitz, M., & Stegun, I.A. (1972). *Handbook of mathematical functions*. New York: Dover.

Agresti, A. (1990). *Categorical data analysis*. New York: Wiley.

Agresti, A. (1996). *An introduction to categorical data analysis*. New York: Wiley.

Agresti, A., & Yang, M.C. (1987). An empirical investigation of some effects of sparseness in contingency tables. *Computational Statistics & Data Analysis*, *5*, 9 - 21.

Aksan, N., Goldsmith, H. H., Smider, N. A., Essex, M. J., Clark, R., Hyde, J. S., Klein, M. H., & Vandell, D. L. (1999). Derivation and prediction of temperamental types among preschoolers. *Developmental Psychology, 35*, 958 - 971.

Anastasi, A. (1994). Geleitwort: Differential psychology: origin and sources. In K. Pawlik (Ed.), *Die differentielle Psychologie in ihren methodischen Grundlagen*. Bern: Verlag Hans Huber.

Anscombe, F. J. (1953). Contribution of discussion of paper by H. Hotelling 'New light on the correlation coefficient and its transform'. *Journal of the Royal Statistical Society, 15*(B), 229 - 230.

Bartholomew, D.J., & Knott, M. (1999). *Latent variable models and factor analysis* (2nd ed). London: Arnold.

Bartoszyk, G.D., & Lienert, G.A. (1978). Konfigurationsanalytische Typisierung von Verlaufskurven. *Zeitschrift für Experimentelle und Angewandte Psychologie*, *XXV*, 1 - 9.

Benjamini, Y., & Hochberg, Y. (1995). Controlling the false discovery rate: A practical and powerful approach to multiple testing. *Journal of the Royal Statistical Society B, 57*, 289 - 300.

Benjamini, Y., & Hochberg, Y. (2000). On the adaptive control of the false discovery rate in multiple testing with independent statistics. *Journal of Educational and Behavioral Statistics, 25*, 60 - 83.

Bergman, L. R. (1996). Studying persons-as-wholes in applied research. *Applied Psychology: An International Review, 45*, 331 - 334.

Bergman, L.R. (2000). I-States as Object Analysis (ISOA) - A way to generate sequences of categories for longitudinal (CFA) analysis. *Psychologische Beiträge*, *42*, 337 - 346.

Bergman, L.R., Cairns, R.B., Nilsson, L.-G., & Nystedt, L. (Eds.).(2000). *Developmental science and the holistic approach*. Mahwah, NJ: Lawrence Erlbaum.

Bergman, L.R. & El-Khouri, B.M. (1998). SLEIPNER - A statistical package for pattern-oriented analysis. University of Stockholm (Sweden): Department of Psychology, statistical software package.

Bergman, L.R., & El-Khouri, B. (1999). Studying individual patterns of development using I-States as Objects Analysis (ISOA). *Biometrical Journal, 41*, 753 - 770.

Bergman, L. R., & Magnusson, D. (1991). Stability and change in patterns of extrinsic adjustment problems. In D. Magnusson, L.R. Bergman, G. Rudinger, & B. Törestad (Eds.), *Problems and methods in longitudinal research* (pp. 323 - 346). Cambridge, UK: Cambridge University Press.

Bergman, L. R., & Magnusson, D. (1997). A person-oriented approach in research on developmental psychopathology. *Development and Psychopathology, 9*, 291 - 319.

Bergman, L.R., Magnusson, D., & El-Khouri, B.M. (2000). *Studying individual development in an interindividual context: A person-oriented approach.* (In preparation).

Bergman, L. R., & von Eye, A. (1987). Normal approximations of exact tests in Configural Frequency Analysis. *Biometrical Journal, 29*, 849 - 855.

Bierschenk, B., & Lienert, G.A. (1977). Simple methods for clustering progiles and learning curves. *Didaktometry, 56*, 1 - 26.

Bishop, Y. M. M., Fienberg, S. E., & Holland, P. W. (1975). *Discrete multivariate analysis*. Cambridge, MA: MIT Press.

Bollen, K.A. (1989). *Structural equations with latent variables*. New York: Wiley.

Bonhoeffer, K. (1917). Die endogenen Reaktionstypen. *Archiv für Psychiatrie und Nervenkrankheiten, 58*, 58 - 70

Bowker, A.H. (1948). A test for symmetry in contingency tables. *Journal of the American Statistical Association, 43*, 572 - 574.

Box, G.E.P., & Tiao, G.C. (1973). *Bayesian inference in statistical analysis*. Reading: Addison-Wesley.

Brandtstädter, J. (1998). Action perspectives on human development. In R.M. Lerner (Ed.), *Handbook of child psychology, Vol. one: Theoretical models of human development* (5th ed., pp. 807 - 863). New York: Wiley.

Cattell, R.B. (1988). The data box. Its ordering of total resources in terms of possible relational systems. In J.R. Nesselroade, & R.B. Cattell (Eds.), *Handbook of multivariate experimental psychology* (2nd ed., pp. 69 - 130). New York: Plenum.

Chipuer, H., & von Eye, A. (1989). Suicide trends in Canada and in

Germany: An application of Configural Frequency Analysis. *Suicide and Life-Threatening Behavior, 19*, 264 - 276.

Christensen, R. (1997). *Log-linear models and logistic regression* (2nd ed.). New York: Springer.

Church, C., & Hanks, P. (1991). Word association norms, mutual information, and lexicography. *Computational Linguistics, 16*, 22 - 29.

Clogg, C.C. (1995). Latent class models. In G. Arminger, C.C. Clogg, & M.E. Sobel (Eds.), *Handbook of statistical modeling for the social and behavioral sciences* (pp. 311 - 359). New York: Plenum.

Clogg, C. C., & Manning, W. D. (1996). Assessing reliability of categorical measurements using latent class models. In A. von Eye & C. C. Clogg (Eds.), *Categorical variables in developmental research. Methods of analysis* (pp. 169 - 182). San Diego: Academic Press.

Clogg, C.C., Petkova, E., & Shihadeh, E.S. (1992). Statistical methods for analyzing collapsibility in regression models. *Journal of Educational Statistics, 17*, 51 - 74.

Cohen, J. (1988). *Statistical power analysis for the behavioral sciences* (2nd ed.). Hillsdale, NJ: Erlbaum.

Cook, T.D., & Campbell, D.T. (1979). *Quasi-experimentation: Design and analysis issues for field settings*. Boston: Houghton Mifflin.

Cribbie, R., Holland, B., & Keselman, H. J. (1999). *Multiple comparisons procedures for large family sizes: Controlling the probability of at least k or more Type I errors*. Montreal: Annual Meeting of the American Educational Research Association.

Darlington, R.B., & Hayes, A.F. (2000). Combining independent *p* values: Extensions of the Stouffer and binomial methods. *Psychological Methods, 5*, 496 - 515.

DuMouchel, W. (1999). Bayesian data mining in large frequency tables, with an application to the FDA spontaneous reporting system. *The American Statistician, 53*, 177 - 190.

Duncan, O.D. (1975). Partitioning polytomous variables in multiway contingency tables. *Social Science Research, 4*, 167 - 182.

Dunkl, E. (2000). A SAS macro to compute Configural Frequency Analysis. *Psychologische Beiträge, 42*, 526 - 535.

Dunnett, C. W., & Tamhane, A. C. (1992). A step-up multiple test procedure. *Journal of the American Statistical Association, 87*, 162 - 170.

Everitt, B. S. (1977). *The analysis of contingency tables*. London:

Chapman & Hall.

Everitt, B. S. (1998). *The Cambridge dictionary of statistics*. Cambridge, UK: Cambridge University Press.

Evers, M., & Namboodiri, N. K. (1978). On the design matrix strategy in the analysis of categorical data. In K. F. Schuessler (Ed.), *Sociological methodology* (pp. 86 - 111). San Francisco: Jossey-Bass.

Feger, H. (1994). *Structure analysis of co-occurrence data*. Aachen: Shaker.

Feller, W. (1957). *Probability theory and its applications*. New York: Wiley.

Fienberg, S. E. (1980). *The analysis of cross-classified categorical data* (2nd ed.). Cambridge, MA: MIT Press.

Finkelstein, J., von Eye, A., & Preece, M. A. (1994). The relationship between aggressive behavior and puberty in normal adolescents: A longitudinal study. *Journal of Adolescent Health, 15*, 319 - 326.

Fisher, R.A., & Yates, F. (1948). *Statistical tables for biological, agricultural, and medical research*. Edinburgh: Oliver & Boyd.

Fleischmann, U. M., & Lienert, G. A. (1982). Die Interaktionsstrukturanalyse als Mittel der Orthogonalitätsbeurteilung faktoriell einfach strukturierter Tests. *Psychologische Beiträge, 24*, 396 - 410.

Fleischmann, U.M., & Lienert, G.A. (1992). A bivariate median test on partial association CFA. *Biometrical Journal*, *6*, 669 - 673.

Funke, W., Funke, J., & Lienert, G. A. (1984). Prädiktionskoeffizienten in der Konfigurationsfrequenzanalyse (Phi-Koeffizienten). *Psychologische Beiträge, 26*, 382 - 392.

Gelman, A., Carlin, J. B., Stern, H. S., & Rubin, D. B. (1995). *Bayesian data analysis*. London: Chapman & Hall.

Glück, J. (1999). *Spatial strategies - Kognitive Strategien bei Raumvorstellungsleistungen* [Spatial strategies - Cognitive strategies for spatial tasks]. Unpublished dissertation, University of Vienna, Austria.

Glück, J., & von Eye, A. (2000). Including covariates in Configural Frequency Analysis. *Psychologische Beiträge*, *42*, 405 - 417.

Goldstein, H.I. (1987). *Multilevel models in educational and social research*. New York: Oxford University Press.

Gonzáles-Debén, A. (1998). *Experiencias con un nuevo índice de falta de adjuste en el análisis de tablas de contingencia*. Unpublished masters thesis, University of Havana, Cuba.

Gonzáles-Debén, A., & Méndez Ramírez, I. (2000). Un nuevo concepto de tipo en el análisis de las frecuencias de las configuraciones de dos muestras. *Multiciênica, 4,* 7 - 17.

Goodman, L. A. (1984). *The analysis of cross-classified data having ordered categories.* Cambridge, MA: Harvard University Press.

Goodman, L.A. (1991). Measures, models, and graphical displays in the analysis of cross-classified data. *Journal of the American Statistical Association, 86,* 1085 - 1111.

Görtelmeyer, R. (1988). *Typologie des Schlafverhaltens.* Regensburg: S. Roderer Verlag.

Görtelmeyer, R. (2000). Veränderungsanalyse in Interventionsstudien mit adjustierter KFA. *Psychologische Beiträge, 42,* 362 - 382.

Görtelmeyer, R. (2001). *Interventions- und Veränderungsanalyse. Ein Vorschlag zum Paradigmenwechsel in der Therapie-Evaluation.* Frankfurt/Main: Peter Lang.

Gottlieb, G. (1992). *Individual development & evolution. The genesis of novel behavior.* New York: Oxford University Press.

Gottlieb, G., Wahisten, D., & Lickliter, R. (1998). The significance of biology for human development: historical and epistemological perspectives. In R.M. Lerner (Ed.), *Handbook of child psychology, vol. one: Theoretical models of human development* (5th ed., pp. 233 - 273). New York: Wiley.

Graham, P. (1995). Modelling covariate effects in observer agreement studies: the case of nominal scale agreement. *Statistics in Medicine, 14,* 299 - 310.

Greenacre, M. J. (1984). *Theory and applications of correspondence analysis.* New York: Academic Press.

Gutiérrez-Peña, E., & von Eye, A. (2000). A Bayesian approach to Configural Frequency Analysis. *Journal of Mathematical Sociology, 24,* 151-174.

Gutiérrez-Peña, E., & von Eye, A. (2000). *The use of prior information in Configural Frequency Analysis - Frequentist and Bayesian approaches.* (in preparation; b)

Haberman, S. J. (1973). The analysis of residuals in cross-classified tables. *Biometrics, 29,* 205 - 220.

Hartigan, J. A. (1975). *Clustering algorithms.* New York: Wiley.

Havránek, T., Kohnen, R., & Lienert, G. A. (1986). Nonparametric evaluation of ANOVA designs by local, regional, and global contingency testing. *Biometrical Journal, 28,* 11 - 21.

Havránek, T., & Lienert, G. A. (1984). Local and regional versus global contingency testing. *Biometrical Journal, 26*, 483 - 494.

Heilmann, W.-R., & Lienert, G. A. (1982). Predictive configural frequency analysis evaluated by simultaneous Berchtold-corrected fourfold X2-tests. *Biometrical Journal, 24*, 723 - 728.

Heilmann, W.-R., Lienert, G. A., & Maly, V. (1979). Prediction models in configural frequency analysis. *Biometrical Journal, 21*, 79 - 86.

Heilmann, W.-R., & Schütt, W. (1985). Tables for binomial testing via the F-distribution in configural frequency analysis. *EDV in Medicine and Biology, 16*, 1 - 7.

Hochberg, Y. (1988). A sharper Bonferroni procedure for multiple tests of significance. *Biometrika, 75*, 800 - 802.

Hoernes, G.E., & Heilweil, M.F. (1964). *Introduction to Boolean algebra and logic design*. New York: McGraw-Hill.

Holland, B. S., & Copenhaver, M. D. (1987). An improved sequentially rejective Bonferroni test procedure. *Biometrics, 43*, 417 - 423.

Holland, P.W. (1986). Statistics and causal inference. *Journal of the American Statistical Association, 81*, 116 - 133.

Holland, P.W. (1988). Causal inference, path analysis, and recursive structural equation models. *Sociological Methodology*, 449 - 493.

Holm, S. (1979). A simple sequentially rejective multiple test procedure. *Scandinavian Journal of Statistics, 6*, 65 - 70.

Hommel, G. (1988). A stagewise rejective multiple test procedure based on a modified Bonferroni test. *Biometrika, 75*, 383 - 386.

Hommel, G. (1989). A comparison of two modified Bonferroni procedures. *Biometrika, 76*, 624 - 625.

Hommel, G., Lehmacher, W., & Perli, H.-G. (1985). Residuenanalyse des Unabhängigkeitsmodells zweier kategorialer Variablen. In J. Jesdinsky & J. Trampisch (Eds.), *Prognose- und Entscheidungsfindung in der Medizin* (pp. 494 - 503). Berlin: Springer.

Hu, T.-C. (1988). A statistical method of approach to Stirling's formula. *The American Statistician, 42*, 204 - 205.

Hütter, U., Müller, U., & Lienert, G. A. (1981). Die Konfigurationsfrequenzanalyse. XIII. Multiple, kanonische und multivariate Prädiktions-KFA und ihre Anwendung in der Medizinsoziologie. *Zeitschrift für Klinische Psychologie und Psychotherapie, 29*, 4 - 13.

Indurkhya, A., & von Eye, A. (2000). The power of tests in Configural Frequency Analysis. *Psychologische Beiträge, 42*, 301 - 308.

Jobson, J. D. (1992). *Applied multivariate data analysis: Vol. 2:*

Categorical and multivariate methods. New York: Springer.

Jöreskog, K., & Sörbom, D. (1993). *LISREL 8 user's reference guide*. Chicago: Scientific Software Inc.

Keenan, D.P., Achterberg, C., AbuShaba, R., Kris-Etherton, P.M., & von Eye, A. (1996). Use of qualitative and quantitative methods to define behavioral fat reduction strategies and their relationship to dietary fat reduction in the Patterns of Dietary Change Study. *Journal of the American Dietetic Association, 96*, 1245 - 1253.

Keselman, H. J., Cribbie, R., & Holland, B. (1999). The pairwise multiple comparison multiplicity problem: an alternative approach to familywise and comparisonwise Type I error control. *Psychological Methods, 4*, 58 - 69.

Keuchel, I., & Lienert, G.A. (1985). Die Konfigurationsfrequenzanalyse. XXIIb. Typen ipsativer Skalenmuster. *Zeitschrift für Klinische Psychologie, Psychopathologie und Psychotherapie, 33*, 232 - 238.

Khamis, H.J. (1996). Application of the multigraph representation of hierarchical log-linear models. In A. von Eye & C.C. Clogg (Eds.), *Categorical variables in developmental research* (pp. 215 - 229). San Diego: Academic Press.

Kieser, M., & Victor, N. (1991). A test procedure for an alternative approach to configural frequency analysis. *Methodika, 5*, 87 - 97.

Kieser, M., & Victor, N. (1999). Configural Frequency Analysis (CFA) revisited - A new look at an old approach. *Biometrical Journal, 41*, 967 - 983.

Kieser, M., & Victor, N. (2000). An alternative approach for the identification of types in contingency tables. *Psychologische Beiträge, 42*, 402 - 404.

Kimball, A. W. (1954). Short cut formulae for the exact partition of chi-square in contingency tables. *Biometrics, 10*, 452 - 458.

Kirk, R.E. (1995). *Experimental design. Procedures for the behavioral sciences* (3rd ed.). Pacific Grove: Brooks/Cole.

Klingenspor, B., Marsiske, M., & von Eye, A. (1993). *Life beyond age 70: Gender-specific differences in social network size*. Unpublished manuscript.

Koehler, K. J., & Larntz, K. (1980). An empirical investigation of goodness-of-fit statistics for sparse multinomials. *Journal of the American Statistical Association, 75*, 336 - 344.

Kohnen, R., & Rudolf, J. (1981). Die Konfigurationsfrequenzanalyse XIVa. Remissionskontrollierte Symptommuster-Abfolgen im Therapie-Wartegruppenvergleich. *Zeitschrift für Klinische*

Psychologie und Psychotherapie, *29*, 110 - 126.

Kotze, P. J. V., & Hawkins, M. M. (1984). The identification of outliers in two-way contingency tables, using 2 x 2 subtables. *Journal of Applied Statistics, 33*, 215 - 223.

Krause, B., & Metzler, P. (1984). *Angewandte Statistik*. Berlin: VEB Deutscher Verlag der Wissenschaften.

Krauth, J. (1973). Nichtparametrische Ansätze zur Auswertung von Verlaufskurven. *Biometrische Zeitschrift*, *15*, 557 - 566.

Krauth, J. (1980a). Nonparametric analysis of response curves. *Journal of Neuroscience Methods*, *2*, 239 - 252.

Krauth, J. (1980b). Ein Vergleich der Konfigurationsfrequenzanalyse mit der Methode der log-linearen Modelle. *Zeitschrift für Sozialpsychologie, 11*, 233 - 247.

Krauth, J. (1993). *Einführung in die Konfigurationsfrequenzanalyse (KFA)*. Weinheim: Beltz. Psychologie Verlags Union.

Krauth, J. (1996a). *Einführung in die Konfigurationsfrequenzanalyse*. Weinheim: Beltz.

Krauth, J. (1996b). Good typal analysis must be based on a precise definition of types. *Applied Psychology: An International Review, 45*, 334 - 337.

Krauth, J., & Lienert, G. A. (1973a). *KFA. Die Konfigurationsfrequenzanalyse und ihre Anwendung in Psychologie und Medizin*. Freiburg: Alber.

Krauth, J., & Lienert, G. A. (1973b). Nichtparametrischer Nachweis von Syndromen durch simultane Binomialtests. *Biometrische Zeitschrift, 15*, 13 - 20.

Krauth, J., & Lienert, G. A. (1974). Zum Nachweis syndromgenerierender Symptominteraktionen in mehrdimensionalen Kontingenztafeln (Interaktionsstrukturanalyse). *Biometrische Zeitschrift, 16*, 203 - 211.

Krauth, J., & Lienert, G.A. (1975). Konfigurationsfrequenzanalytische Auswertung von Verlaufskurven. In W.H. Tack (Ed.), *Bericht über den 29. Kongreß der Deutschen Gesellschaft für Psychologie in Saarbrücken* (pp. 402 - 404). Göttingen: Hogrefe.

Krauth, J., & Lienert, G.A. (1978). Nonparametric two-sample comparison of learning curves based on orthogonal polynomials. *Psychological Research, 40*, 159 - 171.

Krauth, J., & Lienert, G. A. (1982). Die Konfigurationsfrequenzanalyse XVII. Dyslexie-Verdachtstypen bei Jungen und Mädchen. *Zeitschrift für Klinische Psychologie und Psychotherapie, 30*, 196 -

201.

Krebs, H., Ising, M., Janke, W., Macht, M., von Eye, A., Weijers, H.-G., & Weyers, P. (1996). Response curve comparison by pseudo-multivariate two-sample configural frequency analysis. *Biometrical Journal*, *38*, 195 - 201.

Kreppner, K. (1989). Beobachtung und Längsschnitt in der Kleinkindforschung: Überlegungen zur Methodologie und Demonstration eines empirischen Beispiels. In H. Keller (Ed.), *Handbuch der Kleinkindforschung* (pp. 271 - 294). Berlin: Springer.

Kreppner, K., Paulsen, S., & Schütze, Y. (1982). Infant and family development: From dyads to tetrads. *Human Development*, *25*, 373 - 391.

Kristof, W. (1993). Demonstration of metasyndromes in Configuration Frequency Analysis. *Zeitschrift für Klinische Psychologie, Psychiatrie, und Psychotherapie*, *41*, 304 - 306.

Krüger, H.-P., Lienert, G.A., Gebert, A., & von Eye, A. (1979). Eine inferentielle Clusteranalyse für Alternativdaten. *Psychologische Beiträge, 21*, 540 - 553.

Küchenhoff, H. (1986). A note on a continuity correction for testing in three-dimensional Configural Frequency Analysis. *Biometrical Journal, 28*, 465 - 468.

Lange, H.-J., & Vogel, T. (1965). Statistische Analyse von Symptomkorrelationen bei Syndromen. *Methods of Information in Medicine*, *4*, 83 - 89.

Larntz, K. (1978). Small sample comparisons of exact levels for chi-squared goodness-of-fit statistics. *Journal of the American Statistical Association, 73*, 253 - 236.

Lautsch, E., Lienert, G.A., & von Eye, A. (1987). Zur Anwendung der Küchenhoff Stetigkeitskorrektur des Lehmacher KFA-Tests in der Scuhe nach Typen soziogener Neuropathologie. *Zeitschrift für Klinische Psychologie, Psychopathologie und Psychotherapie, 35*, 134 - 140.

Lautsch, E., & von Weber, S. (1995). *Methoden und Anwendungen der Konfigurationsfrequenzanalyse*. Weinheim: Psychologie Verlags Union.

Lautsch, E. (2000). Evaluation von Prädiktionstypen (demonstriert an einem Beispiel aus der kriminologischen Forschung). *Psychologische Beiträge*, *42*, 309 - 326.

Lautsch, E., & Ninke, L. (2000). Kombinierter Einsatz von CHAID und

KFA bei der soziodemographischen Beschreibung von Kriminalitätsfurcht. *Psychologische Beiträge*, *42*, 347 - 361.

Lehmacher, W. (1981). A more powerful simultaneous test procedure in Configural Frequency Analysis. *Biometrical Journal, 23*, 429 - 436.

Lehmacher, W. (2000). Die Konfigurationsfrequenzanalyse als Komplement des log-linearen Modells. *Psychologische Beiträge*, *42*, 418 - 427.

Lehmacher, W., & Lienert, G. A. (1982). Die Konfigurationsfrequenzanalyse XVI. Neue Tests gegen Typen und Syndrome. *Zeitschrift für Klinische Psychologie und Psychotherapie, 30*, 5 - 11.

Lerner, R.M. (Ed.).(1998). *Handbook of child psychology: Vol. one: Theoretical models of human development* (5th ed). New York: Wiley.

Lienert, G.A. (1964). *Belastung und Regression*. Meisenheim am Glan: Hain.

Lienert, G. A. (1968). *Die "Konfigurationsfrequenzanalyse" als Klassifikationsmethode in der klinischen Psychologie*. Paper presented at the 26. Kongress der Deutschen Gesellschaft für Psychologie in Tübingen 1968.

Lienert, G.A. (1969). Die "Konfigurationsfrequenzanalyse" als Klassifikationsmethode in der klinischen Psychologie. In M. Irle (Ed.), *Bericht über den 16. Kongreß der Deutschen Gesellschaft für Psychologie in Tübingen 1968* (pp. 244 - 255). Göttingen: Hogrefe.

Lienert, G.A. (1970). Konfigurationsfrequenzanalsye einiger Lysergsäurediäthylamid-Wirkungen. *Arzneimittelforschung, 20*, 912 - 913.

Lienert, G.A. (1971a). Die Konfigurationsfrequenzanalyse I. Ein neuer Weg zu Typen und Syndromen. *Zeitschrift für Klinische Psychologie und Psychotherapie*, *19*, 99 - 115.

Lienert, G. A. (1971b). Die Konfigurationsfrequenzanalyse III. Zwei- und Mehrstichproben KFA in Diagnostik und Differentialdiagnostik. *Zeitschrift für Klinische Psychologie und Psychotherapie, 19*, 291 - 300.

Lienert, G.A. (1971c). Die Konfigurationsfrequenzanalyse II. Hierarchische und agglutinierende KFA in der klinischen Psychologie. *Zeitschrift für Klinische Psychologie und Psychotherapie*, *19*, 207 - 220.

Lienert, G. A. (1978). *Verteilungsfreie Methoden in der Biostatistik.* (Vol. 2). Meisenheim am Glan: Hain.

Lienert, G.A. (1980). Nonparametric cluster analysis of learning curves based on orthogonal polynomials. In: Hungarian Academy of Sciences (Ed.), *Proceedings of the 4th Meeting of Psychologists from the Danubian Countries* (pp. 595 - 609). Budapest: Akadémiai Kiadó.

Lienert, G.A. (1987). Vergleich unabhängiger Stichproben von qualitativen Variablen mittels geschlossener k-Stichproben-Konfigurationsfrequenzanalyse. In E. Raab & G. Schulter (Eds.), *Perspektiven psychologischer Forschung. Festschrift zum 65. Geburtstag von Erich Mittenecker* (pp. 13 - 24). Wien: Deuticke.

Lienert, G.A. (1988).(Ed.). *Angewandte Konfigurationsfrequenzanalyse.* Frankfurt: Athenäum.

Lienert, G.A. (1989). Victor's alternative approach to configural frequency analysis. In J.A. Keats, R. Taft, R.A. Heath, & S.H. Lovibon (Eds.), *Proceedings of the XXIVth International congress of Psychology: Vol. 4, Mathematical and theoretical systems* (pp. 79 - 97). Amsterdam: Elsevier.

Lienert, G.A., & Barth, A.-R. (1987). Comparing paired samples nonparametrically by Raviv's rank test. *EDV in Medizin und Biologie, 18*, 125 -128.

Lienert, G.A., & Bergman, L.R. (1985). Longisectional Interaction Structure Analysis (LISA) in psychopharmacology and developmental psychopathology. *Neuropsychobiology, 14*, 27 - 34.

Lienert, G.A., Dunkl, E., & von Eye, A. (1990). Kleingruppentests gegen Victor-Typen und -Syndrome. *Zeitschrift für Klinische Psychologie, Psychopathologie und Psychotherapie, 44*, 45 - 51.

Lienert, G. A., & Klauer, K. J. (1983). Kohortenanalyse von Erfolgsbeurteilungen mittels multivariater Prädiktions-KFA. *Zeitschrift für Klinische Psychologie und Psychotherapie, 25*, 297 - 314.

Lienert, G. A., & Krauth, J. (1973a). Die Konfigurationsfrequenzanalyse als Prädiktionsmodell in der angewandten Psychologie. In H. Eckensberger (Ed.), *Bericht über den 28. Kongress der Deutschen Gesellschaft für Psychologie in Saarbrücken 1972* (pp. 219 - 228). Göttingen: Hogrefe.

Lienert, G. A., & Krauth, J. (1973b). Die Konfigurationsfrequenzanalyse V. Kontingenz- und Interaktionsstrukturanalyse multinär skalierter Merkmale. *Zeitschrift für Klinische Psychologie und*

Psychotherapie, 21, 26 - 39.

Lienert, G. A., & Krauth, J. (1973c). Die Konfigurationsfrequenzanalyse VI. Profiländerungen und Symptomverschiebungen. *Zeitschrift für Klinische Psychologie und Psychotherapie, 21*, 100-109.

Lienert, G.A., & Krauth, J. (1973d). Die Konfigurationsfrequenzanalyse VII. Konstellations-, Konstellationsänderungs- und Profilkonstellationstypen. *Zeitschrift für Klinische Psychologie und Psychotherapie, 21*, 197 - 209.

Lienert, G.A., & Krauth, J. (1975). Configural Frequency Analysis as a statistical tool for defining types. *Educational and Psychological Measurement, 35*, 231 - 238.

Lienert, G.A., Ludwig, O., & Rockefeller, K. (1982). Tables of the critical values for simultaneous and sequential Bonferroni z-tests. *Biometrical Journal, 24*, 239 - 255.

Lienert, G.A., & Netter, P. (1985). Die Konfigurationsfrequenzanalyse XXIb. Typenanalyse bivariater Verlaufskurven von Hyper- und Normotonikern. *Zeitschrift für Klinische Psychologie, Psychopathologie und Psychotherapie, 33*, 77 - 88.

Lienert, G.A., & Netter, P. (1986). Nonparametric evaluation of repeated measurement designs by point-symmetry testing. *Biometrical Journal, 28*, 3 - 10.

Lienert, G. A., & Netter, P. (1987). Nonparametric analysis of treatment response tables by bipredictive configural frequency analysis. *Methods of Information in Medicine, 26*, 89 - 92.

Lienert, G.A., Netter, P., & von Eye, A. (1987). Die Konfigurationsfrequenzanalyse XXV. Typen und Syndrome höherer Ordnung. *Zeitschrift für Klinische Psychologie, Psychopathologie und Psychotherapie, 35*, 344 - 352.

Lienert, G. A., & Rey, E.-R. (1982). Die Konfigurationsfrequenzanalyse. XV. Typenexploration und -inferenz (Hybride und agglutinierende Prädiktions-KFA). *Zeitschrift für Klinische Psychologie und Psychotherapie, 30*, 209 - 215.

Lienert, G.A., & Rudolph, J. (1983). Die Konfigurationsfrequenzanalyse. XIX. Remissionskontrollierte Inkrementen-KFA (Zuwachsmuster-Diskriminanztypen) im Therapie-Wartegruppenvergleich. *Zeitschrift für Klinische Psychologie, Psychopathologie und Psychotherapie, 31*, 245 - 253.

Lienert, G.A., & Straube, E. (1980). Die Konfigurationsfrequenzanalyse XI. Strategien des Symptom-Konfigurations-Vergleichs vor und nach einer Therapie. *Zeitschrift für Klinische Psychologie und*

Psychotherapie, 28, 110 - 123.

Lienert, G. A., & von Eye, A. (1984a). Multivariate Änderungsbeurteilung mittels Inkrementen-Konfigurationsclusteranalyse. *Psychologische Beiträge, 26,* 363 - 371.

Lienert, G.A., & von Eye, A. (1984b). Testing for stability and change in multivariate t-point observations by longitudinal configural frequency analysis. *Psychologische Beiträe, 26,* 298 - 308.

Lienert, G. A., & von Eye, A. (1985). Die Konfigurationsclusteranalyse und ihre Anwendung in der klinischen Psychologie. In D. Albert (Ed.), *Bericht über den 34. Kongress der Deutschen Gesellschaft für Psychologie 1984 in Wien* (pp. 167 - 169). Göttingen: Hogrefe.

Lienert, G.A., & von Eye, A. (1986). Nonparametric two-sample CFA of incomplete learning curves. In F. Klix, & H. Hagendorf (Eds.), *Human memory and cognitive capabilities* (pp. 123 - 138). New York: Elsevier.

Lienert, G. A., & von Eye, A. (1987). Nonparametric comparison of longitudinal response patterns from unpaired samples using CFA. *Biometrical Journal, 29,* 675 - 688.

Lienert, G. A., & von Eye, A. (1988). Syndromaufklärung mittels generalisierter Interaktionsstrukturanalyse. *Zeitschrift für Klinische Psychologie, Psychopathologie und Psychotherapie, 36,* 25 - 33.

Lienert, G.A., & von Eye, A. (1989). Die Konfigurationsfrequenzanalyse. XXIV. Konfigurationsclusteranalyse als Alternative zur KFA. *Zeitschrift für Klinische Psychologie, Psychopathologie und Psychotherapie, 36,* 451 - 457.

Lienert, G. A., & Wolfrum, C. (1979). Die Konfigurationsfrequenzanalyse. X. Therapiewirkungsbeurteilung mittels Prädiktions-KFA. *Zeitschrift für Klinische Psychologie und Psychotherapie, 27,* 309 - 316.

Lienert, G.A., & zur Oeveste, H. (1985). Configural Frequency Analysis as a statistical tool for developmental research. *Educational and Psychological Measurement, 45,* 301 - 307.

Lindley, D.V. (2000). The philosophy of statistics. *The Statistician, 49,* 293 - 337.

Lindner, K. (1984). Eine exakte Auswertungsmethode zur Konfigurationsfrequenzanalyse. *Psychologische Beiträge, 26,* 393 - 415.

Ludwig, O., Gottlieb, R., & Lienert, G. A. (1986). Tables of Bonferroni-limits for simultaneous F-tests. *Biometrical Journal, 28,* 25 - 30.

Magnusson, D. (1998). The logic and implications of a person-oriented approach. In R.B. Cairns, L.R. Bergman, & J. Kagan. (Ed.), *Methods and models for studying the individual* (pp. 33-63). Thousand Oaks: Sage.

Magnusson, D., & Bergman, L. R. (2002). Person-centered research. In T. Cook & C. Ragin (Eds.), *International Encyclopedia of the Social and Behavioral Sciences*: *Vol. 8, Logic of inquiry and research design*. Amsterdam: Pergamon. (in press)

Mahoney, J. L. (2000). School extracurricular activity participation as a moderator in the development of antisocial patterns. *Child Development, 71*, 502 - 516.

Marcus, R., Peritz, E., & Gabriel, K. R. (1976). On closed testing procedures with special reference to ordered analysis of variance. *Biometrika, 63*, 655 - 660.

Maxwell, A.E. (1961). *Analyzing qualitative data*. London: Methuen.

McNemar, Q. (1947). Note on the sampling error of the difference between correlated proportions or percentages. *Psychometrika, 12*, 143 - 157.

Meehl, P. E. (1950). Configural scoring. *Journal of Consulting Psychology, 14*, 165 - 171.

Mellenbergh, G. J. (1996). Other null model, other (anti)type. *Applied Psychology: An International Review, 45*, 329 - 330.

Migon, H.S., & Gamerman, D. (1999). *Statistical inference: an integrated approach*. London: Arnold.

Molenaar, W. (1970). *Mathematical Centre tract 31: Approximations to the Poisson, binomial, and hypergeometric distribution functions*. Amsterdam: Mathematisch Centrum.

Müller, M.J., Netter, P., & von Eye, A. (1997). Catecholamine response curves of male hypertensives identified by Lehmacher's two sample Configural Frequency Analysis. *Biometrical Journal, 39*, 29 - 38.

Mun, E.-Y., Fitzgerald, H.E., Puttler, L.I., Zucker, R.A., & von Eye, A. (2001). Early child temperament as predictor of child behavior problems in the context of low and high parental psychopathology. *Infant Mental Health Journal, 22*, 393 - 415.

Mun, E.-Y., von Eye, A., Fitzgerald, H.E., & Zucker, R.A. (2001). Using Mosaic Displays in Configural Frequency Analysis (CFA). *Methods of Psychological Research - Online, 6*, 164 - 196.

Naud, S. J. (1997). *Categorical data analysis: Type I error rate as a function of sampling distribution and hypothesis.* Unpublished paper, Michigan State University, East Lansing.

Naud, S. J. (1999). *Factors influencing Pearson's chi-squared statistic's fit to its asymptotic distributions: implications for sample size guidelines.* Unpublished Doctoral Dissertation, Michigan State University, East Lansing.

Nesselroade, J.R., Pruchno, R., & Jacobs, A. (1986). Reliability and stability in the measurement of psychological states: An illustration with anxiety measures. *Psychologische Beiträge, 28*, 252 - 264.

Neter, J., Kutner, M. H., Nachtsheim, C. J., & Wasserman, W. (1996). *Applied linear statistical models* (4th ed.). Chicago: Irwin.

Netter, P. (1982). Typen sympathomedullärer Aktivität und ihrer psychischen Korrelate. In H. Studt (ed.), *Psychosomatik in Forschung und Praxis* (pp. 216 - 233). München: Urban & Schwarzenberg.

Netter, P. (1996). Prediction CFA as a search for types: History and specifications. *Applied Psychology: An International Review, 45*, 338 - 344.

Netter, P., & Lienert, G.A. (1984). Die Konfigurationsfrequenzanalyse XXIa. Stress-induzierte Katecholamin-Reaktionen bei Hyper- und Normotonikern. *Zeitschrift für Klinische Psychologie, Psychopathologie und Psychotherapie, 32*, 356 - 364.

Netter, P., Toll, C., Rohrmann, S., Hennig, J., & Nyborg, H. (2000). Configural Frequency Analysis of factors associated with testosterone levels in Vietnam veterans. *Psychologische Beiträge, 42*, 504 - 514.

Ohannessian, C.M., Lerner, R.M., Lerner, J.V., & von Eye, A. (1994). A longitudinal study of perceived family adjustment and emotional adjustment in early adolescence. *Journal of Early Adolescence, 14*, 371 - 390.

Olejnik, S., Li, J., Supattathum, S., & Huberty, C. J. (1997). Multiple testing and statistical power with modified Bonferroni procedures. *Journal of Educational and Behavioral Statistics, 22*, 389 - 406.

Osterkorn, K. (1975). Wann kann die Binomial- und Poissonverteilung hinreichend genau durch die Normalverteilung ersetzt werden? *Biometrische Zeitschrift, 17*, 33 - 34.

Overall, J.E., & Gorham, D.R. (1962). The brief psychiatric rating scale. *Psychological Reports, 10*, 799 - 812.

Perli, H.-G. (1984). *Testverfahren in der Konfigurationsfrequenzanalyse bei multinomialem Versuchsschema*. Erlangen: Palm und Enke.

Perli, H.-G., Hommel, G., & Lehmacher, W. (1985). Sequentially rejective test procedures for detecting outlying cells in one- and two-sample multinomial experiments. *Biometrical Journal, 27*, 885 - 893.

Perli, H.-G., Hommel, G., & Lehmacher, W. (1987). Test procedures in Configural Frequency Analysis (CFA) controlling the local and the multiple level. *Biometrical Journal, 29*, 255 - 267.

Pfaundler, H., & von Sehr, L. (1922). Über Syntropie von Krankheitszuständen. *Zeitschrift für Kinderheilkunde, 30*, 100 - 120.

Riley, M.W., Cohn, R., Toby, J., & Riley, J.W., Jr. (1954). Interpersonal orientations in small groups. *American Sociological Review, 19*, 715 - 724.

Rohner,R.P. (1980). *Handbook for the study of parental acceptance and rejection*. Storrs, CT: University of Connecticut.

Rosenthal, R., & Rubin, D.B. (1982). A simple, general purpose display of magnitude of experimental effect. *Journal of Educational Psychology, 74*, 166 - 169.

Rovine, M.J., & von Eye, A. (1997). A 14th way to look at a correlation coefficient: Correlation as the proportion of matches. *The American Statistician, 51*, 42 - 46.

Rudas, T. (1998). *Odds ratios in the analysis of contingency tables*. Thousand Oaks: Sage.

Rudas, T., Clogg, C.C., & Lindsay, B.G. (1994). A new index of fit based on mixture methods for the analysis of contingency tables. *Journal of the Royal Statistical Society, 56*, 623 - 639.

Schneider, J. (2000). 6th annual international picture contest winners. *Popular Photography, 64*, 87 - 123.

Schneider-Düker, M. (1973) *Psychische Leistungsfähigkeit und Ovarialzyklus*. Frankfurt: Lang.

Schuster, C. (1997). *Statistische Beurteilung der Veränderung von Modellparametern in der linearen Regression.* Unpublished Dissertation, Technical University, Berlin.

Schuster, C., & von Eye, A. (2000). Using log-linear modeling to increase power in two-sample Configural Frequency Analysis. *Psychologische Beiträge, 42*, 273 - 284.

Selder, H. (1973). *Einführung in die Numerische Mathematik für Ingenieure*. München: Hanser.

Shaffer, J. P. (1995). Multiple hypothesis testing: A review. *Annual Review of Psychology, 46*, 561 - 584.

Sidak, Z. (1967). Rectangular confidence regions for the means of multivariate normal distributions. *Journal of the American Statistical Association, 62*, 623 - 633.

Simes, R. J. (1986). An improved Bonferroni procedure for multiple tests of significance. *Biometrika, 73*, 751 - 754.

Snedecor, G.W., & Cochran, W.G. (1967). *Statistical methods* (6th ed). Ames, IA: The Iowa State University Press.

Sobel, M.E. (1994). Causal inference in latent variable models. In A. von Eye & C.C. Clogg (Eds.), *Latent variables analysis. Applications for developmental research* (pp. 3 - 35). Newbury Park: Sage.

Sobel, M.E. (1996). Causal inference in the social and behavioral sciences. In G. Arminger, C.C. Clogg, & M.E. Sobel (Eds.), *Handbook of statistical modeling for the social and behavioral sciences* (pp. 1 - 38). New York: Plenum.

Stegmüller, W. (1983). *Erklärung, Begründung, Kausalität.* Berlin: Springer.

Steiger, J. H., Shapiro, A., & Browne, M. W. (1985). On the multivariate asymptotic distribution of sequential chi-square statistics. *Psychometrika, 50*, 253 - 264.

Stemmler, M. (1998). Nonparametric analysis of change patterns in dependent samples. *Methods of Psychological Research - online, 3*, 24 - 38.

Stern, W. (1911). *Die differentielle Psychologie in ihren methodischen Grundlagen* (3rd ed.). Leipzig: Barth.

Stevens, S.S. (1946). On the theory of scales of measurement. *Science, 103*, 677 - 680.

Stevens, S.S. (1951). Mathematics, measurement, and psychophysics. In S.S. Stevens (Ed.), *Handbook of experimental psychology* (pp. 1 - 49). New York: Wiley.

Suppes, P. (1970). *A probabilistic theory of causality.* Amsterdam: North Holland.

Taylor, C.S., Lerner, R.M., Villaruel, F.A., & von Eye, A. (2000). *Annual report for phase II of overcoming the odds: Understanding successful development among African American and Latino male adolescents.* William T. Grant Foundation.

Taylor, C.S., Lerner, R.M., von Eye, A., Sadowski, D., Bilalbegovic, A., & Dowling, E. (2002). Assessing bases of positive individual and social behavior and development among gang and non-gang African American male adolescents. *Journal of Adolescent Research* (in press).

Thompson, K.N., & Schumacker, R.E. (1997). Evaluation of Rosenthal and Rubin's binomial effect size display. *Journal of Educational and Behavioral Statistics, 22*, 109 - 117.

Upton, G.J.G. (1978). *The analysis of cross-tabulated data.* Chichester:

Wiley.

Vargha, A., Rudas, T., Delaney, H.D., & Maxwell, S.E. (1996). Dichotomization, partial correlation, and conditional independence. *Journal of Educational and Behavioral Statistics, 21*, 264 - 282.

Velleman, P.F., & Wilkinson, L. (1993). Nominal, ordinal, interval, and ratio typologies are misleading. *The American Statistician, 47*, 65 -75.

Venables, W.N., & Ripley, B.D. (1994). *Modern applied statistics with S-Plus*. New York: Springer-Verlag.

Victor, N. (1983). An alternative approach to configural frequency analysis. *Methodika, 3*, 61 - 73.

Vogel, F. (1997). *Ein neues Zusammenhangsmaß für nominale Merkmale*. Bamberg: Otto-Friedrich-Universität: Department of Statistics, Arbeiten aus der Statistik.

von Eye, A. (1985). Die Konfigurationsfrequenzanalyse bei gerichteten Variablenbeziehungen (GKFA). *EDV in Medizin und Biologie, 16*, 37 - 51.

von Eye, A. (1986). Strategien der Typen- und Syndromaufklärung mit der Interaktionsstrukturanalyse. *Zeitschrift für Klinische Psychologie, Psychopathologie, und Psychotherapie, 34*, 54 - 68.

von Eye, A. (1988). The general linear model as framework for models in Configural Frequency Analysis. *Biometrical Journal, 30*, 59 - 67.

von Eye, A. (1990). *Introduction to Configural Frequency Analysis. The search for types and antitypes in cross-classifications*. Cambridge, UK: Cambridge University Press.

von Eye, A. (2000). Configural Frequency Analysis - A program for 32 bit Windows operating systems. Manual for program Version 2000. (Version 2). East Lansing, MI.

von Eye, A. (2001). Configural Frequency Analysis - *Version 2000* program for 32 bit operating systems. *Methods of Psychological Research - Online, 6*, 129 - 139.

von Eye, A. (2002). The odds favor antitypes - A comparison of tests for the identification of configural types and antitypes. *Methods of Psychological Research - online* (in press).

von Eye, A. (2002). Configurational analysis. In T. Cook & C. Ragin (Eds.), *International Encyclopedia of the Social and Behavioral Sciences: Vol. 8, Logic of inquiry and research design*. Amsterdam: Pergamon. (in press)

von Eye, A., & Bergman, L. R. (1987). A note on numerical

approximations of the binomial test in Configural Frequency Analysis. *EDP in Medicine and Biology, 17*, 108 - 111.

von Eye, A., & Brandtstädter, J. (1982). Systematization of results of configuration frequency analysis by minimizing Boolean functions. In H. Caussinus, P. Ettinger, & J. R. Mathieu (Eds.), *Compstat 1982, part II: Short communications, summaries of posters* (pp. 91-92). Wien: Physica.

von Eye, A., & Brandtstädter, J. (1988). Application of prediction analysis to cross-classifications of ordinal data. *Biometrical Journal, 30*, 651-655.

von Eye, A., & Brandtstädter, J. (1997). Configural Frequency Analysis as a searching device for possible causal relationships. *Methods of Psychological Research - Online, 2*, 1 - 23.

von Eye, A., & Brandtstädter, J. (1998). The Wedge, the Fork, and the Chain - Modeling dependency concepts using manifest categorical variables. *Psychological Methods, 3*, 169 - 185.

von Eye, A., & Clogg, C.C. (Eds.). (1994). *Latent variable analysis. Applications for developmental research.* Thousand Oaks: Sage.

von Eye, A., & Gutiérrez-Peña, E. (in preparation). *Configural Frequency Analysis of large sparse cross-classifications - frequentist and Bayesian approaches.*

von Eye, A., & Hussy, W. (1980). Zur Verwendung der polynomialen Approximation in der Psychologie. *Psychologische Beiträge, 22*, 208 - 225.

von Eye, A., Indurkhya, A., & Kreppner, K. (2000). CFA as a tool for person-oriented research - Unidimensional and within-individual analyses of nominal level and ordinal data. *Psychologische Beiträge, 42*, 383 - 401.

von Eye, A., Jacobson, L. P., & Wills, S. D. (1990). *Proverbs: Imagery, Interpretation, and Memory.* Paper presented at the 12th West Virginia University Conference on Life-Span Developmental Psychology.

von Eye, A., Kreppner, K., & Weßels, H. (1994). Log-linear modeling of categorical data in developmental research. In D.L. Featherman, R.M. Lerner, & M. Perlmutter (Eds.), *Life-span development and behavior* (Vol. 12, pp. 225 - 248). Hillsdale, NJ: Lawrence Erlbaum.

von Eye, A., Lerner, J.V., & Lerner, R.M. (1999). Modeling reciprocal relations at the level of manifest categorical variables. *Multiciência, 3*, 22 - 51.

von Eye, A., & Lienert, G. A. (1984). Die Konfigurationsfrequenzanalyse XX. Typen und Syndrome zweiter Ordnung. *Zeitschrift für Klinische Psychologie, Psychopathologie und Psychotherapie, 32*, 345 - 355.

von Eye, A., & Lienert, G.A. (1985). Die Konfigurationsfrequenzanalyse. XXIIa. Typen normativer Skalenmuster. *Zeitschrift für Klinische Psychologie, Psychopathologie, und Psychotherapie, 33*, 345 - 355.

von Eye, A., & Lienert, G.A. (1987). Nonparametric comparison of longitudinal response patterns from paired samples using configural frequency analysis. *Biometrical Journal, 29*, 615 - 624.

von Eye, A., Lienert, G.A., & Wertheimer, M. (1991). Syndromkombinaionen als Metasyndrome in der KFA. *Zeitschrift für Klinische Psychologie, Psychopathologie, und Psychotherapie, 39*, 254 - 260.

von Eye, A., & Nesselroade, J.R. (1992). Types of change: Application of Configural Frequency Analysis in repeated measurement designs. *Experimental Aging Research, 18*, 169 - 183.

von Eye, A., & Niedermeier, K. E. (1999). *Statistical analysis of longitudinal categorical data - An introduction with computer illustrations.* Mahwah, NJ: Lawrence Erlbaum.

von Eye, A., & Rovine, M. J. (1988). A comparison of significance tests for Configural Frequency Analysis. *EDP in Medicine and Biology, 19*, 6 - 13.

von Eye, A., Rovine, M.J., & Spiel, C. (1995). Concepts of independence in Configural Frequency Analysis. *Journal of Mathematical Sociology, 20*, 41 - 54.

von Eye, A., & Schuster, C. (1998). On the specification of models for Configural Frequency Analysis - Sampling schemes in Prediction CFA. *Methods of Psychological Research - Online, 3*, 55 - 73.

von Eye, A., & Schuster, C. (1999). Modeling the direction of causal effects using manifest categorical variables. *Multiciência, 3*, 14 - 40.

von Eye, A., Schuster, C., & Gutiérrez-Peña, E. (2000). Configural Frequency Analysis under retrospective and prospective sampling schemes - Frequentist and Bayesian approaches. *Psychologische Beiträge, 42*, 428 - 447.

von Eye, A., & Spiel, C. (1996). Standard and nonstandard log-linear models for measuring change in categorical variables. *The American Statistician, 50*, 300 - 305.

von Eye, A., Spiel, C., & Rovine, M.J. (1995). Concepts of nonindependence in Configural Frequency Analysis. *Journal of Mathematical Sociology, 20*, 41 - 54.

von Eye, A., Spiel, C., & Rovine, M.J. (in press). What goes together and what does not go together - Configural Frequency Analysis in the practice of Neuropsychology. In R.D. Franklin (Ed.), *Prediction in forensic and neuropsychology*. Mahwah, NJ: Erlbaum.

von Eye, A., Spiel, C., & Wood, P.K. (1996a). CFA models, tests, interpretation, and alternatives: A rejoinder. *Applied Psychology: An International Review, 45*, 345 - 352.

von Eye, A., Spiel, C., & Wood, P.K. (1996b). Configural Frequency Analysis in Applied Psychological Research. *Applied Psychology: An International Review, 45*, 301 - 327.

von Neumann, J. (1941). Distribution of the ratio of the mean square successive difference to the variance. *Annals of Mathematical Statistics, 12*, 367 - 395.

Wanberg, K.W., Horn, J.L., & Foster, F.M. (1977). A differential assessment model for alcoholism. The scales of Alcohol Use Inventory. *Journal of Studies on Alcohol*, 38, 512 - 543.

Ward, J.H. (1963). Hierarchical grouping to optimize an objective function. *Journal of the American Statistical Association, 58*, 236 - 244.

Wermuth, N. (1976). Anmerkungen zur Konfigurationsfrequemzanalyse. *Zeitschrift für Klinische Psychologie und Psychotherapie, 23*, 5 - 21.

Wickens, T.D. (1989). *Multiway contingency tables analysis for the social sciences*. Hillsdale, NJ: Erlbaum.

Wilkinson, L. (1999). SYSTAT 9.0. Chicago: SPSS.

Wilkinson, L. (2000). SYSTAT 10.0. Chicago: SPSS.

Williams, V. S. L., Jones, L. V., & Tukey, J. W. (1999). Controlling error in multiple comparisons, with examples from state-to-state differences in educational achievement. *Journal of Educational and Behavioral Statistics, 24*, 42 - 69.

Wise, M. E. (1963). Multinomial probabilities and the chi2 and the X2 distributions. *Biometrika, 50*, 145 - 154.

Wood, P.K. (in preparation). *The search for the syndrome that was there or the variable that wasn't: Configural Frequency Analysis, conditional independence, and Tetrad approaches for categorical data.*

Wood, P.K., Sher, K., & von Eye, A. (1994). Conjugate methods in Configural Frequency Analysis. *Biometrical Journal, 36*, 387 -

410.

Yates, F. (1934). Contingency tables involving small numbers and the X2 test. *Supplement to the Journal of the Royal Statistical Society, 1*, 217 - 235.

Zerbe, G.O. (1979). Randomization analysis of the completely randomized design extended to growth and response curves. *Journal of the American Statistical Association, 74*, 215 - 221.

zur Oeveste, H., & Lienert, G.A. (1984). Methoden der Entwicklings-Konfigurationsfrequenzanalyse. *Psychologische Beiträge, 26*, 372 - 381.

Appendix A

A brief introduction to log-linear modeling

Using the methods of log-linear modeling, researchers attempt to explain the frequency distribution in cross-classifications of categorical variables. To introduce these models, consider the *Generalized Linear Model*,

$$f(Y) = X\beta$$

where Y is the dependent variable, X is the matrix of independent variables, and β is the parameter vector. This vector contains the weights with which the independent variables go into the equation that explains the dependent variable. The function f(Y) is called the *link function*. It describes the transformation performed on the dependent variable.

Using the link function, one can show that the *General Linear Model*, special cases of which include *analysis of variance* (ANOVA) and *regression analysis*, and the *log-linear model*, which is used in this volume to specify most CFA base models, are members of the same family of models that differ, among others, in their link function. Specifically, the link function for the General Linear Model is the *identity function*, called the *identity link*. This function is $f(Y) = Y$. In words, the general linear model uses the dependent variables untransformed. An example of such a model is the multiple regression model of the dependent variable Y and the independent variables X_1, X_2, and X_3,

$$\hat{Y} = \beta_0 + \beta_1 X_1 + \beta_2 X_2 + \beta_3 X_3,$$

where $\hat{Y}$ is the estimate of the dependent measure.

The link function, called the *log-link*, for the log-linear model is the logarithmic function. A Generalized Linear Model that uses the log-link is called a log-linear model. Using again the dependent measure Y and the three independent variables, X_1, X_2, and X_3, one obtains the log-linear model

$$\log(E) = \beta_0 + \beta_1 X_1 + \beta_2 X_2 + \beta_3 X_3,$$

where E is the expected cell frequency in the cross-classification.

For an introduction into log-linear modeling see, for example, Christensen (1997) or Agresti (1996). Von Eye and Niedermeier (1999) use log-linear modeling to specify the base models for CFA and Prediction Analysis of longitudinal data.

There is a number of parallels between the ANOVA and the log-linear models. Specifically, both models are used to explain the response (or its logarithm) that is due to independent variables. Both models use main effects and interactions of the independent variables for explanation. Both models allow the user to include covariates. In addition, the relation of parameters to the matrix X which is also called the *design matrix* or the *indicator matrix* is the same in both models. This relation is

$$\beta = (X^T X)^{-1} X^T m,$$

where $m = log\ E$ (see Chapter 2).

However, there are also several differences between the log-linear and the General Linear models. An obvious difference is that log-linear models are employed to model frequency distributions. The cells of cross-classifications contain frequencies. Thus, each cell frequency y typically describes the responses of y cases rather than the score of an individual case.

The following paragraphs give one complete example of a log-linear model and discuss the relationship between log-linear modeling and CFA.

Log-linear modeling: a complete example. The following example re-analyzes a data set published in the *New York Times* on April 20, 2001. The data describe the number of death penalties issued in a total of 1,521 murder cases in North Carolina in the years 1993 to 1997, depending on the race of the victim and the race of the murderer. These are the murder cases in which a death sentence was possible. Table A1 presents the observed cell frequencies in the cross-classification of the variables Death Penalty (D; yes - no), Race of Murderer (M; nonwhite - white), and Race of Victim (V; nonwhite - white). In addition, Table A1 displays the estimated expected cell frequencies for the log-linear main effect model

$$\log E = \lambda_0 + \lambda_i^D + \lambda_j^M + \lambda_k^V$$

or, in bracket notation [D][M][V], and the standardized residuals which are defined as

$$z = \sqrt{\frac{(N - \hat{E})^2}{\hat{E}}},$$ that is, the square root of the X^2-components, where N is the observed frequency in the cell under scrutiny, and $\hat{E}$ is the estimated expected cell frequency for this cell. The model used here can also be used as a base model for first order global CFA (see Section 5.1).

Table A1: Log-linear main effect model for the cross-classification of the variables Death Penalty, Race of Murderer, and Race of Victim

Configuration DMV	Frequencies observed	Frequencies expected	Standardized Residual
nnn	587	385.03	10.29
nnw	251	456.39	-9.61
nwn	76	265.67	-11.64
nww	508	314.91	10.88
ynn	29	26.81	.42
ynw	33	31.77	.22
ywn	4	18.50	-3.37
yww	33	21.92	2.37

Obviously, the differences between the observed and the estimated expected cell frequencies are large. Accordingly, the $LR\text{-}X^2 = 512.35$ for this model is large and suggests significant model-data discrepancies (df= 4; $p < 0.01$). The standardized residuals in the last column of Table A1 indicate that the discrepancies between the observed and the expected cell frequencies vary over the cells of this table. The parameters of this model cannot be interpreted because the model does not fit.

We now ask whether a more complex model can lead to an improved model fit and to a model that can be retained. Therefore, we now try the model that includes all possible two-way interactions. That is, the model

$$\log E = \lambda_0 + \lambda_i^D + \lambda_j^M + \lambda_k^V$$
$$\lambda_{ij}^{DM} + \lambda_{ik}^{DV} + \lambda_{jk}^{MV},$$

where DM indicates the interaction between Death Penalty and Race of Murderer, DV indicates the interaction between Death Penalty and Race of Victim, and MV indicates the interaction between Race of Murderer and

Race of Victim. This model can also be used as a second order global CFA base model. The standardized residuals for this model are z = {.07, -.11, -.20, .08, -.31, .31, 1.15, -.30}. None of these residuals is large. Accordingly, the overall goodness-of-fit (LR-X^2 = 1.43; df = 1; p = 0.20) suggests excellent model fit. We are now in a position in which we can interpret the parameters. Table A2 displays the parameter estimates, their standard errors, and the z-scores for the null hypotheses that the parameters are equal to zero.

Table A2: Parameter estimates, standard errors, and z-scores for model of all two-way interactions among the death penalty variables

Parameter	estimate	standard error	z
Race of Victim (V)	-.47	.07	-7.25
Race of Murderer (M)	.48	.06	7.50
Penalty (D)	1.41	.06	23.48
M x V	.68	.04	19.31
D x V	.21	.06	3.46
D x M	-.15	.06	-2.43

Table A2 shows that each of the parameters' z-scores exceeds the cutoff of z = 2.0. Thus, each of the parameters is significantly larger than zero.

We now interpret the parameters. The meaning of the parameters results from inserting the indicator matrix in the above equation that describes the relation between the parameters and the matrix X, that is, $\beta = (X^TX)^{-1}X^Tm$. The design matrix X for the present model is

$$X = \begin{bmatrix} 1 & 1 & 1 & 1 & 1 & 1 & 1 \\ 1 & 1 & 1 & -1 & 1 & -1 & -1 \\ 1 & 1 & -1 & 1 & -1 & 1 & -1 \\ 1 & 1 & -1 & -1 & -1 & -1 & 1 \\ 1 & -1 & 1 & 1 & -1 & -1 & 1 \\ 1 & -1 & 1 & -1 & -1 & 1 & -1 \\ 1 & -1 & -1 & 1 & 1 & -1 & -1 \\ 1 & -1 & -1 & -1 & 1 & 1 & 1 \end{bmatrix}$$

From left to right, the columns in this matrix contain vectors for the following effects:

(1) *Constant*: The constant, λ_0, represents the overall mean of the logarithms of the cell frequencies. It can be viewed parallel to the *grand mean* in ANOVA.

(2) *Main effect Death Penalty* (D). The second column vector in X is needed to estimate the main effect parameter, λ_i^D, for the variable Death Penalty. In the vector, the two categories of this variable are contrasted with each other by assigning a 1 to category "death penalty issued" and a -1 to category "other penalty issued."

(3) *Main effect Murderer* (M). The third column vector in X is needed to estimate the main effect parameter, λ_j^M, for the variable Race of Murderer. In this vector, the two categories of this variable are contrasted with each other by assigning a 1 to category "nonwhite" and a -1 to category "white."

(4) *Main effect Victim* (V). The fourth column vector in X is needed to estimate the main effect parameter, λ_k^V, for the variable Race of Victim. As for the Murderer variable, the two categories of this variable are contrasted with each other by assigning a 1 to category "nonwhite" and a -1 to category "white."

(5) *Interaction D x M*. The fifth column vector in X results from element-wise multiplication of the second and the third vectors in X. Based on this vector, the interaction parameter λ_{ij}^{DM} is estimated. The exact interpretation of this parameter follows below.

(6) *Interaction D x V*. The sixth column vector in X results from

element-wise multiplication of the second and the fourth vectors in X. Based on this vector, the interaction parameter λ_{ik}^{DV} is estimated. The exact interpretation of this vector follows below.

(7) *Interaction M x V*. The last column vector in X results from element-wise multiplication of the third and the fourth vectors in X. Based on this vector, the interaction parameter λ_{lk}^{MV} is estimated. The exact interpretation of this vector follows below.

To determine the meaning of the parameters, we insert into the equation for β, and obtain the equations for each parameter. These equations appear in Table A3.

Table A3: Parameter equations for the log-linear model with all two-way interactions in a 2 x 2 x 2 cross-classification, with $m_{ijk} = \log e_{ijk}$

Parameter	Estimate
λ_0	$0.25\Sigma_{ijk}\, m_{ijk}$
λ_i^D	$0.25(m_{111} + m_{112} + m_{121} + m_{122} - m_{211} - m_{212} - m_{221} - m_{222})$
λ_j^M	$0.25(m_{111} + m_{112} - m_{121} - m_{122} + m_{211} + m_{212} - m_{221} - m_{222})$
λ_k^V	$0.25(m_{111} - m_{112} + m_{121} - m_{122} + m_{211} - m_{212} + m_{221} - m_{222})$
λ_{ij}^{DM}	$0.25(m_{111} + m_{112} - m_{121} - m_{122} - m_{211} - m_{212} + m_{221} + m_{222})$
λ_{ik}^{DV}	$0.25(m_{111} - m_{112} + m_{121} - m_{122} - m_{211} + m_{212} - m_{221} + m_{222})$
λ_{lk}^{MV}	$0.25(m_{111} - m_{112} - m_{121} + m_{122} + m_{211} - m_{212} - m_{221} + m_{222})$

Table A3 shows the following characteristics of parameters in log-linear modeling:

(1) In orthogonal designs, that is, in designs in which the correlations among the column vectors in X are zero, the weight with which the

cell frequencies are used in hierarchical models is always equal. The weight can vary in nonstandard designs and in nonorthogonal designs.

(2) **The meaning of a parameter is given by the pattern of signs and by the weights of the cell frequencies** in the equations in the right- hand column in Table A3. For instance, the sign pattern + + + + - - - - for parameter λ_i^D shows that the magnitude of this parameter is the result of the comparison of the first four cells (these are the cells that fall in the first category of the Penalty variable) with the second four cells (these are the cells that fall in the second category of the Penalty variable). This applies accordingly to the other main effect terms. To explain the meaning of the interaction terms, consider, for example, the parameter for the interaction between Penalty and Race of Murderer, λ_{ij}^{DM}. The signs for this parameter are + + - - - - + +. The first four signs are the same as in the vector for the main effect of Murderer, λ_j^M. The second four signs are the inverse of the first four signs. This interaction is thus used to test the hypothesis that the main effect Murderer is the same across the two categories of the variable Death Penalty. Equivalently, one can say that the parameter λ_{ij}^{DM} is used to test whether the main effect Death Penalty is constant across the two categories of the variable Race of Murderer. The parameters for the other two two-way interactions and the three-way interaction (not represented in Table A3) can be interpreted in analogous fashion.

In the above data example, all parameters are significant, thus explaining significant portions of the information in the D x M x V cross-classification given in Table A1. To give an example, let's interpret parameter λ_{ik}^{DV}. The sign pattern for this parameter is + - + - - + - +. The first four of these signs correspond to those for the main effect parameter λ_k^V. The second four are inverted. Thus, using the parameter λ_{ik}^{DV}, one tests the hypothesis that the main effect Race of Victim is the same across the two categories of the variable Death Penalty. As before, one can also say that the parameter λ_{ik}^{DV} is used to test whether the main effect Race of Victim is constant across the categories of the variable Death Penalty. This applies accordingly to interactions of any level.

The relationship between log-linear modeling and CFA. The following brief discussion of the relationship between log-linear modeling and CFA focuses mostly on those cases in which either log-linear models are created using models that could also be used as base models for CFA and methods of residual analysis that are also used in CFA, or vice versa. The fact that (a) log-linear models exist that cannot be CFA base models and (b) methods of calculating expected frequencies exist that are not based on the log-linear model indicate that the two methods overlap only partly.

When comparing the two methods, it must be noted that many CFA base models can be cast in terms of log-linear models. Most of these are hierarchical models. Some include covariates. Accordingly, the methods of estimating expected cell frequencies are the same also. What then is the difference between the two methods? The basic differences lie in the goals of analysis. The method of log-linear modeling, while applicable in the context of person-orientation, is mostly used in the context of variable-centered research (see Section 1.2). Results are typically expressed in terms of variable relationships such as interactions or dependency structures. In contrast, CFA is the prime method of person-centered research. CFA asks whether configurations (e.g., person profiles) occur at rates different than expected, or whether groups of individuals differ significantly in the occurrence rates of particular profiles. Lehmacher (2000) calls CFA a *cell-oriented* method. These diverging goals have one major implication which concerns the role played by the models under scrutiny.

In log-linear modeling, researchers attempt to identify the model that best describes the data. In addition, this model must be parsimonious and there cannot be significant model-data discrepancies. Only then, parameters can be interpreted. In particular when there are significant model-data discrepancies, researchers modify the model, trying to improve model fit. The role played by cell-specific large or small residuals is that of guiding model improvement. This process of model testing and modifying is repeated until an acceptable and interpretable model is found or until the model is rejected. We note that log-linear modeling sometimes implies testing several models before one model is retained.

In contrast, the typical CFA application uses only one base model. When significant model-data discrepancies exist, they are interpreted in terms of types and antitypes. The base model is not changed because of the existence of types and antitypes. If a different base model is considered then either with the goal of identifying the reasons why types and antitypes exist or to test additional hypotheses.

We see from this brief discussion that log-linear modeling and CFA pursue different goals. However, the two methods can also be used in tandem. Here are two sample scenarios.

(1) *Explaining types and antitypes*. The existence of types and antitypes can be explained using substantive arguments. For example, one can explain the antitype that is constituted by the configuration *depressed* + *happy-go-lucky* as logical and confirming these two concepts. In the context of test construction, this antitype could be considered one of the indicators of instrument validity. In addition to substantive arguments, one can ask whether types and antitypes reflect variable interactions. To determine which interactions exist, one can go two routes. The first route involves specifying a different, typically more complex CFA base model. For instance, one can move from a global first order CFA to a global second order CFA. If the new base model makes all types and all antitypes disappear, they can be considered *explained* by the effects included in the base model. It may not always be possible to explain all types and antitypes this way, because the selection of CFA base models underlies restrictions (see Section 2.5) which exclude models that are possible and can be meaningful in the context of a log-linear analysis. The second route involves fitting log-linear models. The result of this effort is a log-linear model that describes the data well, that is, without significant model-data discrepancies. There can be no types or antitypes for a well-fitting model. Regardless of whether the first or the second route is taken, log-linear modeling and CFA complement each other in the sense that log-linear modeling can lead to an explanation of types and antitypes that uses models that do not belong to the class of CFA base models (Lehmacher, 2000).

(2) *Explaining interactions in log-linear models*. Consider a researcher that has found a well- fitting log-linear model. This researcher may then ask whether a finer-grained analysis could help identify the sectors in the cross-classification that carry the effects. One way of answering this question is employing CFA to the model that does not contain the significant effects (if possible, see above). The resulting types and antitypes will tell this researcher where the variable interactions are the strongest (or exist at all).

Conclusions. It seems perfectly all right to only employ log-linear modeling when variable-centered questions need to be answered, and to employ only CFA when the focus of analysis is purely person-centered. However, there are many reasons why methods of analysis can be employed in tandem. This applies to both log-linear modeling and to CFA. In addition, this applies to Bayesian methods of typal analysis and to cell-directed methods of model modification as implemented in SYSTAT. Whatever method of categorical data analysis is employed, other methods can help researchers round out the picture. Thus, variable-centered methods can be used to bolster person- or cell-oriented results in terms of variable relationships. In turn, CFA can be used to add the person perspective to variable-centered analyses.

Appendix B

Table of α*-levels for the Bonferroni and Holm adjustments

t indicates either the total number of cells (for Bonferroni protection of α; see Lienert, Ludwig, & Rockefeller, 1982) or the remaining number of tests (for Holm protection of α).

t	$\alpha^*_{0.01}$	$\alpha^*_{0.05}$
380	0.0001316	0.0000263
379	0.0001319	0.0000264
378	0.0001323	0.0000265
377	0.0001326	0.0000265
376	0.0001330	0.0000266
375	0.0001333	0.0000267
374	0.0001337	0.0000267
373	0.0001340	0.0000268
372	0.0001344	0.0000269
371	0.0001348	0.0000270
370	0.0001351	0.0000270
369	0.0001355	0.0000271
368	0.0001359	0.0000272
367	0.0001362	0.0000272
366	0.0001366	0.0000273
365	0.0001370	0.0000274
364	0.0001374	0.0000275
363	0.0001377	0.0000275
362	0.0001381	0.0000276
361	0.0001385	0.0000277
360	0.0001389	0.0000278
359	0.0001393	0.0000279
358	0.0001397	0.0000279
357	0.0001401	0.0000280
356	0.0001404	0.0000281
355	0.0001408	0.0000282

t	$\alpha^*_{0.01}$	$\alpha^*_{0.05}$
354	0.0001412	0.0000282
353	0.0001416	0.0000283
352	0.0001420	0.0000284
351	0.0001425	0.0000285
350	0.0001429	0.0000286
349	0.0001433	0.0000287
348	0.0001437	0.0000287
347	0.0001441	0.0000288
346	0.0001445	0.0000289
345	0.0001449	0.0000290
344	0.0001453	0.0000291
343	0.0001458	0.0000292
342	0.0001462	0.0000292
341	0.0001466	0.0000293
340	0.0001471	0.0000294
339	0.0001475	0.0000295
338	0.0001479	0.0000296
337	0.0001484	0.0000297
336	0.0001488	0.0000298
335	0.0001493	0.0000299
334	0.0001497	0.0000299
333	0.0001502	0.0000300
332	0.0001506	0.0000301
331	0.0001511	0.0000302
330	0.0001515	0.0000303
329	0.0001520	0.0000304

t	$\alpha^*_{0.01}$	$\alpha^*_{0.05}$
328	0.0001524	0.0000305
327	0.0001529	0.0000306
326	0.0001534	0.0000307
325	0.0001538	0.0000308
324	0.0001543	0.0000309
323	0.0001548	0.0000310
322	0.0001553	0.0000311
321	0.0001558	0.0000312
320	0.0001563	0.0000313
319	0.0001567	0.0000313
318	0.0001572	0.0000314
317	0.0001577	0.0000315
316	0.0001582	0.0000316
315	0.0001587	0.0000317
314	0.0001592	0.0000318
313	0.0001597	0.0000319
312	0.0001603	0.0000321
311	0.0001608	0.0000322
310	0.0001613	0.0000323
309	0.0001618	0.0000324
308	0.0001623	0.0000325
307	0.0001629	0.0000326
306	0.0001634	0.0000327
305	0.0001639	0.0000328
304	0.0001645	0.0000329
303	0.0001650	0.0000330
302	0.0001656	0.0000331
301	0.0001661	0.0000332
300	0.0001667	0.0000333
299	0.0001672	0.0000334
298	0.0001678	0.0000336
297	0.0001684	0.0000337
296	0.0001689	0.0000338
295	0.0001695	0.0000339

t	$\alpha^*_{0.01}$	$\alpha^*_{0.05}$
294	0.0001701	0.0000340
293	0.0001706	0.0000341
292	0.0001712	0.0000342
291	0.0001718	0.0000344
290	0.0001724	0.0000345
289	0.0001730	0.0000346
288	0.0001736	0.0000347
287	0.0001742	0.0000348
286	0.0001748	0.0000350
285	0.0001754	0.0000351
284	0.0001761	0.0000352
283	0.0001767	0.0000353
282	0.0001773	0.0000355
281	0.0001779	0.0000356
280	0.0001786	0.0000357
279	0.0001792	0.0000358
278	0.0001799	0.0000360
277	0.0001805	0.0000361
276	0.0001812	0.0000362
275	0.0001818	0.0000364
274	0.0001825	0.0000365
273	0.0001832	0.0000366
272	0.0001838	0.0000368
271	0.0001845	0.0000369
270	0.0001852	0.0000370
269	0.0001859	0.0000372
268	0.0001866	0.0000373
267	0.0001873	0.0000375
266	0.0001880	0.0000376
265	0.0001887	0.0000377
264	0.0001894	0.0000379
263	0.0001901	0.0000380
262	0.0001908	0.0000382
261	0.0001916	0.0000383

t	$\alpha^*_{0.01}$	$\alpha^*_{0.05}$
260	0.0001923	0.0000385
259	0.0001931	0.0000386
258	0.0001938	0.0000388
257	0.0001946	0.0000389
256	0.0001953	0.0000391
255	0.0001961	0.0000392
254	0.0001969	0.0000394
253	0.0001976	0.0000395
252	0.0001984	0.0000397
251	0.0001992	0.0000398
250	0.0002000	0.0000400
249	0.0002008	0.0000402
248	0.0002016	0.0000403
247	0.0002024	0.0000405
246	0.0002033	0.0000407
245	0.0002041	0.0000408
244	0.0002049	0.0000410
243	0.0002058	0.0000412
242	0.0002066	0.0000413
241	0.0002075	0.0000415
240	0.0002083	0.0000417
239	0.0002092	0.0000418
238	0.0002101	0.0000420
237	0.0002110	0.0000422
236	0.0002119	0.0000424
235	0.0002128	0.0000426
234	0.0002137	0.0000427
233	0.0002146	0.0000429
232	0.0002155	0.0000431
231	0.0002165	0.0000433
230	0.0002174	0.0000435
229	0.0002183	0.0000437
228	0.0002193	0.0000439
227	0.0002203	0.0000441

t	$\alpha^*_{0.01}$	$\alpha^*_{0.05}$
226	0.0002212	0.0000442
225	0.0002222	0.0000444
224	0.0002232	0.0000446
223	0.0002242	0.0000448
222	0.0002252	0.0000450
221	0.0002262	0.0000452
220	0.0002273	0.0000455
219	0.0002283	0.0000457
218	0.0002294	0.0000459
217	0.0002304	0.0000461
216	0.0002315	0.0000463
215	0.0002326	0.0000465
214	0.0002336	0.0000467
213	0.0002347	0.0000469
212	0.0002358	0.0000472
211	0.0002370	0.0000474
210	0.0002381	0.0000476
209	0.0002392	0.0000478
208	0.0002404	0.0000481
207	0.0002415	0.0000483
206	0.0002427	0.0000485
205	0.0002439	0.0000488
204	0.0002451	0.0000490
203	0.0002463	0.0000493
202	0.0002475	0.0000495
201	0.0002488	0.0000498
200	0.0002500	0.0000500
199	0.0002513	0.0000503
198	0.0002525	0.0000505
197	0.0002538	0.0000508
196	0.0002551	0.0000510
195	0.0002564	0.0000513
194	0.0002577	0.0000515
193	0.0002591	0.0000518

t	$\alpha^*_{0.01}$	$\alpha^*_{0.05}$	t	$\alpha^*_{0.01}$	$\alpha^*_{0.05}$
192	0.0002604	0.0000521	158	0.0003165	0.0000633
191	0.0002618	0.0000524	157	0.0003185	0.0000637
190	0.0002632	0.0000526	156	0.0003205	0.0000641
189	0.0002646	0.0000529	155	0.0003226	0.0000645
188	0.0002660	0.0000532	154	0.0003247	0.0000649
187	0.0002674	0.0000535	153	0.0003268	0.0000654
186	0.0002688	0.0000538	152	0.0003289	0.0000658
185	0.0002703	0.0000541	151	0.0003311	0.0000662
184	0.0002717	0.0000543	150	0.0003333	0.0000667
183	0.0002732	0.0000546	149	0.0003356	0.0000671
182	0.0002747	0.0000549	148	0.0003378	0.0000676
181	0.0002762	0.0000552	147	0.0003401	0.0000680
180	0.0002778	0.0000556	146	0.0003425	0.0000685
179	0.0002793	0.0000559	145	0.0003448	0.0000690
178	0.0002809	0.0000562	144	0.0003472	0.0000694
177	0.0002825	0.0000565	143	0.0003497	0.0000699
176	0.0002841	0.0000568	142	0.0003521	0.0000704
175	0.0002857	0.0000571	141	0.0003546	0.0000709
174	0.0002874	0.0000575	140	0.0003571	0.0000714
173	0.0002890	0.0000578	139	0.0003597	0.0000719
172	0.0002907	0.0000581	138	0.0003623	0.0000725
171	0.0002924	0.0000585	137	0.0003650	0.0000730
170	0.0002941	0.0000588	136	0.0003676	0.0000735
169	0.0002959	0.0000592	135	0.0003704	0.0000741
168	0.0002976	0.0000595	134	0.0003731	0.0000746
167	0.0002994	0.0000599	133	0.0003759	0.0000752
166	0.0003012	0.0000602	132	0.0003788	0.0000758
165	0.0003030	0.0000606	131	0.0003817	0.0000763
164	0.0003049	0.0000610	130	0.0003846	0.0000769
163	0.0003067	0.0000613	129	0.0003876	0.0000775
162	0.0003086	0.0000617	128	0.0003906	0.0000781
161	0.0003106	0.0000621	127	0.0003937	0.0000787
160	0.0003125	0.0000625	126	0.0003968	0.0000794
159	0.0003145	0.0000629	125	0.0004000	0.0000800

t	$\alpha^*_{0.01}$	$\alpha^*_{0.05}$
124	0.0004032	0.0000806
123	0.0004065	0.0000813
122	0.0004098	0.0000820
121	0.0004132	0.0000826
120	0.0004167	0.0000833
119	0.0004202	0.0000840
118	0.0004237	0.0000847
117	0.0004274	0.0000855
116	0.0004310	0.0000862
115	0.0004348	0.0000870
114	0.0004386	0.0000877
113	0.0004425	0.0000885
112	0.0004464	0.0000893
111	0.0004505	0.0000901
110	0.0004545	0.0000909
109	0.0004587	0.0000917
108	0.0004630	0.0000926
107	0.0004673	0.0000935
106	0.0004717	0.0000943
105	0.0004762	0.0000952
104	0.0004808	0.0000962
103	0.0004854	0.0000971
102	0.0004902	0.0000980
101	0.0004950	0.0000990
100	0.0005000	0.0001000
99	0.0005051	0.0001010
98	0.0005102	0.0001020
97	0.0005155	0.0001031
96	0.0005208	0.0001042
95	0.0005263	0.0001053
94	0.0005319	0.0001064
93	0.0005376	0.0001075
92	0.0005435	0.0001087
91	0.0005495	0.0001099

t	$\alpha^*_{0.01}$	$\alpha^*_{0.05}$
90	0.0005556	0.0001111
89	0.0005618	0.0001124
88	0.0005682	0.0001136
87	0.0005747	0.0001149
86	0.0005814	0.0001163
85	0.0005882	0.0001176
84	0.0005952	0.0001190
83	0.0006024	0.0001205
82	0.0006098	0.0001220
81	0.0006173	0.0001235
80	0.0006250	0.0001250
79	0.0006329	0.0001266
78	0.0006410	0.0001282
77	0.0006494	0.0001299
76	0.0006579	0.0001316
75	0.0006667	0.0001333
74	0.0006757	0.0001351
73	0.0006849	0.0001370
72	0.0006944	0.0001389
71	0.0007042	0.0001408
70	0.0007143	0.0001429
69	0.0007246	0.0001449
68	0.0007353	0.0001471
67	0.0007463	0.0001493
66	0.0007576	0.0001515
65	0.0007692	0.0001538
64	0.0007813	0.0001563
63	0.0007937	0.0001587
62	0.0008065	0.0001613
61	0.0008197	0.0001639
60	0.0008333	0.0001667
59	0.0008475	0.0001695
58	0.0008621	0.0001724
57	0.0008772	0.0001754

t	$\alpha^*_{0.01}$	$\alpha^*_{0.05}$
56	0.0008929	0.0001786
55	0.0009091	0.0001818
54	0.0009259	0.0001852
53	0.0009434	0.0001887
52	0.0009615	0.0001923
51	0.0009804	0.0001961
50	0.0010000	0.0002000
49	0.0010204	0.0002041
48	0.0010417	0.0002083
47	0.0010638	0.0002128
46	0.0010870	0.0002174
45	0.0011111	0.0002222
44	0.0011364	0.0002273
43	0.0011628	0.0002326
42	0.0011905	0.0002381
41	0.0012195	0.0002439
40	0.0012500	0.0002500
39	0.0012821	0.0002564
38	0.0013158	0.0002632
37	0.0013514	0.0002703
36	0.0013889	0.0002778
35	0.0014286	0.0002857
34	0.0014706	0.0002941
33	0.0015152	0.0003030
32	0.0015625	0.0003125
31	0.0016129	0.0003226
30	0.0016667	0.0003333
29	0.0017241	0.0003448
28	0.0017857	0.0003571
27	0.0018519	0.0003704
26	0.0019231	0.0003846
25	0.0020000	0.0004000
24	0.0020833	0.0004167
23	0.0021739	0.0004348
22	0.0022727	0.0004545
21	0.0023810	0.0004762
20	0.0025000	0.0005000
19	0.0026316	0.0005263
18	0.0027778	0.0005556
17	0.0029412	0.0005882
16	0.0031250	0.0006250
15	0.0033333	0.0006667
14	0.0035714	0.0007143
13	0.0038462	0.0007692
12	0.0041667	0.0008333
11	0.0045455	0.0009091
10	0.005	0.001
9	0.0555556	0.0011111
8	0.00625	0.00125
7	0.0071438	0.0014286
6	0.0083333	0.0016667
5	0.01	0.002
4	0.0125	0.0025
3	0.0166667	0.0033333
2	0.025	0.005

Author Index

Subject Index